I0063750

Causes and Consequences of Species Diversity in Forest Ecosystems

Causes and Consequences of Species Diversity in Forest Ecosystems

Special Issue Editors

Aaron M. Ellison
Frank S. Gilliam

MDPI • Basel • Beijing • Wuhan • Barcelona • Belgrade

MDPI

Special Issue Editors

Aaron M. Ellison
Harvard University
USA

Frank S. Gilliam
University of West Florida
USA

Editorial Office
MDPI
St. Alban-Anlage 66
4052 Basel, Switzerland

This is a reprint of articles from the Special Issue published online in the open access journal *Forests* (ISSN 1999-4907) from 2018 to 2019 (available at: https://www.mdpi.com/journal/forests/special_issues/causes_consequences_diversity)

For citation purposes, cite each article independently as indicated on the article page online and as indicated below:

LastName, A.A.; LastName, B.B.; LastName, C.C. Article Title. *Journal Name* **Year**, *Article Number, Page Range.*

ISBN 978-3-03921-309-2 (Pbk)
ISBN 978-3-03921-310-8 (PDF)

Cover image courtesy of Aaron M. Ellison.
Paramachaerium gruberi Brizicky, an endangered canopy tree of Central American rainforests, shown here growing on the Osa Peninsula of Costa Rica.

© 2019 by the authors. Articles in this book are Open Access and distributed under the Creative Commons Attribution (CC BY) license, which allows users to download, copy and build upon published articles, as long as the author and publisher are properly credited, which ensures maximum dissemination and a wider impact of our publications.
The book as a whole is distributed by MDPI under the terms and conditions of the Creative Commons license CC BY-NC-ND.

Contents

About the Special Issue Editors

Aaron M. Ellison is the Deputy Director of, and a senior ecologist at, the Harvard Forest, the Senior Research Fellow in Ecology at Harvard University in the Department of Organismic and Evolutionary Biology, and a semi-professional photographer and writer. He received his B.A. in East Asian philosophy from Yale University in 1982 and his Ph.D. in evolutionary ecology from Brown University in 1986. After post-doctoral positions at Cornell and with the Organization for Tropical Studies in Costa Rica, Dr. Ellison taught for a year at Swarthmore College before moving to Mount Holyoke College in 1990. There, he was the Marjorie Fisher Assistant, Associate, and Full Professor, founding director of Mount Holyoke's Center for Environmental Literacy, and Associate Dean for Science, and he taught biology, environmental studies, and statistics until 2001. Following a sabbatical year at Harvard in 2001–2002, Dr. Ellison assumed his current position at the Harvard Forest—Harvard's 1500-hectare outdoor classroom and laboratory for ecological research. Dr. Ellison works in wetlands and forests to study the disintegration and reassembly of ecosystems following natural and anthropogenic disturbances; the evolutionary ecology of carnivorous plants; the response of plants and ants to global climate change; the application of Bayesian statistical inference to ecological research and environmental decision-making; and the critical reaction of ecology to Modernism. He has authored or co-authored over 200 scientific papers, dozens of book reviews and software reviews, and the books *A Primer of Ecological Statistics* (2004; 2nd edition 2012), *A Field Guide to the Ants of New England* (2012), *Stepping in the Same River Twice: Replication in Biological Research* (2017), *Carnivorous Plants: Physiology, Ecology, and Evolution* (2018), and *Vanishing Point* (2017)—a collection of photographs and poetry from the Pacific Northwest. From 2010–2015, Dr. Ellison was the Editor-in-Chief of *Ecological Monographs*, in 2012 he was elected a Fellow of the Ecological Society of America, and he is currently a Senior Editor of *Methods in Ecology & Evolution*.

Frank S. Gilliam is a professor of biology at the University of West Florida and a professor emeritus at Marshall University. He completed his B.S. in biology at Vanderbilt University and received a Ph.D. in plant ecology at Duke University, studying the fire ecology of southeastern coastal plain pine forests. Following post-doctoral appointments at Kansas State University to study the fire ecology of tallgrass prairie and at the University of Virginia to study hardwood forest canopy/atmosphere interactions, Dr. Gilliam began his 28-year tenure on the faculty of Marshall University in 1990. His research lies primarily at the conceptual boundary between terrestrial plant communities and ecosystems, including the movement and cycling of plant nutrients, especially nitrogen (N). These interests extend to fire ecology and the effects of fire on nutrient cycling, plants, and soils in fire-prone ecosystems. Additionally, related to his ecosystem approach to ecological research is an interest in atmospheric deposition and precipitation chemistry, leading to the study of pollutant conditions (acid deposition, excess N, ozone) in forested areas. Other work includes secondary succession and the species dynamics of the herbaceous layer of forests, as well as the variety of biotic and abiotic factors that influence species composition and change within this vegetation stratum. Ongoing work includes vegetation dynamics in forest ecosystems, N cycling in forest ecosystems, and species composition and stand structure in longleaf pine forests. Dr. Gilliam currently serves as Associate Editor for *Journal of Ecology* and *Journal of Plant Ecology*. He has authored or co-authored more than 100 peer-reviewed articles, in addition to book chapters and reviews of books, current scientific articles, and software, and has authored/co-authored the books *Terrestrial Plant Ecology*, 3rd Edition (1999) and *The Herbaceous Layer in Forests of Eastern North America* (2003; 2nd Edition 2014). He is the grateful husband of Laura P. Gilliam and father of Rachel M. Gilliam, M.Div., and Ian S. Gilliam LTJG USN.

Preface to "Causes and Consequences of Species Diversity in Forest Ecosystems"

Forests have the highest plant diversity among terrestrial ecosystems. Among forests, tree species diversity tends to be highest in the tropics and at low elevations and is positively associated with increasing precipitation and resource availability. Within forests, trees themselves create physical structures and habitats for other species. Stratified "layers" consist of species of similar life forms. The lowest, herbaceous layer is a mixture of resident species (e.g., mosses, liverworts, ferns, flowering herbs) and transient seedlings of trees and shrubs that eventually grow into higher strata. Shrub, subcanopy, and canopy layers, in turn, are dominated increasingly by woody shrubs and trees. Epiphytes, epiphylls, herbaceous vines, and lianas depend on trees for support, while arthropods, birds, and other animals use them for food and shelter.

Considerable research in recent decades has yielded new insights into the mostly positive relationships between species diversity and ecosystem processes. There has been a concomitant escalation of concern that declines in biodiversity of forests caused by increasing human population size and land-use intensity, together with shifts in biodiversity caused by rapid climatic change and new disturbance regimes, will compromise the ecosystem "services" forests provide to human society. Although forests have always been seen as dynamic systems—Henry Chandler Cowles described ecological succession over a century ago as "a variable approaching a variable, not a constant"—the rapid increase in atmospheric concentrations of carbon dioxide and other "greenhouse gases" and frequencies of extreme floods, droughts, fires, and catastrophic cyclonic windstorms presage extensive changes and rearrangement of forests worldwide. Thus, research on the causes and consequences of biodiversity in forests now intersects with the anxieties of the Anthropocene.

The 14 papers in this book, reprinted from a 2018–2019 Special Issue of the journal *Forests*, illustrate these intersections. The first four papers document patterns of diversity at different temporal and spatial scales. Dexter et al. place North American forest diversity in a phylogenetic context and highlight that species diversity is not limited to modern forests but has its roots in evolutionary processes and deep time. W. Li et al. examine diversity in a single family (Salicaceae) in China, using species distribution models to explore its climate-driven changes in diversity from the Last Glacial Maximum (22,000 years ago) through the present and into the late 21st century. Wang et al. zero in on patterns of diversity of woody Ericaceae in China's Yunnan Province, while Nguyen et al. focus on spatial-scale-dependent diversity patterns of trees within individual 2-hectare plots in north-central Vietnam.

The next four papers investigate drivers of diversity in unmanaged forests. Ellison et al. use data from large (≥15-hectare) forest plots in the Western Hemisphere to look for statistical "fingerprints" of foundation tree species—those species that control the biodiversity and ecosystem dynamics of the forests they define and structure. Van Tatenhove et al. document the complex interaction of climatic factors on changing elevational distributions of forest birds and small mammals in New Hampshire's White Mountains. Peterson develops a new index of "damage diversity" and shows how it is related to climatic drivers (wind disturbance) and in turn, influences structure diversity and complexity in the forests of eastern North America. Gilliam reviews experimental studies of a quarter-century of nitrogen addition on an experimental forest in West Virginia. This work has revealed that nitrogen addition at levels similar to those coming from atmospheric deposition leads to a decline in species diversity of herbaceous species in the forest understory and a greater sensitivity of the remaining species to changes in light availability defined by the woody overstory.

The final six papers place forest diversity squarely in the context of human impacts and management. Belote confirms the expected pattern that water availability and soil fertility positively affect species diversity and productivity in the United States. He goes on to show—perhaps unsurprisingly but rarely documented—that people are more likely to manage and modify highly productive, species rich forests but conserve forests that have fewer species and lower productivity. In a curious parallel, Tadesse et al. find that home gardens in western Ethiopia have a nearly threefold higher tree diversity than nearby "natural" parklands. People manage for productivity and diversity. They cultivate many non-native tree species that provide food, fiber, and lumber, whereas parklands have primarily native species that provide similar, albeit less-productive, ecosystem services. Heinrichs et al. illustrate that managing forests in Germany as mixtures rather than monocultures increases local- and landscape-scale diversity of vascular plants, bryophytes, and lichens. Y. Li et al. examine diversity of soil arthropods in monoculture plantations of poplar (*Populus deltoides*) in eastern China, finding generally higher species richness in older (21-year-old) stands but also temporal shifts in species composition. Bolton and D'Amato show that managing with disturbance (harvest gaps) increases diversity of both native and non-native understory plants in silvicultural systems in Minnesota (USA). Lastly, Oldfield and Peterson draw the useful distinction between diversity and species composition in their report that salvage logging following wind disturbance has little effect on diversity (as number of species) but substantial effects on species composition in forests of north Georgia (USA).

Taken together, the papers in this book cover a broad range of forest types across four continents and examine a wide range of topics relevant to understanding the causes and consequences of forest diversity. They also illustrate the central importance on this human-dominated planet of managing for, and with, species diversity in forests.

Aaron M. Ellison, Frank S. Gilliam
Special Issue Editors

forests

MDPI

Article

Exploring the Concept of Lineage Diversity across North American Forests

Kyle G. Dexter [1,2,*], Ricardo A. Segovia [1,3] and Andy R. Griffiths [1]

[1] School of GeoSciences, University of Edinburgh, Edinburgh EH9 3FF, UK; segoviacortes@gmail.com (R.A.S.); andy.griffiths@ed.ac.uk (A.R.G.)
[2] Royal Botanic Garden Edinburgh, Edinburgh EH3 5LR, UK
[3] Instituto de Ecología y Biodiversidad, Santiago 7800003, Chile
*   Correspondence: kyle.dexter@ed.ac.uk; Tel.: +44-(0)-131-650-7439

Received: 30 May 2019; Accepted: 16 June 2019; Published: 22 June 2019

Abstract: Lineage diversity can refer to the number of genetic lineages within species or to the number of deeper evolutionary lineages, such as genera or families, within a community or assemblage of species. Here, we study the latter, which we refer to as assemblage lineage diversity (ALD), focusing in particular on its richness dimension. ALD is of interest to ecologists, evolutionary biologists, biogeographers, and those setting conservation priorities, but despite its relevance, it is not clear how to best quantify it. With North American tree assemblages as an example, we explore and compare different metrics that can quantify ALD. We show that both taxonomic measures (e.g., family richness) and Faith's phylogenetic diversity (PD) are strongly correlated with the number of lineages in recent evolutionary time, but have weaker correlations with the number of lineages deeper in the evolutionary history of an assemblage. We develop a new metric, time integrated lineage diversity (TILD), which serves as a useful complement to PD, by giving equal weight to old and recent lineage diversity. In mapping different ALD metrics across the contiguous United States, both PD and TILD reveal high ALD across large areas of the eastern United States, but TILD gives greater value to the southeast Coastal Plain, southern Rocky Mountains and Pacific Northwest, while PD gives relatively greater value to the southern Appalachians and Midwest. Our results demonstrate the value of using multiple metrics to quantify ALD, in order to highlight areas of both recent and older evolutionary diversity.

Keywords: temperate forests; species richness; assemblage lineage diversity; phylogenetic diversity; evolutionary diversity; United States; trees; TILD

---

1. Introduction

The evolutionary lineage is a fundamental concept in biology, denoting a group of organisms connected by ancestor-descendent relationships [1]. Evolutionary lineages are hierarchically structured; multiple younger evolutionary lineages can be nested within an overarching older lineage, or clade. Thus, multiple genetically diverged lineages can exist within a single taxonomic species, and multiple species can belong to older evolutionary lineages, such as genera, families or orders. Knowing the number of lineages in different ecological assemblages and biogeographic regions gives insights into evolutionary process, biogeographic history, and conservation priorities. For example, an assemblage or region that houses many lineages can be interpreted as having a richer evolutionary history, and therefore may be a greater priority for conservation than one that houses few. However, the conservation value of lineage diversity has yet to be fully, and persuasively, communicated [2–4]. Providing clear and accurate quantification of lineage diversity may assist its integration into conservation practice.

In its most basic form, quantifying the number of lineages in assemblages could consist of counting the number of species. However, the term lineage diversity is generally applied when the units are not species, but a shallower or deeper evolutionary level, i.e., within or above the species taxonomic rank (see [5–9] for examples below species rank; see [10–14] for examples above species rank). In this paper, we focus on lineage diversity above the species rank. Employing tree assemblages in the contiguous United States, we explore various metrics by which assemblage lineage diversity (hereafter ALD) might be quantified, using taxonomic and phylogenetic approaches. Given its pertinence to conservation prioritisation, we focus specifically on the richness dimension of ALD.

Taxonomy is a hierarchical system for organising biological diversity. As such, it provides an apparently straightforward means of quantifying ALD at different evolutionary depths, for example by tallying the number of genera, families or orders in assemblages. However, Linnean taxonomic ranks are not 'natural' in the sense that they do not directly correlate to any precise evolutionary age. Some clades of a given taxonomic rank may actually be younger than clades of a putatively lower taxonomic rank. For example, the genus *Pinus* (Pinaceae) may be as old as 100 million years [15], which is older than most angiosperm families [16]. If one were to compare an assemblage of four *Pinus* species with an assemblage of four angiosperm species belonging to different genera in the same family, and ALD were estimated as the number of genera in each assemblage, the angiosperm assemblage would appear to have 4x higher ALD. However, all four species in the assemblage of *Pinus* may have diverged from each other prior to the age of the most recent common ancestor of the four species in the angiosperm assemblage (similar to mock assemblages B and C in Figure 1), which could mean that the assemblage of *Pinus* has greater conservation value because it encompasses greater total evolutionary history, in terms of time or branch lengths.

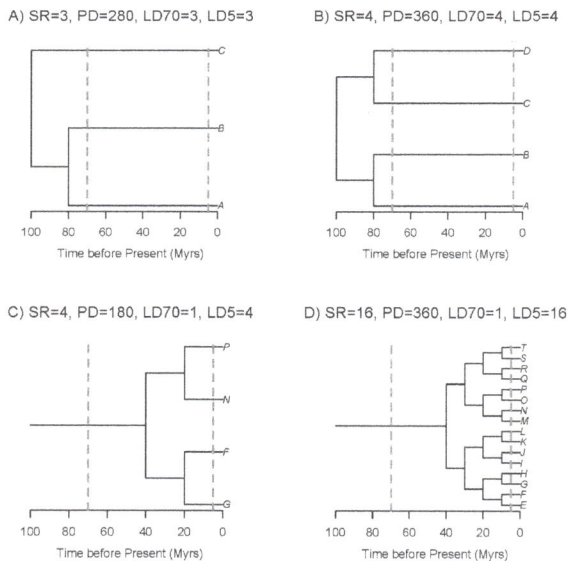

Figure 1. Example phylogenies for four mock assemblages (**A–D**) with contrasting species richness (SR), phylogenetic diversity (PD) and phylogenetic structure (LD70 = number of lineages 70 Ma; LD5 = number of lineages 5 Ma).

The advent of molecular phylogenetics has allowed researchers to move past taxonomic approaches to quantifying ALD. Using a temporally calibrated phylogeny, one can choose a certain evolutionary age–say X millions of years (Myrs)–and then readily estimate the number of lineages at

X million years ago (Ma) in an assemblage of species. Further, one could examine how the number of lineages varies at different time slices across a set of assemblages, or geographic space (*sensu* Jønsson et al., 2011). This is directly analogous to constructing a lineage through time plot for a given evolutionary clade [17], and indeed, studies have proposed constructing lineage through time plots for individual communities or assemblages [18]. However, it is not clear at which evolutionary age, or phylogenetic depth, one should be counting lineages. An assemblage that has more lineages than another assemblage at one, deeper time slice might have fewer lineages at a more recent time slice (compare assemblages B vs. D in Figure 1), which could be driven by variation in diversification histories, community assembly, or numerous other processes. It would be ideal to have metrics for ALD that integrate over the evolutionary history of the clade being studied.

Faith (1992) [19] developed a simple metric, phylogenetic diversity (PD), to estimate the evolutionary history present in communities or assemblages of species, which is calculated by summing the length of all branches in a phylogeny that includes all taxa present in an assemblage, and only those taxa. While this metric is related to the number and age of evolutionary lineages present in an assemblage, and thus may serve as a proxy for ALD, Figure 1 demonstrates that inferences of ALD based on calculating PD may not always be straightforward. In this contrived scenario, it seems clear that Assemblage A has less ALD than Assemblage B and that Assemblage C has less ALD than Assemblage D. The calculations of PD, and even species richness, would support this visual observation. Further, it seems plausible that Assemblage A has more lineage diversity than Assemblage C, even though Assemblage C has more species. However, do Assemblages B and D really have identical ALD even though they have such a discrepancy in species richness? Comparing Assemblages B and D is challenging because they have such different phylogenetic structures. Assemblage B has 4x as many lineages at 70 Ma, while Assemblage D has 4x as many lineages at 5 Ma. For this reason, researchers have suggested that the amount of PD an assemblage contains above or below that expected given its SR is a better measure of ALD [12,13]. However, if we were to follow that approach, then Assemblage C might be considered to have more ALD than Assemblage D (its ratio of PD:SR is twice that of Assemblage D), even though at all phylogenetic depths Assemblage D has the same or more lineages than Assemblage C. Clearly, more work is needed to determine which metrics derived from phylogenies may provide the best measures of ALD that integrate over evolutionary timescales.

The overarching goal of the present manuscript is to explore the behaviour of different metrics that may potentially be used to quantify ALD. As our empirical example, we focus on tree assemblages in the contiguous United States. These tree assemblages provide an ideal system for such an empirical study, as over 150,000 forest inventory plots have been sampled in a standardised way by the U.S. National Forest Service, and existing time-calibrated phylogenies encompass nearly all species present in the plots. We use this large dataset to (1) quantify the ALD using different taxonomic and phylogenetic metrics; (2) assess the relationship of different metrics with each other and with the number of lineages at different evolutionary depths; and (3) map variation in ALD across the contiguous United States. To give context to our results, we conduct a clustering analysis of assemblages based on their shared evolutionary history, thereby determining the main evolutionary groups of tree assemblages in the contiguous United States.

2. Materials and Methods

2.1. Data Sources

We accessed compositional data from 177,549 plots sampled across the contiguous United States by the Forest Inventory and Analysis (FIA) Program of the U.S. Forest Service [20], via the BIEN package [21] for the R Statistical Environment [22]. The FIA protocol records trees $\geq$12.7 cm diameter at breast (dbh) in four 168.3 m^2 subplots that are 36.6 m apart. The main evident spatial data gaps in this dataset are the state of Louisiana and the eastern part of the state of Kentucky.

In order to obtain a phylogeny that covered all species in the FIA tree plot inventory dataset, we combined the temporally calibrated ultrametric phylogenies for North American gymnosperm and angiosperm trees from Ma et al. (2016) [23] (see Figure 2). We set the age of the split between angiosperms and gymnosperms at 350 Ma [24]. After resolving synonyms according to The Plant List (2013), version 1.1 (http://www.theplantlist.org/, accessed in December 2018), we manually added the tree species present in the FIA dataset, but absent in the phylogeny. Their exact placement was based on consultations of the systematics literature (see Table S1 for species added and associated literature reference), with the added taxon being placed halfway along the branch leading to its sister species or clade in the phylogeny. The branch length leading to the added taxon was set to a value such that the tree remained ultrametric. The phylogeny file used in this study is available in Appendix B.

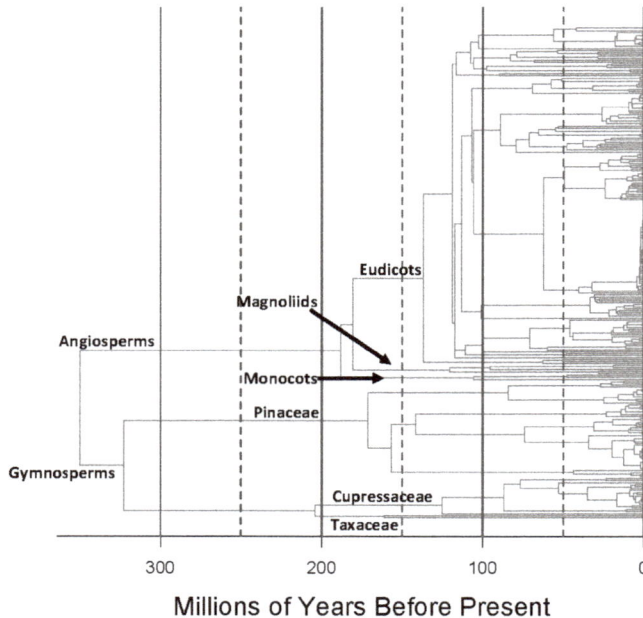

Figure 2. Phylogeny of all tree species present in the contiguous United States in the US Forest Inventory and Analysis (FIA) dataset, based on the phylogenies for gymnosperms and angiosperms in Ma et al. (2016).

2.2. Assemblage Lineage Diversity (ALD) Metrics

2.2.1. Taxonomic Measures

In the absence of phylogenetic data, the number of supraspecific lineages in assemblages can be calculated as the number of taxa of a higher taxonomic rank. Classification systems are consistent across angiosperms and gymnosperms up to the order level, and we therefore tabulated the following taxonomic measures of lineage diversity for assemblages: **number of genera, number of families and number of orders**. The taxonomy table is available in Appendix C.

2.2.2. Phylogenetic Measures

Since the advent of molecular phylogenetics, diverse metrics have been developed and implemented to quantify the lineage, or evolutionary, diversity of assemblages from phylogenies (e.g., References [25–27]). We focus here on metrics that aim to quantify the 'richness dimension of phylogenetic diversity' [28], as our interest is in 'how much' lineage diversity is in assemblages, not how diverged lineages are from each other (e.g., as quantified by mean pairwise phylogenetic distance) or how evenly lineages are represented (e.g., as quantified by phylogenetic species evenness; [29]). In addition, conservation prioritisation is generally based on which species are present, not their relative abundance (which could reflect disturbance histories or other idiosyncratic processes), and we therefore focus on presence/absence metrics. This also increases the general utility of our approach, as abundance information is not available for many datasets.

We started by calculating the most basic metric of ALD, **phylogenetic diversity**, or **PD** [19], which is the sum of all branch lengths in each assemblage, including the branch that goes to the root of all seed plants. We also include its estimate standardised for variation in species richness. This is accomplished by calculating the first two moments of the null expectation for PD, given the number of species in the assemblage, and using them to calculate a standardised effect size. The moments of the null distribution can be calculated by randomly shuffling the tips of the phylogeny many times, but there is an analytical expectation for these moments, which is the approach we used [30]. We refer to this metric as the **standardised phylogenetic diversity**, or **sPD**.

We also calculated two additional proposed measures of the richness dimension of phylogenetic diversity, the **phylogenetic species richness**, or **PSR** [29] and the sum of **evolutionary distinctiveness**, or **sumED** [31]. PSR can essentially be considered a measure of species richness that takes into account the phylogenetic relatedness of taxa in an assemblage. If the assemblage is composed entirely of closely related species, this will produce a lower value of PSR than if the assemblage were composed of distantly related taxa. In practice, this measure is obtained by multiplying the mean pairwise phylogenetic distance between species in an assemblage by its species richness (and dividing by two, so that it represents distance to tips from the most recent common ancestor for each pair of species). For sumED, we first calculated the evolutionary distinctiveness of each species in our dataset, based on the entire phylogeny representing all species, following the fair proportions approach of Isaac et al., (2007) [32]. This is essentially a measure of how phylogenetically isolated each species is, relative to the given phylogeny. We then summed the evolutionary distinctiveness values for the species in each assemblage, following Safi et al., (2013) [31].

As our overarching goal in this study was to quantify ALD over the full evolutionary time of the clade of study (here, seed plants), we developed an additional metric that may better capture this, which we term **time integrated lineage diversity**, or **TILD**. If one constructs a lineage through time (LTT) plot for each assemblage (*sensu* Yguel et al., 2016) [18], one can simply integrate the area under this curve as a measure of the total lineage diversity of the assemblage over time. This integral is mathematically identical to the phylogenetic diversity of the assemblage, when including the root branch. However, in considering an LTT plot built from extant species, as LTT plots for extant assemblages are, they necessarily monotonically increase towards the present and under a constant diversification rate, this increase is exponential. The integral therefore is necessarily weighted towards the number of lineages in recent evolutionary time compared to the number of lineages in deeper evolutionary time. In order to downweight the number of recent lineages when calculating TILD, we log-transformed the y-axis (i.e., the number of lineages at each point in time) prior to taking the integral.

2.3. Statistical Analyses

As the individual FIA plots are quite small in scale, we combined all plots within 0.2° grid cells prior to calculating ALD metrics (n = 13,207 grid cells). In order to determine the main evolutionary groups of tree assemblages across grid cells, we used k-means clustering of assemblages based on their shared phylogenetic branch length, as quantified by the Phylosorensen Index [33]. An elbow analysis suggested that 14 groups was a parsimonious number that minimised within group variation in phylogenetic composition (Figure S1). Preliminary analyses of the distribution of these groups over geographic and climatic space showed that several pairs of groups overlapped both in geographic location and climatic environment. These pairs of groups were combined to give nine total evolutionary groups that were geographically and climatically cohesive. A silhouette analysis [34] was then run for these nine evolutionary groups in order to determine if any individual sites were closer in their evolutionary composition to the medoid value of another group than the group to which they were originally assigned (as measured by the Phylosorensen Index). If such was found, these sites were then reassigned to the group to which they were more similar in evolutionary composition.

In order to visualise the compositional relationships of these different groups, we ordinated assemblages based on the presence versus absence of evolutionary lineages, as quantified by the occurrence of individual nodes in the phylogeny in each assemblage. We specifically used the evolutionary principal component analysis developed by Pavoine (2016) [35], with the occurrence data Hellinger transformed prior to ordination [36]. This approach also allows identification of the evolutionary lineages that are associated with different components of the ordination space. We determined the lineages that are most strongly correlated with the first two principal components. In order to further characterise the composition of the evolutionary groups, we conducted a standard indicator analysis to determine the species most strongly associated with each group [37]. Lastly, to further characterise the evolutionary groups, we mapped where they occur in geographic and climatic space. In order to better visualise how the groups occupy geographic and climatic space, we generated 95% kernel density estimates [38] of the distribution of each group over two climatic dimensions, mean annual temperature and precipitation, and two geographic dimensions, elevation and latitude.

There is wide variation in the number of individual trees sampled across the combined plots in each grid cell (887 ± 1204 inds, mean ± s.d.; range 2–17,307 inds), and all of the ALD metrics that we calculated, except sPD, were positively correlated with the number of individuals sampled (Pearson's r = 0.60−0.76). In order to obtain comparable estimates of ALD, we rarefied grid cells to the same number of individuals. While rarefaction can be problematic because it excludes assemblages from analysis below the abundance threshold used and introduces heteroscedasticity in the diversity estimate that is related to the number of individuals sampled [39], we do not know of any estimates of the richness dimension of ALD or phylogenetic diversity that are robust to variation in sample size (in terms of number of individuals sampled). While Rao's quadratic entropy has been proposed as an estimate of phylogenetic diversity that is robust to variation in sample size, it measures the divergence dimension of phylogenetic diversity, not the richness dimension [28], and was therefore not of interest to us here.

In order to determine the number of individuals to select in rarefactions, we first selected the subset of assemblages that have at least 1000 individuals (3660 grid cell assemblages). We estimated the species richness of these assemblages when rarefied to 1000 individuals (i.e., expected number of species per 1000 stems). We then rarefied these assemblages to smaller numbers of individuals, and observed how the richness estimate for a smaller number of stems correlated with the richness estimate per 1000 stems. Once assemblages were rarefied to less than 50 stems, the correlation (pearson's r) between the two richness estimates dropped below 0.95. We therefore chose 50 individuals as the size for rarefied assemblages. We repeated rarefactions 100 times, and calculated the average of each ALD metric over these 100 rarefactions.

In order to assess the general behaviour of ALD metrics, we calculated the spearman's rank correlation (rho) between a given ALD metric and the number of lineages at different phylogenetic depths (in intervals of 1 Myrs between the present and the root of the seed plant phylogeny at 350 Ma). We used spearman's rank correlation because these relationships are not necessarily linear, and because our goal is to evaluate if assemblages would be ranked similarly, e.g., for conservation prioritisation, if counting the number of lineages at a particular time slice vs. using a given ALD metric. In order to obtain an overall measure of the behaviour of a lineage diversity metric, we then obtained the mean of the spearman's rho values across all phylogenetic depths. All analyses were carried out in the R Statistical Environment [22] using functions in the ape [40], picante [41], vegan [42], cluster [43], adiv [44] and PhyloMeasures [30] packages. The analysis script is available in Appendix A.

3. Results

Clustering analyses based on shared evolutionary history resulted in nine major evolutionary groups of assemblages, which vary in their geographic (Figure 3), elevational and climatic distributions (Figure 4). The west coast of the United States is dominated by a single group, but as one moves inland there are four different evolutionary groups that are spatially mixed across much of the western United States. They occupy relatively distinct regions of climatic space, and their spatial interdigitation likely results from environmental variation generated by topographic heterogeneity. In contrast, the four groups east of the Mississippi are clearly arranged in a latitudinal manner, reflecting the fact that environmental gradients are more gradual in the eastern United States (Figure 3). These groups clearly replace each other along a temperature gradient from colder to warmer sites (Figure 4B). There are two groups in the centre of the United States, one of which is most predominant in Texas, but also extends in a scattered distribution further north in the Great Plains and westwards into the southern Rockies. The other central group is scattered through the more eastern, wetter portions of the Great Plains and also in the Midwest, with its core extent in the northern Great Plains. More detailed comments regarding the groups can be found in Table S2.

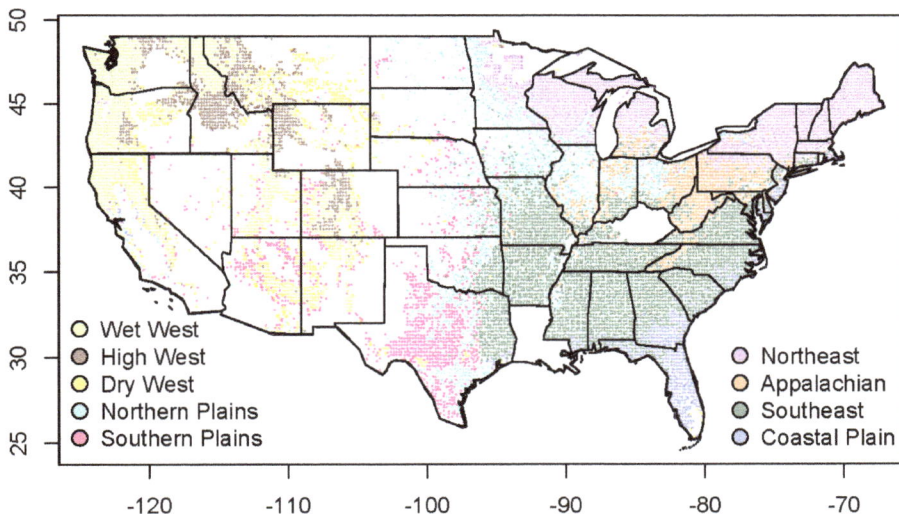

Figure 3. Map of tree assemblages included in this study, coloured by evolutionary group following a clustering analysis based on shared evolutionary history. Names for groups were chosen based on their geographic and/or climatic characteristics. Each assemblage consists of all FIA plots within a 0.2° by 0.2° grid cell.

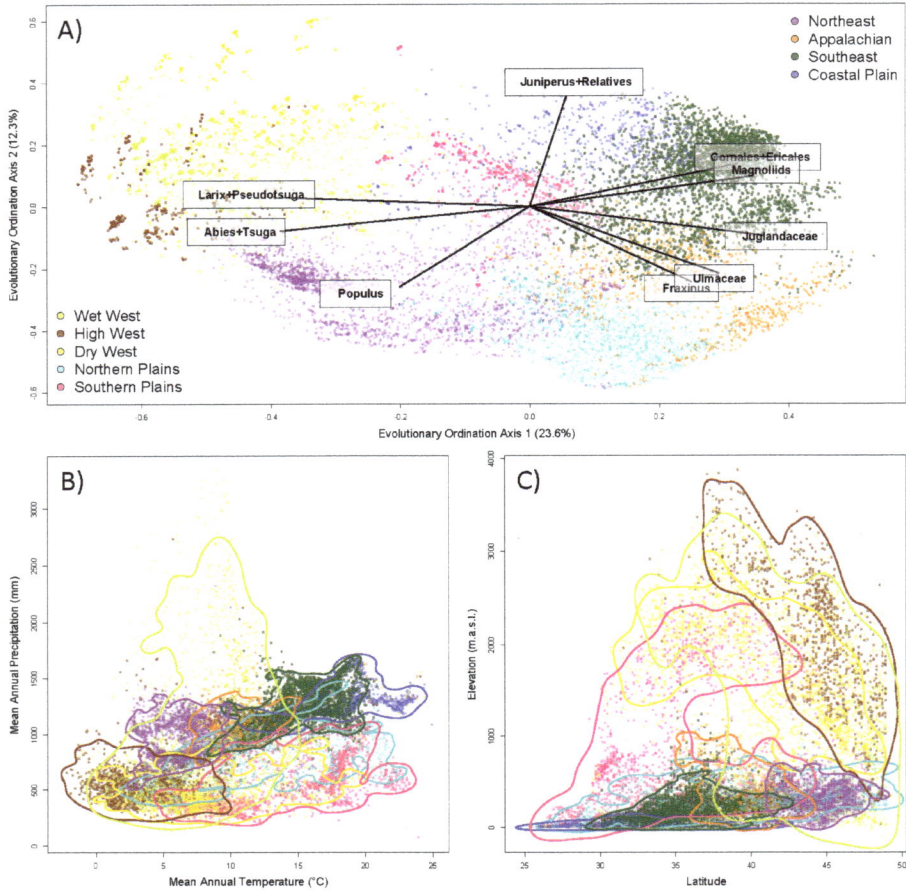

Figure 4. Distribution of nine main evolutionary groups of tree assemblages in the contiguous United States across: (**A**) an ordination space based on shared evolutionary history, with the influence of the key clades highlighted; (**B**) climatic space; and (**C**) elevation and latitude.

The majority of grid cell assemblages sampled at least 50 individuals (11,547 of 13,207 assemblages or 87%), and were therefore included in calculations of assemblage lineage diversity (ALD). We estimated ALD metrics for each assemblage, including constructing lineage through time (LTT) plots for each to calculate the time integrated lineage diversity (TILD). We show a sample of these LTT plots for each evolutionary group in Figure 5. There is clear variation across groups in when they accumulate lineage diversity. The Northern Plains group is composed entirely of eudicot angiosperms, and therefore most assemblages only have a single lineage (a log value of 0 on y-axis) until the eudicots begin to diversify ~120 Ma. An entirely contrasting pattern can be found in assemblages of the Dry West group, which have multiple lineages of gymnosperms, and thus have high lineage diversity deep in time. However, these assemblages are relatively poor in angiosperms and so do not achieve the same number of lineages in recent time periods as the Northern Plains group. The eastern groups have the highest number of lineages in recent time slices and also tend to have high lineage diversity deep in time, except for the Appalachian group which is relatively poor in gymnosperms.

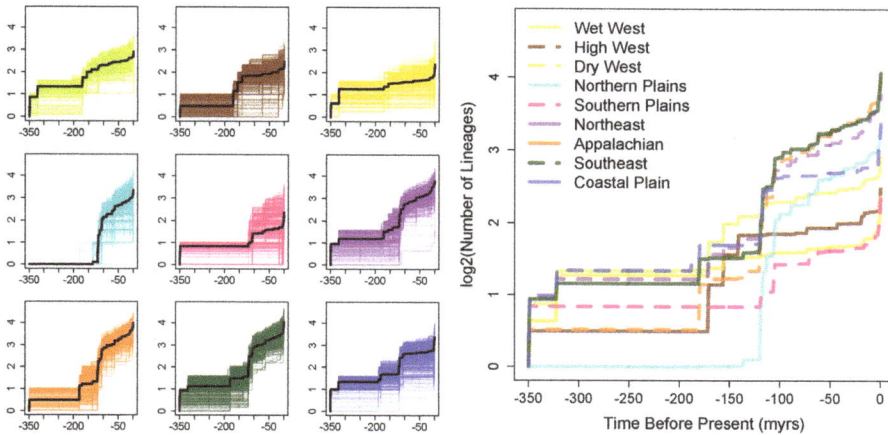

Figure 5. (**Left**) Lineage through time (LTT) plots for a sample of 500 assemblages from each evolutionary group, with each coloured line representing the average values over 100 rarefactions of the given assemblage to 50 individuals. Meanwhile, the thick black line gives the mean value at each evolutionary depth across all assemblages for a given evolutionary group. (**Right**) The mean number of lineages at each evolutionary depth across all assemblages in each group (thick black line from left panel), arrayed on one plot for direct comparison. Note log2 transformation of y-axes and that the number of lineages at the right-hand side of each graph represents the (mean) species richness of the assemblage(s).

Except for standardised phylogenetic diversity (sPD), the different ALD metrics we calculated are highly correlated with each other (Pearson's r = 0.60−0.95; Figure 6) and with the number of species in assemblages (Pearson's r = 0.57−0.93; Figure 6). The correlation of species richness (SR) with the number of lineages declines with increasing phylogenetic depth, dropping to very low values prior to the radiation of the Eudicots at 120 Ma (Figure 7). The other taxonomic measures of ALD all show a similar pattern to SR; i.e., none show a strong correlation with number of lineages prior to ∼120 Ma. Because of this, none of the taxonomic measures of lineage diversity show a mean correlation over all evolutionary depths greater than 0.5 (Figure 7).

The phylogenetically-derived metrics of ALD vary in their pattern of correlation with number of lineages over different evolutionary depths (Figure 7). Neither sPD or sum of evolutionary distinctiveness (sumED) show high mean correlations (mean rho = 0.30 and 0.46 respectively), and these two metrics show contrasting patterns over phylogenetic depth. sPD is more strongly correlated with the number of lineages deep in evolutionary time, while sumED shows a pattern more similar to taxonomic measures of lineage diversity. Time integrated lineage diversity (TILD) shows the highest mean correlation with number of lineages across all phylogenetic depths (mean rho = 0.76), but phylogenetic diversity (PD) and phylogenetic species richness (PSR) also showed relatively high mean correlations (mean rho = 0.67 and 0.66 respectively). PD and PSR show stronger correlations with the number of lineages in recent evolutionary time, while TILD shows stronger correlations with the number of lineages in deeper evolutionary time (Figure 7). Given that other taxonomic measures of ALD are strongly correlated with SR, that sPD and sumED show low mean correlations with number of lineages across most phylogenetic depths and that PSR does not show a different pattern from PD, with which it is highly correlated (r = 0.95; Figure 6), we focus below on patterns with respect to SR, PD and TILD.

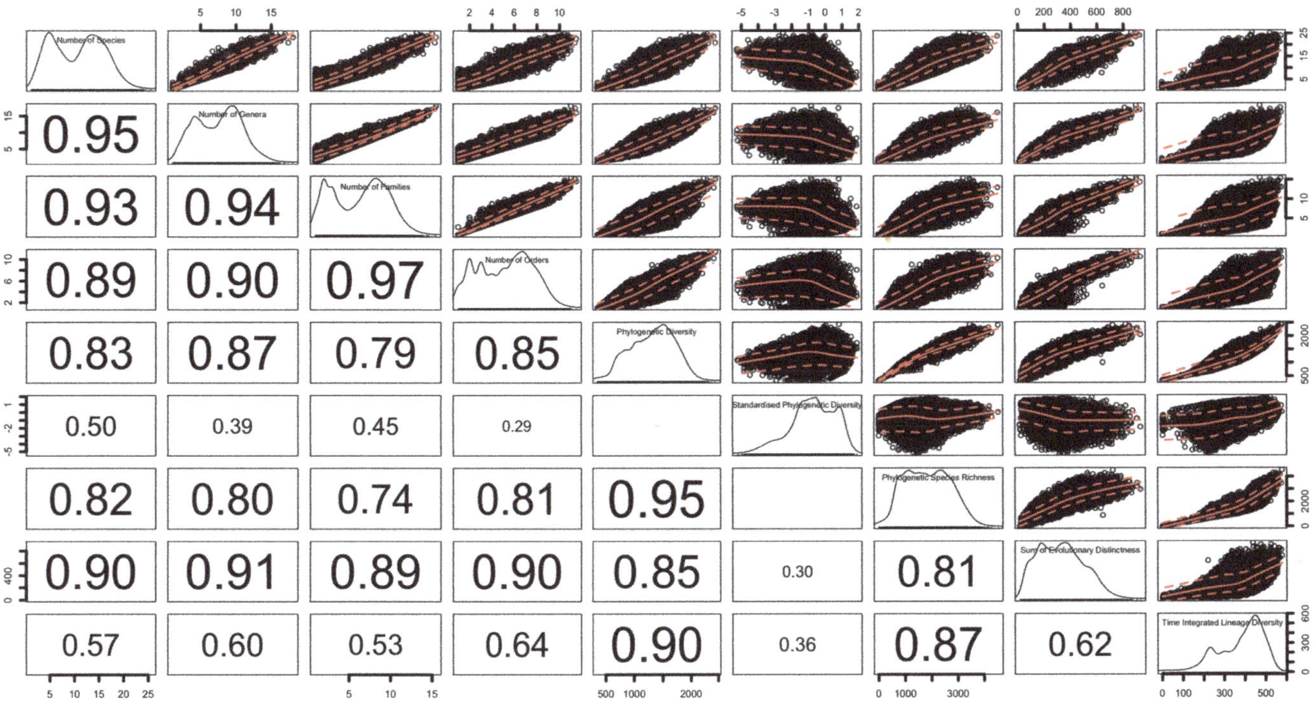

Figure 6. Pairwise relationships between lineage diversity metrics, with Pearson's correlation coefficient given on the lower half of the matrix. The solid red line represents a loess, moving average regression and the dashed red lines represent the conditional variability over the range of the x-axis. The diagonal gives probability density plots for each metric.

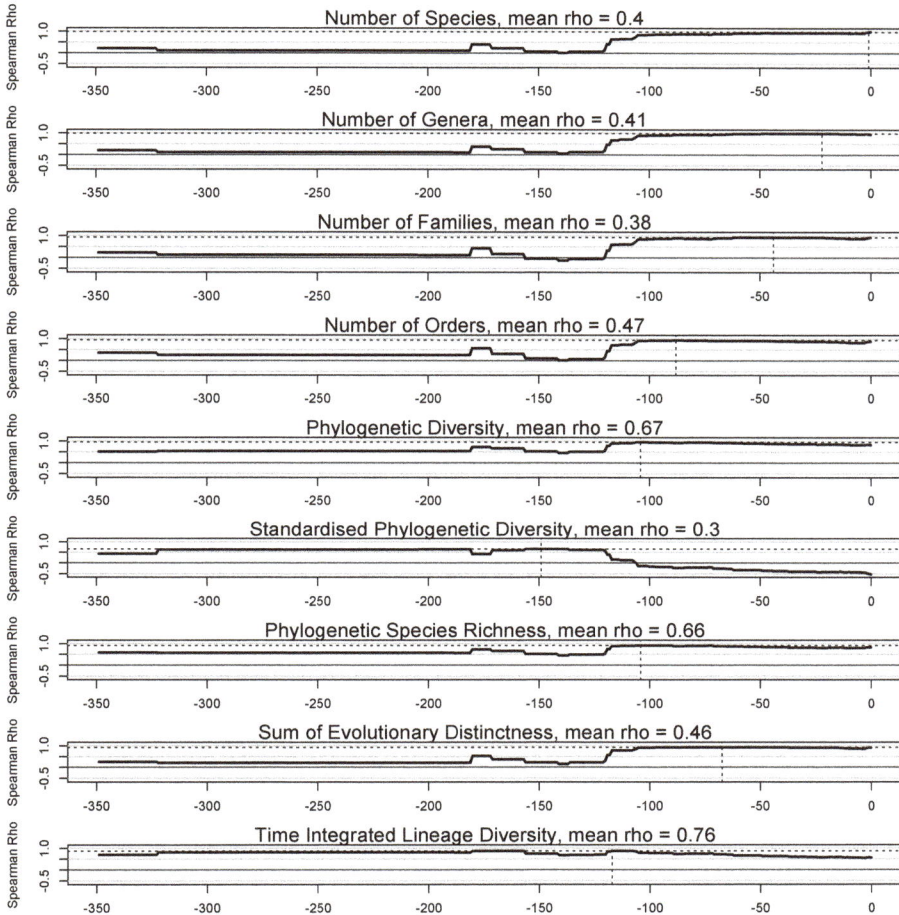

Figure 7. Spearman's rank correlations between different synthetic measures of assemblage lineage diversity and the number of lineages at different phylogenetic depths. The mean value of spearman's rho across all depths (excluding t = 0 and 350) is given above each plot. The phylogenetic depth at which the maximum correlation is found is marked with dashed lines going to the x and y-axes.

Species richness of assemblages, quantified as the number of species per 50 trees, shows clear spatial patterns across the contiguous United States (Figure 8). Low values are generally observed in the western half of the United States, while in the eastern half, low values are observed in Florida and parts of the Northeast. Among the western groups, the highest SR is found in assemblages in the Wet West group, while in the east, the highest values are in the Southeast group. PD shows similar patterns to SR (Figure 8), although it gives higher values on average for the Northeast group than the Appalachian group, while the opposite holds for SR. Also for PD, the Wet West group approaches values observed in the eastern groups, while that is not the case for SR. The TILD metric shows patterns that contrast with those for SR and PD (Figure 8). TILD gives higher values on average for the Southern Plains group than the Northern Plains group, while PD finds the opposite relative ranking. TILD also gives the Wet West group equal value to that for eastern groups. Within the east, TILD gives values for the Coastal Plain group equal to that for the Northeast and Southeast groups, while it gives relatively lower values for the Appalachian group. Overall, for most groups, PD shows a pattern for groups that is intermediate between that observed for SR and TILD. In analysing the deviation of TILD from the expectation given PD (based on the residuals of a regression of TILD on the logarithm of PD), we see that TILD gives higher values than expected for the Wet West, Dry West, Southern Plains and the Coastal Plain, and lower values than expected for the Northern Plains and Appalachians (Figure 8, bottom row).

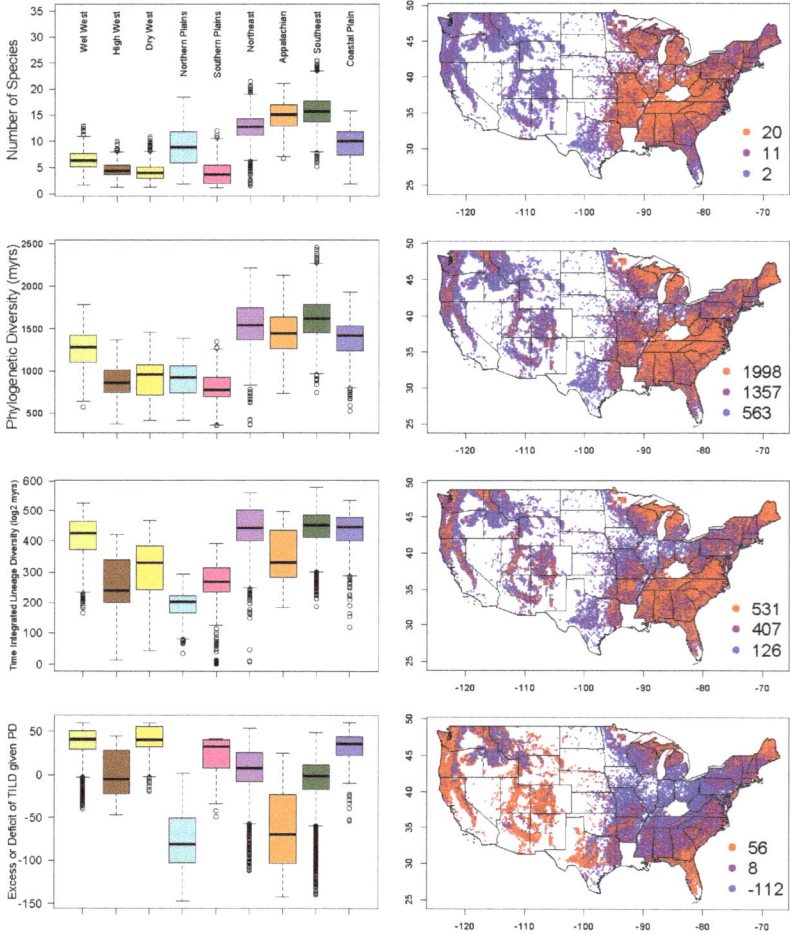

Figure 8. Left: Variation in species richness, phylogenetic diversity (PD), time integrated lineage diversity (TILD) and the excess or deficit of TILD given PD for tree assemblages, separated by major evolutionary group. The latter represent the residuals from a regression of TILD on the logarithm of PD. Values for each assemblage represent the average across 100 rarefactions to 50 tree individuals. **Right:** Maps of the same metrics, per assemblage, across the contiguous United States, with the colours for the upper 95% quantile, median and lower 95% quantile for a given metric in the lower right-hand corner of each map.

4. Discussion

Many different metrics could potentially be used to quantify the lineage diversity of organismal assemblages. For tree assemblages across the contiguous United States, we find that two metrics, which can be derived from temporally calibrated phylogenies, show the greatest average correlation with number of lineages over the full evolutionary history of seed plants, and thus seem best suited to quantify assemblage lineage diversity (ALD). These are Faith's phylogenetic diversity (PD) and a metric that is newly derived here, time integrated lineage diversity (TILD). Other metrics derived from molecular phylogenies either showed lower average correlations with number of lineages in assemblages, or were strongly correlated with PD. Meanwhile, taxonomic metrics, including species richness (SR), failed to correlate with the number of lineages deep in evolutionary time, specifically, prior to the origin of the Eudicots. This is because the high number of species, genera, families and orders of Eudicots in assemblages in the eastern United States drive the pattern of variation in taxonomic metrics. If prioritisation schemes were to be based solely on SR, or other taxonomic richness measures of ALD, the entire western half of the US would receive less conservation attention than the eastern half. Yet, western US tree assemblages, dominated by older, relatively species-poor gymnosperm lineages, can still represent substantial reservoirs of evolutionary history, as reflected in TILD values comparable to the most lineage-diverse tree assemblages in the eastern US.

4.1. Taxonomic Measures of Lineage Diversity

In many studies [12,45], species richness has been found to be strongly correlated with phylogenetic diversity (PD), and has therefore been suggested as a suitable proxy for ALD [46]. Our study suggests that, at least for tree assemblages in the contiguous US, this is not the case. Higher-level taxonomic measures that we explored, specifically the numbers of genera, families and orders in assemblages, do not perform much better. As expected, as higher taxonomic ranks are used, strong correlations with number of lineages persist deeper into evolutionary time (compare the x-intercept of highest correlation for different taxonomic ranks in Figure 7), but none of the taxonomic measures provide a high correlation with number of lineages prior to ∼120 Ma. This is perhaps unsurprising as the majority of lineages deeper in evolutionary time are gymnosperms (Figure 2), and all the gymnosperms in our dataset come from a single order, three families and 15 genera, while angiosperms dominate the variation in taxonomic measures of ALD with 18 orders, 35 families and 68 genera. Thus, for clades with highly imbalanced phylogenies, like seed plants, taxonomic measures of lineage diversity are not likely to provide an adequate, synthetic measure of ALD [47].

4.2. Phylogenetic Measures of Lineage Diversity

Time integrated lineage diversity (TILD) represents the area beneath a lineage through time plot where the number of lineages per time slice has been log-transformed. TILD is mathematically related to PD, which is identical to the area beneath a raw (i.e., non-log-transformed) lineage through time plot. PD was originally conceived as a metric to aid conservation prioritisation [19], and it has always been properly interpreted as a measure of the total evolutionary diversity in assemblages, which is certainly worth quantifying. But, it is strongly skewed towards the number of lineages present in recent evolutionary time, downweighting older evolutionary divergences. We suggest that researchers may use PD to quantify ALD more recently in evolutionary time, and complementarily, TILD may be more suitable to obtain a measure of ALD that gives equal weight to deeper evolutionary time as recent evolutionary time. While phylogenetic species richness (PSR; [29]) is strongly correlated with PD and could represent an alternative to it, we suggest researchers continue to use PD, because of its historical precedence and because, as with TILD, it is directly interpretable in terms of numbers of lineages.

For this dataset, the standardised phylogenetic diversity (sPD) correlates well with the number of lineages deep in evolutionary time, but not with numbers of lineages in recent evolutionary time. In fact, sPD is negatively correlated with numbers of lineages less than 70 million years old. We suggest that sPD may better serve as a metric of phylogenetic community structure, which is interesting in its own right [48], but that it should not be used as a measure of the richness dimension of lineage diversity in assemblages. In contrast to sPD, the sum of evolutionary distinctiveness in assemblages, sumED, showed weaker correlations deeper in evolutionary time and stronger relationships in recent evolutionary time (Figure 7). In fact, sumED showed a very similar pattern to taxonomic metrics of lineage diversity (Figure 7), and of the phylogenetically-derived metrics in this study, it shows the strongest correlation with taxonomic metrics (Figure 6). As a conservation prioritisation metric, sumED has a clear intuitive value, since it represents the totality of phylogenetic diversity in a given assemblage that is rare in the entire dataset, but values for assemblages will be sensitive to phylogenetic taxon sampling in the overall dataset, even if a given assemblage itself is fully phylogenetically sampled (see Isaac et al., (2007) [32] for full explanation of how ED is derived for each species). Conversely, PD and TILD will only vary based on sampling within quantified assemblages, and are therefore more straightforward to apply.

4.3. Tree Diversity Patterns across the Contiguous United States

Consistent with previous assessments [49,50], the most evident spatial contrast in the species richness pattern is between the eastern and western United States. To a very coarse approximation, this reflects the dominance of gymnosperms in the western United States, the dominance of angiosperms in the eastern United States, and the fact that angiosperms are a much more diverse clade than gymnosperms (even when focusing only on trees). Previous studies have identified the high plateau south of the Appalachian Mountains [49], and the Florida panhandle, Alabama/Georgia border region [50], as areas of maximal tree species richness in the United States. In contrast we found the highest tree species richness in a region centred on Kentucky and Tennessee. This contrast in results could be due to the different spatial grain size of analysis, our use of plot data rather than overlap in range maps or the fact that we only include taxa larger than 12.7 cm dbh and thus exclude small tree taxa. The forest region centred on Kentucky and Tennessee corresponds to the mixed mesophytic forest region [51]. The first part of the name reflects that there are no particularly dominant tree species in the region and most forest stands have a mix of dominant species. Braun (1950) [51] recognised the exceptional tree species richness of this region and characterised it as "the association of the Deciduous Forest which occupies the area of optimum moisture and temperature conditions of North America" (p. 42). Indeed, moisture stress for plants is lower in this region of the US compared to regions southeast of the Appalachians or the entire western US, while temperatures do not reach the extreme lows that occur in the northern parts of the contiguous US. Meanwhile, as the second part of the name, mesophytic, implies, the forests in this area are also found on more fertile soils compared to other forests in the US. Thus, the high alpha diversity of tree assemblages in this region may reflect an environment that is the most benign for the majority of tree species occurring in the contiguous United States. This is similar to the pattern found in another large biogeographic region, the Amazon, where the most species-rich tree assemblages are found in the western Amazon, which has relatively fertile soils and is subject to less moisture stress than the southern or eastern Amazon [12].

The spatial patterns of PD and TILD for tree assemblages show several evident contrasts with the spatial pattern of tree species richness (SR). In general, the western US shows almost uniformly low values of SR, at least in comparison to the eastern US, but the PD and TILD metrics show much greater variation. In particular, the TILD metric shows values for assemblages in the Wet West group that are comparable to values for assemblages in the eastern United States. One evident hotspot of lineage diversity in the west is the temperate rain forest region of the Pacific northwest, with high PD and TILD values. This temperate rain forest region includes the 'Miracle Mile' in the Klamath mountains of northern California which holds 18 species of conifers [52], albeit not that many occur in any of the assemblages derived from FIA plots. PD and TILD plummet as one crosses eastward over the Cascade mountain range or the Sierra Nevada mountain range. Presumably the arid conditions on the eastern side of these mountain ranges limit tree lineage diversity. Though species richness minima to the east of the Sierra Nevada (as well as the Rocky Mountains) have previously been noted [49], this contrast across mountain ranges is most evident when considering PD, and particularly TILD. Other regions of notable ALD in the western US include scattered areas in the southern Rocky Mountains and northern Idaho.

In the eastern United States, the spatial pattern of PD, and particularly TILD, also contrasts with that of species richness, although there are areas that are high for all three of these diversity measures (e.g., much of the Southeast). The most north-eastern state in the contiguous US (Maine) as well as the area around the Great Lakes emerge as regions of high PD and TILD, which is presumably due to the increasing prevalence of conifers in the far north and their deep evolutionary heritage. Meanwhile, the mixed mesophytic forest region that has the highest tree species richness values does not show the highest PD and TILD values. Higher PD and TILD are found to the south and east of the mixed mesophytic forest region. The south-eastern United States was highlighted as a region of high angiosperm tree PD in a previous study based on range maps [23], and our results show this is consistent when using inventory data, and when incorporating gymnosperms into the quantification of overall seed plant PD.

The Coastal Plain and Southern Plains groups are both notable in showing substantially higher TILD than SR values relative to other groups, which may be due to the incursion of tropical angiosperm lineages into these southern areas. For example, the Coastal Plain group is home to *Sabal* species [Arecaeae], *Persea borbonia* [Lauraceae], *Annona glabra* [Annonaceae], among other species, which belong to old, largely tropical families. Indeed, the average age of angiosperm families in the coastal plain is higher than anywhere else in the contiguous US [53]. Meanwhile, the Southern Plains group has among the lowest values for SR, but shows intermediate values for TILD, and includes represents of tropical dry region genera such as *Prosopis* and *Vachellia* (Fabaceae), which do not occur elsewhere in the contiguous United States.

5. Conclusions

Our study has explored the concept of lineage diversity, and how it might be quantified, for tree assemblages across the contiguous United States. We have shown how temporally calibrated molecular phylogenies can be used to quantify assemblage lineage diversity (ALD) and aid conservation prioritisation [19]. As might be expected, metrics derived from a molecular phylogeny showed stronger correlations than taxonomic metrics with the numbers of lineages over the evolutionary history of the focal clade (seed plants). As for specific recommended metrics, we suggest that phylogenetic diversity (PD) and time integrated lineage diversity (TILD) metrics both be used. PD has precedence in the literature and is useful for comparison with previous studies. We do stress however that PD is skewed towards the number of lineages in recent evolutionary time, while the newly-derived TILD metric is shown to better represent the entire evolutionary history of the clade of interest, and should therefore also be used.

We employed an empirical dataset on tree assemblages of the contiguous United States to explore these different metrics of ALD. We found that the spatial patterns of PD and TILD differ in important ways from the spatial patterns of species richness, for example by highlighting the high conservation value of temperate rainforests in the Pacific Northwest. PD and TILD also give somewhat contrasting results, with the former giving relatively higher values for tree assemblages in the northern Great Plains, Midwest and high elevation areas of the Appalachians, and the latter giving relatively higher values for some dry areas in the western US and the southeast Coastal Plain. However, it would be naïve to suggest that conservationists in the United States are unaware of the high conservation value of these different forests. Indeed, the tree flora of the United States is likely well enough known, such that good awareness already exists regarding which areas have particularly high or low conservation value with respect to tree species composition and lineage diversity. Where these metrics may be particularly useful is in less well known floras, such as in many tropical biogeographic regions. There has been one study of variation in lineage diversity across ~300 sites in the rain forests of the Amazon basin [12], but we know of no such similar study outside of tropical moist forest, or in other tropical regions.

Supplementary Materials: The following are available online at http://www.mdpi.com/1999-4907/10/6/520/s1, Table S1: Species added to the molecular phylogeny, Table S2: Descriptions of the major evolutionary groups in the tree flora of the contiguous United States, Figure S1: Clustering validation (Elbow plot).

Author Contributions: K.G.D. and R.A.S. conceived the manuscript and led analyses. A.R.G. contributed to analyses. All authors contributed to writing.

Funding: This research received no external funding.

Acknowledgments: We thank two anonymous reviewers for their thorough and insightful reviews, which greatly improved the quality of the manuscript. K.G.D. was supported by a Leverhulme International Academic Fellowship during the time this research was completed. R.A.S. is supported by Newton International Fellowship from The Royal Society, Conicyt PFCHA/Postdoctorado Becas Chile/2017 N° 3140189 and CONICYT PIA APOYO CCTE AFB170008. A.R.G. is supported by a NERC studentship (NE/L002558/1).

Conflicts of Interest: The authors have no conflicts of interest to report.

Appendix A

R Codes to execute analyses.

Appendix B

Combined molecular phylogeny derived from combining the angiosperm and gymnosperm phylogenies in Ma et al., 2016, with the addition of unsampled species.

Appendix C

Taxonomy table used to determine taxonomic richness of tree assemblages.

References

1. De Queiroz, K. The general lineage concept of species, species criteria, and the process of speciation. In *Endless Forms: Species and Speciation*; Howard, D.J., Berlocher, S., Eds.; Oxford University Press: New York, NY, USA, 1998; pp. 57–75.
2. Winter, M.; Devictor, V.; Schweiger, O. Phylogenetic diversity and nature conservation: Where are we? *Trends Ecol. Evol.* **2013**, *28*, 199–204. [CrossRef] [PubMed]
3. Geeta, R.; Lohmann, L.G.; Magallón, S.; Faith, D.P.; Hendry, A.; Crandall, K.; De Meester, L.; Webb, C.; Prieur-Richard, A.H.; Mimura, M.; et al. Biodiversity only makes sense in the light of evolution. *J. Biosci.* **2014**, *39*, 333–337. [CrossRef] [PubMed]
4. Forest, F.; Crandall, K.A.; Chase, M.W.; Faith, D.P. Phylogeny, extinction and conservation: Embracing uncertainties in a time of urgency. *Philos. Trans. R. Soc. B* **2015**, *370*, 20140002.10.1098/rstb.2014.0002. [CrossRef] [PubMed]
5. Kozak, K.H.; Blaine, R.A.; Larson, A. Gene lineages and eastern North American palaeodrainage basins: Phylogeography and speciation in salamanders of the *Eurycea bislineata* species complex. *Mol. Ecol.* **2006**, *15*, 191–207. [CrossRef] [PubMed]
6. Oliver, P.; Keogh, J.S.; Moritz, C. New approaches to cataloguing and understanding evolutionary diversity: A perspective from Australian herpetology. *Aust. J. Zool.* **2015**, *62*, 417–430. [CrossRef]
7. Milián-García, Y.; Jensen, E.L.; Madsen, J.; Álvarez Alonso, S.; Serrano Rodríguez, A.; Espinosa López, G.; Russello, M.A. Founded: Genetic Reconstruction of Lineage Diversity and Kinship Informs *Ex situ* Conservation of Cuban Amazon Parrots (*Amazona leucocephala*). *J. Hered.* **2015**, *106*, 573–579. [CrossRef] [PubMed]
8. Becking, L.E.; de Leeuw, C.A.; Knegt, B.; Maas, D.L.; De Voogd, N.J.; Suyatna, I.; Peijnenburg, K.T. Highly divergent mussel lineages in isolated Indonesian marine lakes. *PeerJ* **2016**, *4*, e2496. [CrossRef] [PubMed]
9. Moritz, C.C.; Pratt, R.C.; Bank, S.; Bourke, G.; Bragg, J.G.; Doughty, P.; Keogh, J.S.; Laver, R.J.; Potter, S.; Teasdale, L.C.; et al. Cryptic lineage diversity, body size divergence, and sympatry in a species complex of Australian lizards (*Gehyra*). *Evolution* **2018**, *72*, 54–66. [CrossRef]
10. Linder, H.P.; Eldenas, P.; Briggs, B.G. Contrasting patterns of radiation in African and Australian Restionaceae. *Evolution* **2003**, *57*, 2688–2702. [CrossRef]

11. Jønsson, K.A.; Fabre, P.H.; Ricklefs, R.E.; Fjeldså, J. Major global radiation of corvoid birds originated in the proto-Papuan archipelago. *Proc. Natl. Acad. Sci. USA* **2011**, *108*, 2328–2333. [CrossRef]

12. Honorio-Coronado, E.N.; Dexter, K.G.; Pennington, R.T.; Chave, J.; Lewis, S.L.; Alexiades, M.N.; Alvarez, E.; Alves de Oliveira, A.; Amaral, I.L.; Araujo-Murakami, A.; et al. Phylogenetic diversity of Amazonian tree communities. *Divers. Distrib.* **2015**, *21*, 1295–1307. [CrossRef]

13. Rezende, V.L.; Dexter, K.G.; Pennington, R.T.; Oliveira-Filho, A.T. Geographical variation in the evolutionary diversity of tree communities across southern South America. *J. Biogeogr.* **2017**, *44*, 2365–2375. [CrossRef]

14. Law, C.J.; Slater, G.J.; Mehta, R.S. Lineage diversity and size disparity in Musteloidea: Testing patterns of adaptive radiation using molecular and fossil-based methods. *Syst. Biol.* **2017**, *67*, 127–144. [CrossRef] [PubMed]

15. Leslie, A.B.; Beaulieu, J.; Holman, G.; Campbell, C.S.; Mei, W.; Raubeson, L.R.; Mathews, S. An overview of extant conifer evolution from the perspective of the fossil record. *Am. J. Bot.* **2018**, *105*, 1531–1544. [CrossRef] [PubMed]

16. Magallón, S.; Gómez-Acevedo, S.; Sánchez-Reyes, L.L.; Hernández-Hernández, T. A metacalibrated time-tree documents the early rise of flowering plant phylogenetic diversity. *New Phytol.* **2015**, *207*, 437–453. [CrossRef] [PubMed]

17. Raup, D.M.; Gould, S.J.; Schopf, T.J.; Simberloff, D.S. Stochastic models of phylogeny and the evolution of diversity. *J. Geol.* **1973**, *81*, 525–542. [CrossRef]

18. Yguel, B.; Jactel, H.; Pearse, I.S.; Moen, D.; Winter, M.; Hortal, J.; Helmus, M.R.; Kühn, I.; Pavoine, S.; Purschke, O.; et al. The evolutionary legacy of diversification predicts ecosystem function. *Am. Nat.* **2016**, *188*, 398–410. [CrossRef]

19. Faith, D.P. Conservation evaluation and phylogenetic diversity. *Biol. Conserv.* **1992**, *61*, 1–10. [CrossRef]

20. Burrill, E.; Wilson, A.; Turner, J.; Pugh, S.; Menlove, J.; Christiansen, G.; Conkling, B.; David, W. *The Forest Inventory and Analysis Database: Database Description and User Guide version 8.0 for Phase 2*; U.S. Department of Agriculture, Forest Service: Washington, DC, USA, 2018. Available online: www.fia.fs.fed.us/library/database-documentation (accessed on 18 December 2018).

21. Maitner, B.S.; Boyle, B.; Casler, N.; Condit, R.; Donoghue, J.; Durán, S.M.; Guaderrama, D.; Hinchliff, C.E.; Jørgensen, P.M.; Kraft, N.J.; et al. The bien r package: A tool to access the Botanical Information and Ecology Network (BIEN) database. *Methods Ecol. Evol.* **2018**, *9*, 373–379. [CrossRef]

22. R Core Team. *R: A Language and Environment for Statistical Computing*; R Foundation for Statistical Computing: Vienna, Austria, 2018.

23. Ma, Z.; Sandel, B.; Svenning, J.C. Phylogenetic assemblage structure of North American trees is more strongly shaped by glacial–interglacial climate variability in gymnosperms than in angiosperms. *Ecol. Evol.* **2016**, *6*, 3092–3106. [CrossRef]

24. Harris, L.W.; Davies, T.J. A Complete Fossil-Calibrated Phylogeny of Seed Plant Families as a Tool for Comparative Analyses: Testing the 'Time for Speciation' Hypothesis. *PLoS ONE* **2016**, *11*, e0162907. [CrossRef] [PubMed]

25. Nunn, C.L.; Altizer, S.; Sechrest, W.; Jones, K.E.; Barton, R.A.; Gittleman, J.L. Parasites and the evolutionary diversification of primate clades. *Am. Nat.* **2004**, *164*, S90–S103. [CrossRef] [PubMed]

26. Mazel, F.; Guilhaumon, F.; Mouquet, N.; Devictor, V.; Gravel, D.; Renaud, J.; Cianciaruso, M.V.; Loyola, R.; Diniz-Filho, J.A.F.; Mouillot, D.; et al. Multifaceted diversity–area relationships reveal global hotspots of mammalian species, trait and lineage diversity. *Glob. Ecol. Biogeogr.* **2014**, *23*, 836–847. [CrossRef] [PubMed]

27. Blaimer, B.B.; Brady, S.G.; Schultz, T.R.; Fisher, B.L. Functional and phylogenetic approaches reveal the evolution of diversity in a hyper diverse biota. *Ecography* **2015**, *38*, 901–912. [CrossRef]

28. Tucker, C.M.; Cadotte, M.W.; Carvalho, S.B.; Davies, T.J.; Ferrier, S.; Fritz, S.A.; Grenyer, R.; Helmus, M.R.; Jin, L.S.; Mooers, A.O.; et al. A guide to phylogenetic metrics for conservation, community ecology and macroecology. *Biol. Rev.* **2017**, *92*, 698–715. [CrossRef] [PubMed]

29. Helmus, M.R.; Bland, T.J.; Williams, C.K.; Ives, A.R. Phylogenetic measures of biodiversity. *Am. Nat.* **2007**, *169*, E68–E83. [CrossRef]

30. Tsirogiannis, C.; Sandel, B. PhyloMeasures: A package for computing phylogenetic biodiversity measures and their statistical moments. *Ecography* **2016**, *39*, 709–714. [CrossRef]

31. Safi, K.; Armour-Marshall, K.; Baillie, J.E.; Isaac, N.J. Global patterns of evolutionary distinct and globally endangered amphibians and mammals. *PLoS ONE* **2013**, *8*, e63582. [CrossRef]

32. Isaac, N.J.; Turvey, S.T.; Collen, B.; Waterman, C.; Baillie, J.E. Mammals on the EDGE: Conservation priorities based on threat and phylogeny. *PLoS ONE* **2007**, *2*, e296. [CrossRef]
33. Bryant, J.A.; Lamanna, C.; Morlon, H.; Kerkhoff, A.J.; Enquist, B.J.; Green, J.L. Microbes on mountainsides: Contrasting elevational patterns of bacterial and plant diversity. *Proc. Natl. Acad. Sci. USA* **2008**, *105*, 11505–11511. [CrossRef]
34. Rousseeuw, P.J. Silhouettes: A graphical aid to the interpretation and validation of cluster analysis. *J. Comput. Appl. Math.* **1987**, *20*, 53–65. [CrossRef]
35. Pavoine, S. A guide through a family of phylogenetic dissimilarity measures among sites. *Oikos* **2016**, *125*, 1719–1732. [CrossRef]
36. Legendre, P.; Gallagher, E.D. Ecologically meaningful transformations for ordination of species data. *Oecologia* **2001**, *129*, 271–280. [CrossRef] [PubMed]
37. Dufrêne, M.; Legendre, P. Species assemblages and indicator species: The need for a flexible asymmetrical approach. *Ecol. Monogr.* **1997**, *67*, 345–366. [CrossRef]
38. Duong, T. ks: Kernel density estimation and kernel discriminant analysis for multivariate data in R. *J. Stat. Softw.* **2007**, *21*, 1–16. [CrossRef]
39. McMurdie, P.J.; Holmes, S. Waste Not, Want Not: Why Rarefying Microbiome Data Is Inadmissible. *PLoS Comput. Biol.* **2014**, *10*, 1–12. [CrossRef] [PubMed]
40. Paradis, E.; Claude, J.; Strimmer, K. APE: Analyses of phylogenetics and evolution in R language. *Bioinformatics* **2004**, *20*, 289–290. [CrossRef]
41. Kembel, S.W.; Cowan, P.D.; Helmus, M.R.; Cornwell, W.K.; Morlon, H.; Ackerly, D.D.; Blomberg, S.P.; Webb, C.O. Picante: R tools for integrating phylogenies and ecology. *Bioinformatics* **2010**, *26*, 1463–1464. [CrossRef]
42. Oksanen, J.; Blanchet, F.G.; Kindt, R.; Legendre, P.; Minchin, P.R.; O'hara, R.; Simpson, G.L.; Solymos, P.; Stevens, M.H.H.; Wagner, H.; et al. Package 'Vegan': Community Ecollogy Package. 2018. Available online: https://cran.r-project.org/web/packages/vegan/index.html (accessed on 22 June 2019).
43. Maechler, M.; Rousseeuw, P.; Struyf, A.; Hubert, M.; Hornik, K. Package 'cluster'. 2018. Available online: https://cran.r-project.org/web/packages/cluster/index.html (accessed on 22 June 2019).
44. Pavoine, S. Package adiv: Analysis of diversity. 2018. Available online: https://cran.r-project.org/web/packages/adiv/index.html (accessed on 22 June 2019).
45. Forest, F.; Grenyer, R.; Rouget, M.; Davies, T.; Cowling, R.; Faith, D.; Balmford, A.; Manning, J.; Proches, S.; van der Bank, M. Preserving the evolutionary potential of floras in biodiversity hotspots. *Nature* **2007**, *445*, 757–760. [CrossRef]
46. Rapacciuolo, G.; Graham, C.H.; Marin, J.; Behm, J.E.; Costa, G.C.; Hedges, S.B.; Helmus, M.R.; Radeloff, V.C.; Young, B.E.; Brooks, T.M. Species diversity as a surrogate for conservation of phylogenetic and functional diversity in terrestrial vertebrates across the Americas. *Nat. Ecol. Evol.* **2018**, *3*, 1. [CrossRef]
47. Miller, J.T.; Jolley-Rogers, G.; Mishler, B.D.; Thornhill, A.H. Phylogenetic diversity is a better measure of biodiversity than taxon counting. *J. Syst. Evol.* **2018**, *56*, 663–667. [CrossRef]
48. Webb, C. Phylogenies and community ecology. *Annu. Rev. Ecol. Evol. S* **2002**, *33*, 475–505. [CrossRef]
49. Currie, D.J.; Paquin, V. Large-scale biogeographical patterns of species richness of trees. *Nature* **1987**, *329*, 326–327. [CrossRef]
50. Jenkins, C.N.; Van Houtan, K.S.; Pimm, S.L.; Sexton, J.O. US protected lands mismatch biodiversity priorities. *Proc. Natl. Acad. Sci. USA* **2015**, *112*, 5081–5086. [CrossRef] [PubMed]
51. Braun, E.L. Deciduous Forests of Eastern North America. *Soil Sci.* **1950**, *71*, 259–279. [CrossRef]
52. DeSiervo, M.H.; Jules, E.S.; Kauffmann, M.E.; Bost, D.S.; Butz, R.J. Revisting john sawyer and dale thornburgh's 1969 vegetation plots in the russian wilderness: A legacy continued. *Fremontia* **2016**, *44*, 20.
53. Hawkins, B.A.; Rueda, M.; Rangel, T.F.; Field, R.; Diniz-Filho, J.A.F. Community phylogenetics at the biogeographical scale: cold tolerance, niche conservatism and the structure of North American forests. *J. Biogeogr.* **2014**, *41*, 23–38. [CrossRef]

© 2019 by the authors. Licensee MDPI, Basel, Switzerland. This article is an open access article distributed under the terms and conditions of the Creative Commons Attribution (CC BY) license (http://creativecommons.org/licenses/by/4.0/).

forests

MDPI

Article

Climatic Change Can Influence Species Diversity Patterns and Potential Habitats of Salicaceae Plants in China

Wenqing Li [1], Mingming Shi [1], Yuan Huang [2], Kaiyun Chen [1], Hang Sun [1] and Jiahui Chen [1,*]

[1] CAS Key Laboratory for Plant Diversity and Biogeography of East Asia, Kunming Institute of Botany, Chinese Academy of Sciences, Kunming 650201, Yunnan, China; liwenqing@mail.kib.ac.cn (W.L.); shimingming@mail.kib.ac.cn (M.S.); chenkaiyun@mail.kib.ac.cn (K.C.); sunhang@mail.kib.ac.cn (H.S.)
[2] School of Life Sciences, Yunnan Normal University, Kunming 650092, Yunnan, China; huangyuan@mail.kib.ac.cn
* Correspondence: chenjh@mail.kib.ac.cn

Received: 18 January 2019; Accepted: 25 February 2019; Published: 1 March 2019

Abstract: Salicaceae is a family of temperate woody plants in the Northern Hemisphere that are highly valued, both ecologically and economically. China contains the highest species diversity of these plants. Despite their widespread human use, how the species diversity patterns of Salicaceae plants formed remains mostly unknown, and these may be significantly affected by global climate warming. Using past, present, and future environmental data and 2673 georeferenced specimen records, we first simulated the dynamic changes in suitable habitats and population structures of Salicaceae. Based on this, we next identified those areas at high risk of habitat loss and population declines under different climate change scenarios/years. We also mapped the patterns of species diversity by constructing niche models for 215 Salicaceae species, and assessed the driving factors affecting their current diversity patterns. The niche models showed Salicaceae family underwent extensive population expansion during the Last Inter Glacial period but retreated to lower latitudes during and since the period of the Last Glacial Maximum. Looking ahead, as climate warming intensifies, suitable habitats will shift to higher latitudes and those at lower latitudes will become less abundant. Finally, the western regions of China harbor the greatest endemism and species diversity of Salicaceae, which are significantly influenced by annual precipitation and mean temperature, ultraviolet-B (UV-B) radiation, and the anomaly of precipitation seasonality. From these results, we infer water–energy dynamic equilibrium and historical climate change are both the main factors likely regulating contemporary species diversity and distribution patterns. Nevertheless, this work also suggests that other, possibly interacting, factors (ambient energy, disturbance history, soil condition) influence the large-scale pattern of Salicaceae species diversity in China, making a simple explanation for it unlikely. Because Southwest China likely served as a refuge for Salicaceae species during the Last Glacial Maximum, it is a current hotspot for endemisms. Under predicted climate change, Salicaceae plants may well face higher risks to their persistence in southwest China, so efforts to support their in-situ conservation there are urgently needed.

Keywords: Climatic change; species diversity; potential habitats; China; Maxent; Salicaceae

1. Introduction

Understanding geographical variation in biological diversity and its underlying mechanisms has re-emerged as a research hotspot in ecology and biogeography [1–3]. With accelerating specimen digitization and more high-resolution environmental data, biodiversity can now be gleaned from online databases (e.g., Global Biodiversity Information Facility, GBIF). Over the last 20 years, such data has been instrumental for revealing species richness patterns and been used to test hypotheses about its spatial dynamics, thereby advancing macroecology [4,5]. The formation of species richness patterns is quite complicated, however. Many contentious hypotheses try to explain how their large-scale patterning arises mechanistically, primarily through historical geological processes or contemporary environments and their biophysical constraints [6–9]. The former view holds that regional differences in current biodiversity mainly depend on sudden events in geological history, such as periodic variation of glacial-interglacial periods [10,11], with historical climate change in particular having shaped species' recent distributions and spatial patterns [12,13]. This would suggest the number of species found in a region is the net result of species formation, extinction, migration, and dispersal rates over evolutionary time. The other prominent view is that contemporary energy and water affect species richness by (1) determining an ecosystem's primary productivity, which shapes its food chain and drives other indirect effects (i.e., "primary productivity hypothesis"), and (2) by the extent to which an organism's tolerance to one or more environmental factors determines its species distribution (i.e., "physiological tolerance hypothesis") [14,15]. Hence, the water–energy dynamics hypothesis presumes the environment influences the allocation of abiotic resources (e.g., water, temperature, ultraviolet-B (UV-B) among different species and that this is what primarily determines the distribution patterns of regional biodiversity [16]. Its basic tenet is that water (annual precipitation typically) and energy (annual temperature and UV-B radiation) together determine the large-scale patterns of species diversity [11,17].

Changes in climate could also influence species distribution, community composition, and ecosystem structure [18–20]. Global climate change is no longer a matter of debate, although many scientists may not agree on the exact predictions from different models [21], and resolving the relationships between vegetation and climate dynamics is an outstanding critical issue of current research [22,23]. By the end of 21st century, the average temperature on the planet will have risen by 0.3–4.5 °C [24], with the Arctic having undergone especially rapid warming over recent decades. Since Salicaceae species are a key component of pan-Arctic vegetation, climate change will likely continue to influence their species' geographic distributions [25–27]. In this context, studying Salicaceae's species distribution and biodiversity patterns is invaluable for insight into environmental science, natural resource management, and biodiversity conservation.

Members of the Salicaceae s. str. (hereinafter inclusive) family are the main woody plant component of northern temperate forests, having considerable ecological and economic value in timber and ornamental applications, and now serving as the most important woody bioenergy crop [28–30]. Salicaceae consists of the genera *Populus* and *Salix*. The species richness of Salicaceae is unevenly distributed, however: most of its species occur in the temperate zone of the Northern Hemisphere, with East Asia and the mountains of southwest China especially rich in species diversity [28,29]. China alone contains some 347 species (ca. 56% of Salicaceae)—the most of any country in the world—of which 236 are endemic. Despite their widespread human use, how the current distribution patterns of Salicaceae plants formed remains mostly unknown [31], and these may be significantly affected by global climate warming [32].

Although the ecological niche modeling (ENM) [33] is now a common technique, our study's novelty lies in applying this ecological modeling to the family taxonomic rank as a whole, and to single species as typically done. Importantly, we used ENM to simulate the potential distribution of Salicaceae given the following key considerations: (1) The accuracy of distribution data is crucial for robust niche modeling and will affect the simulation results. Generally, species' distribution data is likely biased by wrongly identified species, whereas the identification of organisms at the family level is far less prone to such errors. Hence, the accuracy of family-level distribution data recorded for specimens exceeds that of one specie. (2) Different species may have different ecological niches, yet the Salicaceae is a typical temperate family belonging to pan-Arctic flora and most its member species prefer humid environments. An important premise of ENM is that the ecological requirements and distributions of species are in equilibrium [34]. Although Salicaceae plants are widely distributed in China, their distribution also shows strong regularity, and they share a similar climatic niche on a larger spatial scale [29]. (3) We integrated the distribution data of all extant species of Salicaceae in China, so the simulation results were able to fully capture and convey this family's nationwide geographical distribution patterns [35,36]. In summary, the above three considerations validate our use of ENM to simulate the potential distribution of Salicaceae in China at different times (past, present, and future).

Here, we took advantage of a comprehensive georeferenced occurrence dataset of Salicaceae in China coupled to high-resolution environmental data, to which we applied Maxent algorithms to build species distribution models. We hypothesized that dry and cold climatic conditions during the glacial period would have decreased the total suitable distribution area for Salicaceae plants and that with more climatic warming in the future, suitable habitats in the present would also face the risk of being lost. Based on these presumptions, we had three goals: (1) to project and quantify the Salicaceae species' changes in the extent of their potential suitable distribution under a variety of past, present, and future climate scenarios/years; (2) to map the spatial patterns of species diversity and endemism for Salicaceae members; (3) to determine which factors play key roles in shaping the patterns of Salicaceae plant species richness, weighted endemism, and corrected-weighted endemism.

2. Materials and Methods

2.1. Spatial Data

Species distribution data relating to the occurrence of Salicaceae plants were extracted from China National Specimen Information Infrastructure (NSII) (http://www.nsii.org.cn/), the Chinese Virtual Herbarium (CVH) (http://www.cvh.org.cn/), and the Global Biodiversity Information Facility (GBIF) (https://www.gbif.org/). All the species names were verified and rectified by using Taxonomic Name Resolution Service (TNRS) and by taxonomic experts. We obtained 10,322 Salicaceae specimens in total, of which 2107 were *Populus* and 8215 were *Salix*; the latter included specimens of 5 subgenera: 2848 of *Chamaetia*, 42 of *Chosenia*, 2120 of *Salix*, 61 of *Triandra*, and 3144 of *Vetrix*. However, because most specimens we collected lacked information on their latitude and longitude, we georeferenced these records based on their described location information. When collecting plant specimens, more intensive sampling is typically carried out in one area or region over others [37], and this sampling bias will skew the data representativeness for the species distributions vis-à-vis climate variables. To reduce and correct this spatial bias, we applied a systematic sampling method [38] using a resolution of 1×1 km, due to the spatial error in georeferenced specimen records and to match the resolution of environmental data (WorldClim). Our study region was divided into a grid of equally-sized square areas of 1 km^2, which ensures that each grid cell has only one distributed record. After filtering, 2673 uniquely located records remained (Table S1), each with an accurate location, were used to simulate potential distributions of the Salicaceae family: 1025 *Populus*, 2306 *Salix*, 760 *Chamaetia*, 29 *Chosenia*, 1136 *subgenus Salix*, 43 *Triandra*, and 1114 *Vetrix* records. To obtain the species richness pattern for Salicaceae, we simulated the potential distribution of each species. Species currently occupying less than 5 grid cells were removed from our analyses [39,40]. A total of 6678 georeferenced specimen records belonging to the 5 *Salix* subgenera representing 215 species (Table S2) were used for these niche models. To avoid too much data stifling the model runs, the simulation results were resampled with a spatial resolution of 20×20 km for analyzing the relationships between species diversity patterns and environmental variables.

2.2. Environmental Parameters

The growth and distribution of Salicaceae species are susceptible to various interacting factors, namely climate (e.g., temperature and precipitation), UV-B radiation, and soil [41–43]. For example, UV-B radiation and temperature caused gender imbalance in Salicaceae plants, driving divergent evolutionary trends in *Populus* and *Salix* genera and changes in their population numbers [44,45]. Hageer et al. [46] and Chen et al. [47] studies showed that in addition to climate variables, the effects of soil variables should also be taken into account in the simulation of the potential distribution of species. Hence, for our modeling, we chose four types of environmental variables: climate, UV-B radiation, soil, and topography.

Initially, 39 environmental variables were selected to build a species distribution model. These consisted of elevation, 19 bioclimatic variables, 15 Food and Agriculture Organization (FAO) soil variables, and 4 UV-B variables. Before any modeling, to avoid multicollinearity of variables—this leads to over-fitted species distribution models, with the highly correlated variables affecting model accuracy—we calculated Spearman's rank correlation coefficients among them, selecting those variables with r_s-values < 0.75 for inclusion (Tables S3 and S4). This key filtering step resulted in 15 environmental variables in total used to predict the patterns of Salicaceae (5 bioclimatic, 1 topographic, 5 soil, and 4 UV-B factors; as listed in Table 1).

Table 1. Environmental variables used to predict the geographical distributions of Salicaceae plants.

Code	Environment Variables	Resolution	Unit
Bio1	Annual mean temperature	30″	°C × 10
Bio3	Isothermality (BIO2/BIO7) ∗ 100	30″	–
Bio7	Temperature annual range	30″	°C × 10
Bio12	Annual precipitation	30″	mm
Bio15	Precipitation seasonality (Coefficient of variation)	30″	–
Elevation	Elevation	30″	m
S-CE	Cation exchange capacity (CEC) clay subsoil	30″	–
T-BS	Base saturation% topsoil	30″	%
T-C	Organic carbon pool topsoil	30″	–
T-N	Nitrogen % topsoil	30″	%
Drain	Soil drainage class	30″	–
UV-B1	Annual mean UV-B	15′	$J/m^2/day$
UV-B2	Ultraviolet-B (UV-B) seasonality	15′	$J/m^2/day$
UV-B3	Mean UV-B of lightest month	15′	$J/m^2/day$
UV-B4	Mean UV-B of lowest month	15′	$J/m^2/day$

2.2.1. Current Environment Variables

Data on the 5 bioclimatic factors and elevation were obtained from the WorldClim website (https://www.worldclim.org/), while those for the 5 soil factors were obtained from the China Soil Map Based Harmonized World Soil Database (v1.1) (http://westdc.westgis.ac.cn/). Data on the 4 UV-B factors were taken from the gIUV database (http://www.ufz.de/gluv/). All data were resampled to a spatial resolution of 1 km^2.

2.2.2. Historical and Future Climate Scenarios

The Fifth Assessment Report (AR5) of the Intergovernmental Panel on Climate Change (IPCC) shows that, over the past 100 years (1880–2012), the average annual temperature at the Earth's surface has increased by 0.85 °C, most of which (0.072 °C rise) has occurred in the last 60 years (1951–2012) [24]. The IPCC's AR5 adopted a climate-coupled model of the international coupling model for phase 5 (CCMIP5) and discussed four representative greenhouse gas emission scenarios: RCP2.6, RCP4.5, RCP6.0, and RCP8.5. RCP stands for "representative concentration pathway" (i.e., the typical concentration trajectory); its assigned number conveys the amount of radiative forcing (RF, in W/m^{-2}) in 2100 relative to its value in 1750. We selected the two most typical concentration trajectories, RCP4.5 and RCP8.5, since they respectively represent an intermediate stable trajectory and a highly concentrated one.

Because climate change impact on ecological phenomena and processes has both occurred and is ongoing, we selected six climatic scenarios: 2 in the past and 4 in the future (climatic scenarios/years). The former were the Last Inter Glacial period (LIG, ~120,000–140,000 years) and the Last Glacial Maximum (LGM, ~22,000 years ago), and latter were RCP4.5-2050 (averaged for the years 2041–2060 under scenario RCP4.5), RCP4.5-2070 (averaged for the years 2061–2080 under scenario RCP4.5), RCP8.5-2050 (averaged for the years 2041–2060 under scenario RCP8.5), and RCP8.5-2070 (averaged for the years 2061–2080 under scenario RCP8.5). These past and future climate scenarios were adopted following the BCC-CSM model, developed by the National Center for Atmospheric Research (NCAR), from the WorldClim database (https://www.worldclim.org/). From it only 5 bioclimatic variables (Bio1, Bio3, Bio7, Bio12, and Bio15) under past and future scenarios are currently available for use in the analyses, whereas these were lacking for the 10 environmental variables (i.e., Alt, S-CE, T-BS, T-C, T-N, Drain, UVB1, UVB2, UVB3, and UVB4); hence, the latter did not change over time.

To better capture the changes in historical climate, we calculated the difference between temperature and precipitation during the present and LGM (i.e., bio1-anomaly = $Bio1_{present} - Bio1_{LGM}$; Table 3).

2.3. Building the Species Distribution Model (SDM)

The species distribution model (SDM) is widely used to simulate the suitable distribution pattern of species under climate change scenarios [48–50]. The maximum entropy (Maxent) model is a type of machine-learning algorithm model. Specifically, its algorithm adopts the Maxent principle, of not constraining any unknown distribution information yet also preserving the information constraint of the distributed environment variable data, to predict scientifically the potential distribution of one or more species. A logistically attractive feature of Maxent is that it can predict species' potential distributions by their presence-only data and environmental variables. Compared with other SDMs, the predictions of Maxent are better when the number of distribution points is uncertain and the correlation between different climatic factors is unclear [51]. In recent years, SDMs using the maximum entropy model have played a major contributing role to species diversity conservation in the future [52].

Using the environmental variables at present associated with the distribution points of Salicaceae species, we constructed the model based on the maximum entropy theory to simulate this family's realized niche. Next, these simulated real ecological niches were projected to different times (i.e., past or future) for calculation to build the model [53]. After modeling the current suitable habitat area for Salicaceae with current climate data, we conducted modeling projections for past (i.e., LIG and LGM) and future climatic scenarios (i.e., RCP4.5-2050, RCP4.5-2070, RCP8.5-2050, and RCP8.5-2070) to predict their respective availability of suitable habitat. These SDM projections were made in Maxent software. Based on the 15 environmental variables and our species distribution data, we built an SDM model for the Salicaceae family for past, present, and future climate scenarios. For 215 species, we also simulated the potential distribution of each by using the maximum entropy model in MaxEnt v3.3 software (http://biodiversityinformatics.amnh.org/open_source/maxent/). Only those species with at least 5 distribution records were used to establish the species distribution model, as anything less is considered insufficient for model predictive ability [40]. MaxEnt was run with these rules: use only linear features for SDM with distribution data of <5 records; add quadratic features when occurrence records consisted of ≥5 and <15 records; add hinge features for ≥15 but <80 records; include product and threshold features when species records exceeded 80 [54]. In China, Salicaceae species have rather specific environmental niches and geographical distribution ranges; hence, we applied the clamping function of Maxent. We calculated the contribution of each environmental variable via Jackknife testing and the set cross-validation function, with logistic outputs. This operation was repeated 10 times, with 75% of species occurrence records used for model training and the remainder (25%) for model testing. To reduce model over-fitting to low levels, we set the regularized multiplier to a value of 2.00 and the number of background points to 10,000 [55]. Next, an ASCII grid layer was produced with the largest obtained AUC (area under the receiver operating characteristic (ROC) curve). The ROC curve was employed to test the model's predictive accuracy by judging the AUC value (range of 0–1): a prediction was perfect if its AUC value = 1. This rarely happens, yet when AUC > 0.7 the simulation results are considered valid [56,57].

Next, the arithmetic results from MaxEnt were loaded into ArcGIS v10.2 to carry out a fitness classification and visualization expression, which enabled us to generate the potential distributions of Salicaceae plants. The source of the map was the National Catalogue Service for Geographic Information (http://www.webmap.cn/main.do?method=index). It was critical to choose an appropriate probability threshold of species existence when converting the continuous suitability index maps to binary habitats. Extensive research has demonstrated that a "maximum training sensitivity-plus-specificity" approach leads to highly accurate predictions and is thus superior to other threshold division methods [58,59]. Therefore, to demarcate the habitat and non-habitat of Salicaceae species, we used a "maximum training sensitivity-plus-specificity" threshold in our study.

2.4. Biodiversity Pattern Indices

We used three common biodiversity metrics—species richness, weighted endemism, and corrected-weighted endemism—to quantify the species diversity patterns. Once the probability threshold of species existence had been determined, the matrix of all presence/absence layers was generated. Its rows represented the 26,137 grid cells covering China's landmass while its columns indicated the presence of the 215 Salicaceae species we modeled.

Species richness (SR) was defined as the sum of unique species occurring per cell. It is one of the most commonly used indice in large-scale biodiversity conservation research, characterized by its simplicity and intuition. Endemism is the limitation of a biological group unit (species, genus, or family) to a particular geographical area. In this respect, it is of great value to understand the nature, occurrence, and evolution of a regional flora. Weighted endemism (WE), the sum of the reciprocal total number of cells each species in a grid cell is found in, emphasizes areas having a high proportion of species with restricted ranges. Corrected-weighted endemism (CWE) is simply the weighted endemism divided by the total number of species in a cell [60]. Hence, the CWE emphasizes areas that have a high proportion of species with restricted ranges, yet these are not necessarily species-rich areas. Both the WE and CWE can be used to reveal the distribution center of endemism phenomena. The spatial distribution pattern of the WE value will depends largely on the species richness pattern. The CWE fully embodies the spatial distribution characteristics of endemic species, but it lacks objectivity because it uses thean artificial boundary. Therefore, these three indices complement each other. SR, WE, and CWE were mapped in ArcGIS v10.2, and calculated as follows:

$$SR = K, \tag{1}$$

$$WE = \Sigma\, 1/C, \tag{2}$$

$$CWE = WE/K, \tag{3}$$

where C is the number of grid cells in which each endemic species occurs, and K is the total number of species in a grid cell.

2.5. Changes in Core Distribution Centers

To examine changing trends from the past to the future, we tested and compared the changes in potential habitat area and distribution centers of the past, present, and future suitable areas by using the SDM toolbox [61]. This tool calculates the distributional changes between two binary SDMs (i.e., past vs. present SDMs), providing a table output depicting the predicted contraction, expansion, and areas of no change in a given species distribution [62]. The tool was also used to reduce Salicaceae' distributions to a centroid and to form a vector file to describe the changes in a centroid's range and direction.

2.6. Statistical Analysis

To explain the species diversity patterns of Salicaceae, we conducted variation partitioning analysis. It is useful to explore the relative explanatory power of independent variables, especially between two groups of environmental factors—i.e., contemporary environments vs. historical climate change—in forming species diversity patterns. This explanatory power from environmental variables amounted to 60.9% (Figure S2), so their selection should help to better explain biodiversity patterns. Generalized linear models (GLM) were used to explore the relationships between the species diversity indices and environmental variables. We built three models: (i) species richness and environmental variables model; (ii) weighted endemism and environmental variable model; and (iii) corrected-weighted endemism model. However, spatial autocorrelation in species diversity patterns might inflate Type I error rates [63]. To eliminate this influence on our significance testing, we used modified the t test applied to all the models' coefficients [64,65]. The geographical coordinate system used for the central point of each grid cell was WGS1984. All statistical analyses were done in R v3.5.1 (http://www.r-project.org/).

3. Results

3.1. Species Distribution Model and Its Accuracy

China was divided into 26,137 grid cells, with 2673 of them (10.22%) having exact geographic coordinates. Models for the Salicaceae family with a cross-validation AUC close to 0.75 were considered useful in our study. The training and testing of the AUC value for the 215 species models was >0.8, which indicated these models were non-random and could be reliably used for modeling the relationship between diversity patterns and environmental factors. Potential suitable habitat areas are currently widely distributed in China's subtropical and warm temperate zones, but occurred mainly in its central and southwestern regions. Highly suitable areas were found in the Hengduan Mountains, Daba Mountains, and Qinling Mountains (Figure 1).

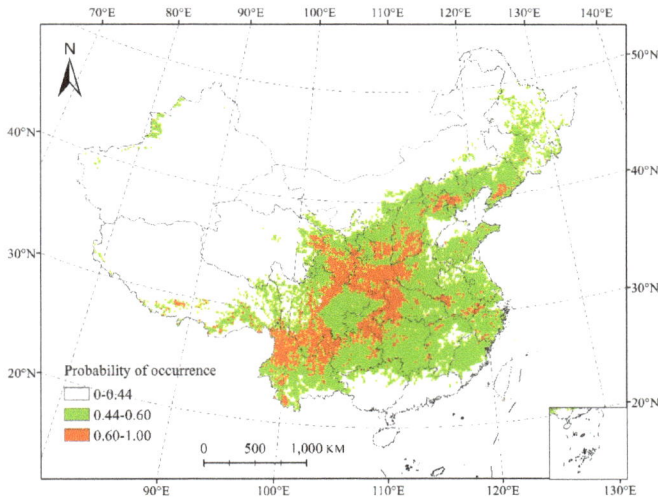

Figure 1. Present potential distribution pattern of Salicaceae plants in China, using the Albers projection. The map was plotted in ArcGIS v10.2.

3.2. Changes in the Potential Range of Salicaceae

Compared with the LIG, the niche model predicted losses in current suitable habitat areas mainly distributed in the middle temperate zone of northern China, namely Inner Mongolia, Heilongjiang, Jilin, Xinjiang, Qinghai, and Gansu (Figure 2A). The projected area of reduction amounted to 139.72×10^4 km^2, which accounts for 38.67% of the currently suitable area (Table 2). This range contraction was mainly concentrated in high latitude areas. Conversely, the gains in suitable range area totaled 62.79×10^4 km^2 (17.38% of the currently suitable area), mostly distributed in provinces of Ningxia, Shaanxi, Gansu, Henan, and Hebei. Expansion occurred mainly in the mid-latitudes. Overall, from the last glacial period to the present, we saw a wide range of contractions, with the total suitable habitats reduced by ca. 76.92×10^4 km^2 (Figure 2A, Table 2). In contrast to the LIG, the suitable habitats increased significantly—the total suitable habitat area increased by ca. 69.63×10^4 km^2—from the LGM to the present (Figure 2B, Table 2). These gains amounted to 106.88×10^4 km^2, or 29.58% of the currently suitable area, and were concentrated in parts of Tibet, Sichuan, Qinghai, Gansu, Ningxia, Inner Mongolia, Shannxi, Hebei, Liaoning, Beijing, Tianjin, Jilin, and Heilongjiang. Expansion in habitats increased with increasing latitude. The predicted losses occurred mainly in the subtropics of southern China, namely Guangxi, Guangdong, and Fujian, covering an area of ca. 37.25×10^4 km^2 (10.31% of the currently suitable area; Figure 2B, Table 2).

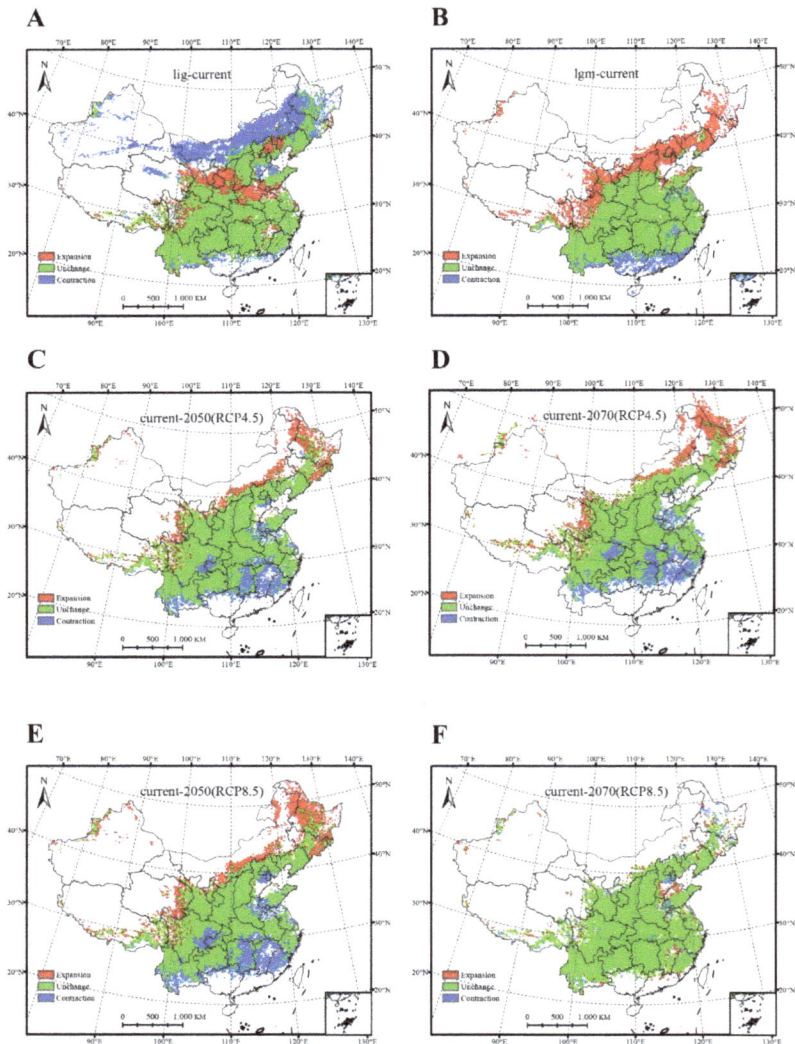

Figure 2. Spatial changes of Salicaceae plants in China under different combinations of climate scenario/years (Albers projection). Comparisons between the species distribution model (SDM) for Salicaceae in the present and (**A**) the SDM under the Last Inter Glacial ("lig"); (**B**) the SDM under the Last Glacial Maximum ("lgm"); (**C**) the SDM under future climate scenario RCP4.5 in 2050; (**D**) the SDM under future climate scenario RCP4.5 in 2070; (**E**) the SDM under future climate scenario RCP8.5 in 2050; (**F**) the SDM under future climate scenario RCP8.5 in 2070. The map was plotted in ArcGIS v10.2.

Table 2. Dynamic changes in the suitable area for Salicaceae plants under different combination of climate scenario/years.

Climate Scenario/Year	Area ($\times 10^4$ km^2)				Proportion of Area (%)			
	Contraction	Expansion	Unchanged	Total	Contraction	Expansion	Unchanged	Total
LIG [a]	139.72	62.79	298.40	−76.92	38.67	17.38	82.58	−21.29
LGM [b]	37.25	106.88	254.34	69.63	10.31	29.58	70.39	19.27
RCP4.5-2050 [c]	65.63	59.09	295.56	−6.54	18.16	16.35	81.80	−1.81
RCP4.5-2070 [d]	73.18	67.82	288.01	−5.36	20.25	18.77	79.71	−1.48
RCP8.5-2050 [e]	82.06	78.06	279.09	−4.00	22.71	21.60	77.24	−1.11
RCP8.5-2070 [f]	20.21	20.18	341.05	−0.04	5.59	5.58	94.39	−0.01

Negative values indicate contractions in suitable habitat area. Comparisons between the suitable area for Salicaceae plants in the present and [a] the suitable area under the Last Inter Glacial ("LIG"); [b] the suitable area under the Last Glacial Maximum ("LGM"); [c] the suitable area under future climate scenario RCP4.5 in 2050; [d] the suitable area under future climate scenario RCP4.5 in 2070; [e] the suitable area under future climate scenario RCP8.5 in 2050; [f] the suitable area under future climate scenario RCP8.5 in 2070.

Under all four future climate scenario/year combinations of RCP4.5-2050, RCP4.5-2070, RCP8.5-2050, and RCP8.5-2070, the predicted area of suitable habitats shrunk (Table 2). The parts of Salicaceae's range with high risks of habitat loss were distributed in subtropical zones, including southern and eastern regions of China, Jiangxi, Fujian, eastern Hunan, southwestern Yunnan, northern Guangxi and Guangdong, southeast Hubei, the border of Chongqing and Sichuan, the junction of Anhui, Henan, Jiangsu, and Shandong, as well in limited regions of Hebei, Tianjin, and Beijing (Figure 2C–F). The Maxent model predicted total losses of 20.21×10^4 km^2–82.06×10^4 km^2 (5.59%~22.71% of the currently suitable area; Table 2). These losses were mostly concentrated at low elevation and the contraction generally increased at lower latitudes. Most of the predicted gains in suitable habitat area were restricted to Salicaceae's range in the north of China's warm temperate zone (Tibet, Sichuan, Qinghai, Inner Mongolia, Heilongjiang, and Jilin), which amounted to ca. 20.18×10^4 km^2–78.06×10^4 km^2 (5.58%~21.60% of the currently suitable area; Table 2), all located in high latitude regions. The total potential suitable habitat area decreased under the four future climatic scenario/years, with contractions in Jiangxi, Fujian, and eastern Hunan most apparent (Figure 2C–F).

3.3. Core Distributional Shifts

The center of predicted suitable areas in the present was located in the Hubei province, with geographical coordinates of 110.62 E and 32.79 N (Figure 3). The centroid of potential suitable habitats in the LIG shifted to northern regions (111.11 E, 35.12 N) and high latitudes. By contrast, the centroid in the LGM period shifted to southern regions and low latitude (110.28 E, 29.52 N). Under the future climate scenario/years combinations, the center of potential suitable habitats was located at more northerly areas with higher latitudes, except under RCP8.5-2070. Evidently, the distribution center of Salicaceae plants is likely to shift to high latitudes from the LGM into the future (Figure 3).

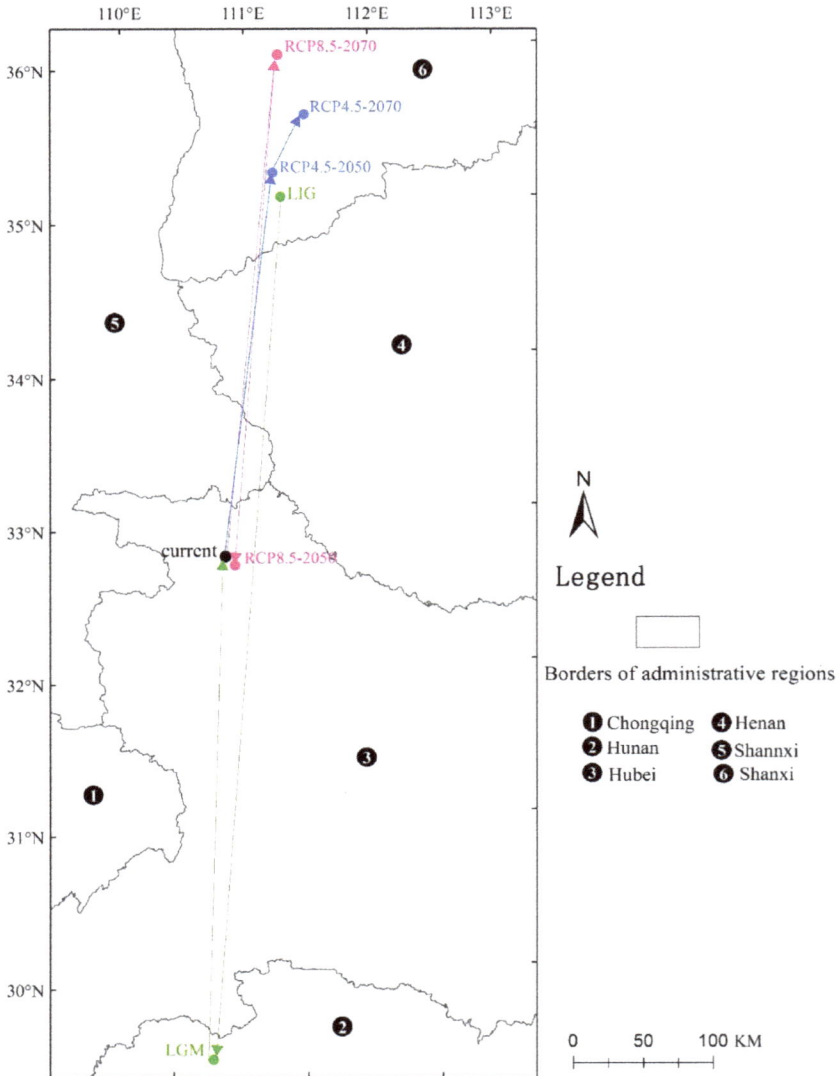

Figure 3. Core distributional shifts under different climate scenario/years for Salicaceae plants in China (Albers projection). LIG, Last Inter Glacial; LGM, Last Glacial Maximum. RCP4.5-2050, future climate scenario RCP4.5 in 2050. RCP4.5-2070, future climate scenario RCP4.5 in 2070. RCP8.5-2050, future climate scenario RCP8.5 in 2050. RCP8.5-2070, future climate scenario RCP8.5 in 2070. The green, black, purple, and blue dots respectively indicate the geometric centers of past (two glacial periods), current and future (four scenarios) suitable areas, with the arrows depicting the magnitude and direction of predicted change through time. The map was plotted using ArcGIS v10.2.

3.4. Relationships between Species Richness and Environmental Parameters

The pattern of Salicaceae species richness was obtained from 215 species potential distribution model layers. Botanical species richness ranged from 0 to 115 in each grid cell in China, with an average of 35 per cell. The southwest and central regions of China showed the highest Salicaceae species diversity, which included the eastern Tibet Plateau and the Hengduan Mountains, Qinling Mountains, Daba Mountains, Dalou Mountains, Wumeng Mountains, Wulian Mountains, Longmen Mountains, Daxue Mountains, Liupan Mountains, and Kunji Mountains. In short, high species diversity was mainly concentrated in alpine areas. The northeast region of China ranked second in species diversity. By contrast, species diversity of northwest China, south China, and the western Tibet Plateau were relatively low (Figure 4).

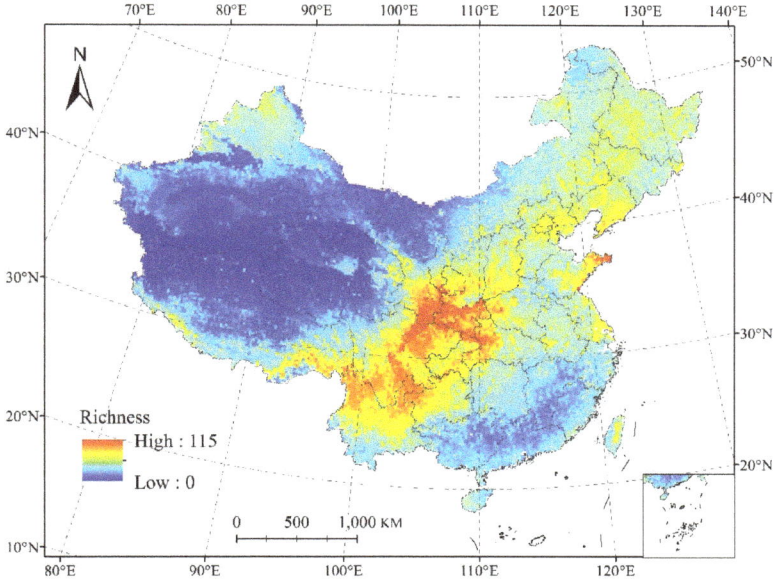

Figure 4. Species richness pattern of Salicaceae plants in China (using 20 × 20 km). The map was plotted using ArcGIS v10.2.

The GLMs showed that the variables related to contemporary energy had stronger explanatory ability than contemporary water availability, soil conditions, heterogeneity, and historical climate change. Specifically, UV-B radiation was the strongest variable, as it explained 35% ($p < 0.05$). Species richness decreased as UV-B factors and annual mean temperature increased, indicating that areas with high species diversity are warm with low UV-B radiation. Elevation and anomaly of precipitation seasonality both adversely affected species diversity patterning, while annual precipitation and nitrogen content of the topsoil were positively associated with it. Those places featuring a climate warmer and wetter than in the LGM tended to have more Salicaceae species than other regions (Table 3).

Table 3. Explanatory power (R^2) of the environmental variables for Salicaceae species diversity patterns evaluated by the generalized linear models (GLMs). Modified t-tests were used to examine the significance.

	SR	WE	CWE
Contemporary energy			
Bio1 (°C)	5.26%(+) ***	3.73%(+) **	1.00%(+)
Bio7 (°C)	5.85%(−)	9.94%(−)	6.53%(−)
Bio3	1.20%(+)	0.92%(+)	6.40%(+)
UVB1 (J m^{-2}·day^{-1})	4.80%(−) **	0.08%(−) ***	1.90%(−) ***
UVB2 (J m^{-2}·day^{-1})	18.20(−) ***	8.75%(−)	1.20%(−)
UVB3 (J m^{-2}·day^{-1})	11.85%(−) ***	2.72%(−) ***	0.11%(−) **
UVB4 (J m^{-2}·day^{-1})	0.03%(−)	2.33%(−) ***	6.8%(+) ***
Contemporary water availability			
Bio12 (mm)	10.22%(+) **	10.07%(+) **	5.90%(+) ***
Bio15	2.04%(−) **	2.21%(−)	3.81%(−)
Contemporary soil conditions			
S-CE	9.71%(+)	3.74%(+)	0.01%(+)
T-N	4.25%(+) ***	8.01%(+) ***	11.06%(+) ***
Drain	0.12%(+)	0.21%(−)	0.02%(−)
T-BS	0.10%(+)	0.18%(+)	1.95%(+)
Heterogeneity			
Elevation	6.59%(−) **	1.31%(−)	0.04%(−) **
Historical climate change			
Bio1-Ano	0.18%(−) ***	0.45%(−)	0.05%(−)
Bio3-Ano	2.92%(+) **	3.18%(+) ***	4.19%(+) **
Bio7-Ano	4.58%(−) **	1.22%(−)	0.09%(−)
Bio12-Ano	13.17%(+)	6.30%(+) ***	1.33%(+) ***
Bio15-Ano	19.97%(−) **	12.63%(−) ***	3.80%(−) ***

** $p < 0.05$; *** $p < 0.01$; +, positive effects; −, negative effects. Refer to Table 1 for the abbreviations for environmental variables. SR: species richness, WE: weighted endemism, CWE: Corrected-weighted endemism, Bio1-Ano: Bio1 anomaly, Bio3-Ano: Bio3 anomaly, Bio7-Ano: Bio7 anomaly, Bio12-Ano: Bio12 anomaly, Bio15-Ano: Bio15 anomaly.

3.5. Relationships between Weighted Endemism (WE), Corrected-Weighted Endemism (CWE), and Environmental Factors

Plant-weighted endemism ranged from 0 to 0.0295 in each grid cell in China, with an average value of 0.0068, and 1.61% of the total grids had more values >0.0200 (Figure 5A). Weighted-endemism followed a similar pattern as did that found for species richness. The Himalayas, and the mountains of Hengduan, Qinling, Daba, Longmen, Wumeng, Wulian, Yunling, and Kunji all featured high values of weighted endemism. The corrected-weighted endemism ranged from 0 to 0.0004 (Figure 5B), with the high values concentrated in northwest China, including the Himalayas and Hengduan Mountains. The latter is where endemic species of Salicaceae had their ranges most narrowly distributed.

The GLMs showed that both historical climate change and contemporary water availability were the main factors shaping the patterns of weighted endemism (Table 3). The anomalies of precipitation seasonality and annual precipitation explained 22% of its variation. While nitrogen content of topsoil and UV-B radiation also had significant positive effects on the corrected-weighted endemism pattern. Elevation was a significant factor in the three types of pattern, but it showed weaker explanatory power for the weighted endemism and corrected-weighted endemism patterns. The relationships between nitrogen content of topsoil and patterns of weighted endemism and corrected-weighted endemism were significant, showing strong positive correlations (Table 3).

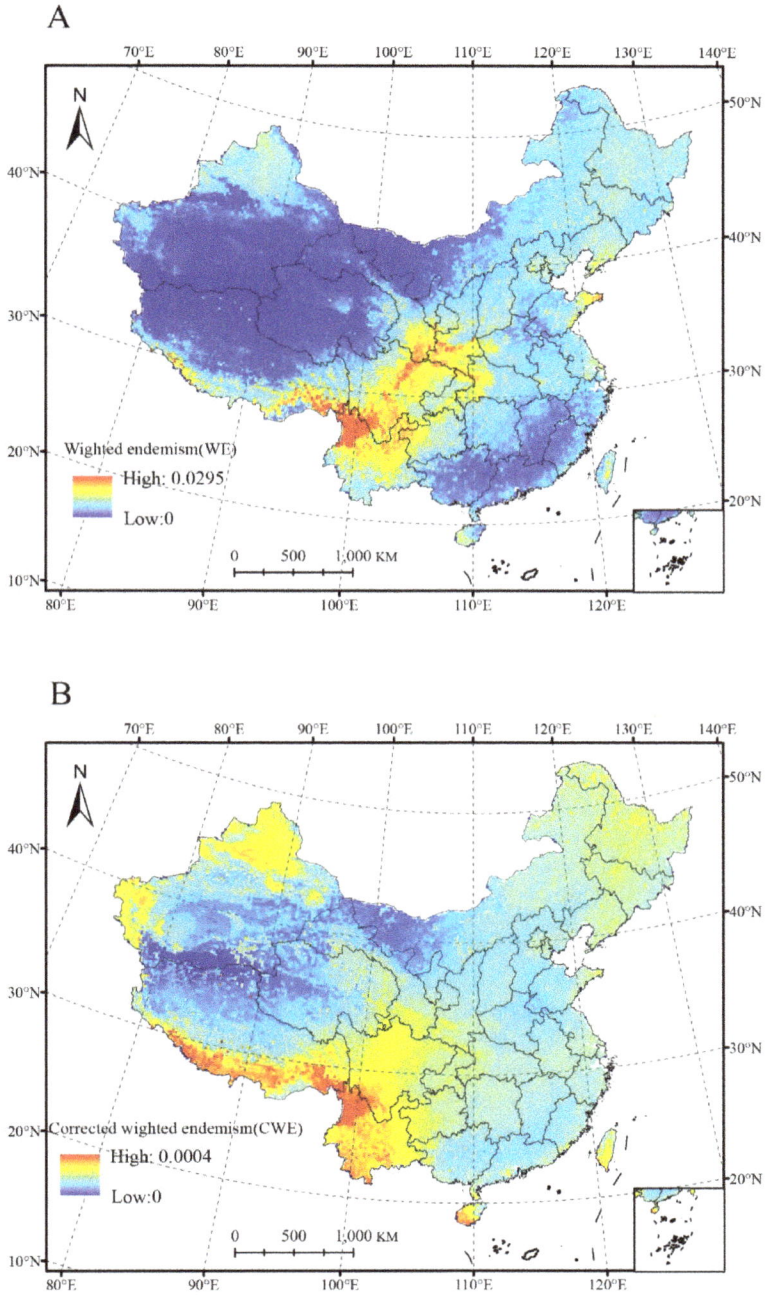

Figure 5. Patterns of weighted endemism (**A**) and corrected-weighted endemism (**B**) for Salicaceae plants in China (Albers projection). The map was plotted using ArcGIS v10.2.

4. Discussion

4.1. Changes in the Potential Range of Salicaceae in China

Using the georeferenced occurrence data of Salicaceae, our results showed that its predicted potential distribution area covers most regions of China. The predicted results are consistent with the actual distribution of Salicaceae plants in China, which features strong regularity and forms part of the temperate deciduous forest community. The similarity of all the plant species—they are deciduous, water demanding, and require a temperate bioclimate, to name a few key traits—and the model's effective prediction accuracy (>0.7) support the generalization of our results. Second, by simulating Salicaceae species' ranges and patterns in different periods we could infer their historical and future population dynamics (Figure S1). Salicaceae may have experienced a significant and extensive expansion from south to north during the Last Inter Glacial period (LIG). This strongly suggests Salicaceae can track suitable niches during the glacial and interglacial cycles, in that its distribution shifted southward in the glacial period but shifted back northward after the glacial period. Salicaceae is a group of plants whose ecological amplitude is large and they can adapt to many different ecological environments, from those ranging from high (~20 °C) to low (~3 °C) temperatures, and from swamp to sand habitats [66,67]. The temperature (average: ~5 °C) and rainfall (average: ~557 mm) in the LIG were within the tolerance levels of Salicaceae species. However, with significant reductions in temperature and rainfall, the climate became drier and colder in the Last Glacial Maximum (LGM) [68,69]. In response, because those areas with suitable habitat were generally greatly reduced, Salicaceae's distribution range retreated to areas low in elevation [70]. Therefore, from these results we infer this plant family was able to spread rapidly in the glacial-interglacial cycle [11]. Most of its species (i.e., members of *Salix* subgenera *Chamaetia* and *Vetrix*, represent ca. 75% species of Salicaceae) are Arctic-alpine taxa well adapted to cold and hostile environments [71]. Therefore, the considerable northward expansion in China of the Salicaceae distribution area is both plausible and reasonable. In sum, Quaternary climate change profoundly influenced the historical spatial population dynamics of ancient Chinese Salicaceae species [72].

As the climate warms, extreme weather events are becoming more frequent and intense [73–75]. Climate research has made much progress in recent years, but most of it is based on qualitative rather than quantitative analyses [76–79]. In our study, we conducted a detailed quantitative analysis of the potential distribution of Salicaceae in the future. Nevertheless, simulation results are expected to vary between different climate models [80,81]. Salicaceae are widely distributed, occurring in Africa, Europe, Asia, South America, and North America from ca. 82° N to 52° S [82], but their habitats are often fragmented. East Asia is the present distribution center of modern species of Salicaceae, and China harbors the most species numerically.

Our modeling study predicted ca. 361.33×10^4 km^2 for Salicaceae species under the present climate scenario with its core distributional area being Hubei Province, which is broadly within the distribution area of extant Salicaceae. In our study's projections, the centroid of suitable distribution shifted northerly to higher latitude areas under the future climate scenarios, and this trend agrees with other works [83–86]. Our results also show that if global warming becomes more intense, the suitable range of Salicaceae faces the risk of reduction under the medium and high carbon dioxide emission scenarios (RCP4.5 and RCP8.5), a result that is also in line with previous studies [79,87]. Species of Salicaceae are, therefore, clearly vulnerable to climate change effects, with its range contractions mainly concentrated at low latitudes [88–90]. Nevertheless, since the model involved all species of the family these results are not applicable to endangered species or any other particular species. Earlier studies by Bomhard et al. [91] and Midgley et al. [92] suggested that future climate change and land-use transformations would have synergistic effects on species habitats, giving rise to unsuitable habitats for their persistence. In this context, we found that Salicaceae species do display strong adaptability under the present and future climate scenarios; however, some of the suitable range may become unsuitable due to intense disturbance by human activities and land-use changes. For example, much of the suitable range for Salicaceae plants in China has been converted into farmland or construction land [93]. Clearly, the effects from future land-use changes should also be taken into account in species distribution models [94]. We suggest that local protection measures (Figure S3) can be adopted to reduce anthropocentric-driven damage to suitable habitat. Meanwhile, we can also try to implement *ex situ* protection projects, such as transplanting species from a threatened area to a site with non- or less-degraded habitat, or by planting them in multiple botanical gardens.

4.2. Species Richness and Endemism Patterns of Salicaceae in China

Using the distributions of 215 Salicaceae species, we explored their large-scale variation in species richness and endemism and the environment variables likely determining these patterns. China is famous for its high species diversity; better documenting and understanding of its diversity and endemic patterns will inform and strengthen biological conservation, by providing baselines for biodiversity management and relevant policy-making. China's southwest, lying south of the Hengduan Mountains, is well known for its high suitability for and diversity of Salicaceae species (Figures 2, 4 and 5). Our variance partition analyses revealed pure effects of contemporary environments and historical climate change that respectively accounted for 47.90% and 18.10% of the variance in Salicaceae diversity patterns across China (Figure S2). Consistent with other research [14,95], our study showed the climate exerts a important influence on Salicaceae species richness, weighted endemism, and corrected-weighted endemism (Table 3 and Figure 2 and Figure S2). For all three, the most important factor was the contemporary environment (energy, water availability and soil conditions), since it had the strongest explanatory power (Table 3). This indicates high species diversity of Salicaceae mostly occurs in areas with relatively warm and high humidity [96]. High temperatures at low latitudes may curtail breeding times and accelerate species formation, favoring the emergence of new plant species [97]. This is consistent with evidence that imposed warming considerably increased the root, stem, and leaf biomass of Salicaceae plants [98] with drought strongly impeding their growth [99,100]. Meanwhile, considering that Salicaceae species originated in the temperate zone [101] and most of them prefer wet environments [102], the positive association between precipitation, temperature and species diversity lends supports to the water–energy dynamic equilibrium hypothesis as a species richness formation mechanism. It has been shown that water–energy dynamics (precipitation and coldest monthly evapotranspiration) were the main factors affecting the diversity of South African trees [103–105]. Other studies have found biodiversity patterns that were also related to water availability [106,107]. Temperature and liquid water are not only vital for plants to absorb and transport nutrients but also mediate key physiological activities (e.g., photosynthesis) [108,109].

Besides water and temperature, our results suggested UV-B radiation also contributes to the biodiversity pattern of Salicaceae plants. The negative associations indicate that where UV-B radiation is intense, both species richness and endemism are lower. UV-B radiation is a factor known to limit the distribution range of terrestrial and marine life: its negative relationship with biodiversity patterns arises from its adverse effects on the protective mechanisms and organs of plants [110]. Related research has also found that high-intensity UV-B radiation can reduce the photosynthetic performance of Salicaceae species [43,111,112], in addition to diminishing the biomass accumulation [16], seed germination [113] and membrane structure integrity [44] of plants. Therefore, we propose that, generally, intensive UV-B radiation reduces Salicaceae plant biodiversity. Compared with other environmental variables, the association between soil fertility and biodiversity pattern was relatively weak, which is in line with other research [114]. Nevertheless, it worth emphasizing that topsoil nitrogen content makes a non-trivial contribution to the diversity patterns of Salicaceae species. This may be linked to how N can markedly influence gender differences and competitiveness between the two sexes in these plants [47]. Some studies do show that temperature stability [115] and historical climatic stability [116] also affect species diversity patterns. We found historical geological processes also played an essential role by altering the range (Figure 2A,B), diversity and endemism (Table 3, Figure S2) of Salicaceae species during the glacial period. Thus, while contemporary water-energy and historical climate change both have significant impacts on Salicaceae, other factors (such as disturbance history) warrant quantitative study in future research [117]. It is well known that the ozone layer protects the forests and other life forms on the Earth's surface from the sun's harmful ultraviolet radiation [118]. During the last few decades, the stratospheric ozone layer has been depleted by anthropogenic pollutants such as chlorofluorocarbons (CFCS), leading to an increase in surface-level ultraviolet radiation. As a result of the 1987 Montreal protocol and its amendments, the anthropogenic loading of ozone-depleting substances in the atmosphere is being reduced. Nevertheless, future climate change could cause extreme dynamic events [119], volcanic eruptions [120] and irregular changes in solar flux [121], all of which are likely to damage the ozone layer. Caldwell's results showed that for every 1% reduction in ozone, the surface's UV-B (280–320 mm) radiation increases by 2% [122]. With greater UV-B radiation possible in the future, the geographical distribution and species diversity of Salicaceae plants may become threatened.

5. Conclusions

Based on niche model, we propose the Salicaceae family underwent an extensive population expansion of its species during the LIG, but retreated to low latitudes during the LGM. As climate warming intensifies, suitable habitats will shift to higher latitudes while those at lower latitudes will contract in area and abundance. Habitats of Salicaceae in China are most likely to be lost in the future, especially in Jiangxi, Fujian, and eastern Hunan. The western and central regions of China currently show the highest diversity of Salicaceae species. Because they are typical pioneer plants, with a strong ability to disperse and colonize, they could fill new or open habitats easily; these traits make it difficult to resort to a single hypothesis to reveal their pattern of species diversity. Nevertheless, our results point to the water–energy dynamic equilibrium and historical geological processes as being key drivers of Salicaceae diversity patterns in China. Our study could thus assist in the conservation and sustainable use of Salicaceae, and our results provide a solid baseline for biodiversity management and relevant policy-making for these important woody plants.

Supplementary Materials: The following are available online at http://www.mdpi.com/1999-4907/10/3/220/s1, Table S1: The filtered occurrence data of Salicaceae; Table S2: The 215 species catalogue used to construct the species diversity pattern of Salicaceae; Table S3: Spearman's rank correlation coefficients (r_s) for the 5 bioclimatic variables and elevation. Bio1: Annual mean temperature, Bio3: Isothermality (Bio2/Bio7) (*100), Bio7: Temperature annual range (Bio5-Bio6), Bio12: Annual precipitation, Bio15: Precipitation seasonality (coefficient of variation); Table: S4 Spearman's rank correlation coefficients (r_s) for the 5 soil variables. S-CE: Cation exchange capacity (CEC) clay subsoil, T-BS: Base saturation% topsoil, T-C: Organic carbon pool topsoil, T-N: Nitrogen % topsoil, Drain: Soil drainage class; Figure S1: The potential distribution pattern of Salicaceae plants in China under different combination of climate scenario/years, using the Albers projection. (A) the potential distribution pattern under the Last Inter Glacial; (B) the potential distribution pattern under the Last Glacial Maximum; (C) the potential distribution pattern under future climate scenario RCP4.5 in 2050; (D) the potential distribution pattern under future climate scenario RCP4.5 in 2070; (E) the potential distribution pattern under future climate scenario RCP8.5 in 2050; (F) the potential distribution pattern under future climate scenario RCP8.5 in 2070; Figure S2: The pure and combined effect of contemporary environments and historical climate changes in shaping *Salix* species diversity; Figure S3: The spatial pattern of nature reserves and the shrinking distribution range under the future climate scenarios.

Author Contributions: Conceptualization, W.L., M.S., Y.H., H.S. and J.C.; Data curation, W.L., M.S. and K.C.; Formal analysis, W.L., Y.H., H.S. and J.C.; Funding acquisition, H.S. and J.C.; Investigation, W.L., Y.H. and K.C.; Methodology, W.L., M.S. and K.C.; Project administration, W.L., H.S. and J.C.; Resources, W.L., M.S., Y.H. and J.C.; Software, W.L., M.S. and Y.H.; Supervision, Y.H., H.S. and J.C.; Validation, W.L., M.S., K.C. and H.S.; Visualization, W.L. and K.C.; Writing—original draft, W.L. and J.C.; Writing—review and editing, W.L., H.S. and J.C.

Funding: This study was supported by grants from the National Natural Science Foundation of China (NSFC 31590823 to HS, 31670198 to JC, 31560062 to YH); the Science and Technology Research Program of Kunming Institute of Botany, the Chinese Academy of Sciences, Grant NO. KIB2016005; and the Youth Innovation Promotion Association, the Chinese Academy of Sciences.

Conflicts of Interest: The authors declare that they have no competing interests.

References

1. Hong, Q.; Ricklefs, R.E. Large–scale processes and the asian bias in species diversity of temperate plants. *Nature* **2000**, *407*, 180–182.
2. Buckley, L.B.; Davies, T.J.; Ackerly, D.D.; Kraft, N.J.; Harrison, S.P.; Anacker, B.L.; Cornell, H.V.; Damschen, E.I.; Grytnes, J.A.; Hawkins, B.A.; et al. Phylogeny, niche conservatism and the latitudinal diversity gradient in mammals. *Proc. R. Soc.* **2010**, *277*, 2131–2138. [CrossRef] [PubMed]
3. Brown, J.H. Why are there so many species in the tropics? *J. Biogeogr.* **2014**, *41*, 8–22. [CrossRef] [PubMed]
4. Beck, J.; Ballesteros-Mejia, L.; Buchmann, C.M.; Dengler, J.; Fritz, S.A.; Gruber, B.; Hof, C.; Jansen, F.; Knapp, S.; Kreft, H.; et al. What's on the horizon for macroecology? *Ecography* **2012**, *35*, 673–683. [CrossRef]
5. Hawkins, B.A. Invited views in basic and applied ecology: Are we making progress toward understanding the global diversity gradient? *Basic Appl. Ecol.* **2004**, *5*, 1–3. [CrossRef]
6. Brown, J.H.; Gillooly, J.F.; Allen, A.P.; Savage, V.M.; West, G.B. Toward a metabolic theory of ecology. *Ecology* **2004**, *85*, 1771–1789. [CrossRef]
7. Colwell, R.K.; Rahbek, C.; Gotelli, N.J. The mid-domain effect and species richness patterns: What have we learned so far? *Am. Nat.* **2004**, *163*, E1–E23. [CrossRef] [PubMed]
8. Joy, J.B. The global diversity of birds in space and time. *Nature* **2012**, *491*, 444–448.
9. Jansson, R. Global variation in diversification rates of flowering plants: Energy vs. Climate change. *Ecol. Lett.* **2008**, *11*, 173–183. [CrossRef] [PubMed]
10. Mittelbach, G.G.; Schemske, D.W.; Cornell, H.V.; Allen, A.P.; Brown, J.M.; Bush, M.B.; Harrison, S.P.; Hurlbert, A.H.; Knowlton, N.; Lessios, H.A.; et al. Evolution and the latitudinal diversity gradient: Speciation, extinction and biogeography. *Ecol. Lett.* **2007**, *10*, 315–331. [CrossRef] [PubMed]
11. Levsen, N.; Tiffin, P.; Olson, M. Pleistocene speciation in the genus populus (salicaceae). *Syst. Biol.* **2012**, *61*, 401–412. [CrossRef] [PubMed]
12. Lee, C.B.; Chun, J.H. Retracted article: Habitat heterogeneity and climate explain plant diversity patterns along an extensive environmental gradient in the temperate forests of south korea. *Folia Geobot.* **2016**, *1*. [CrossRef]
13. Veloz, S.D.; Williams, J.W.; Blois, J.L.; He, F.; Otto-Bliesner, B.; Liu, Z. No–analog climates and shifting realized niches during the late quaternary: Implications for 21st–century predictions by species distribution models. *Glob. Chang. Biol.* **2012**, *18*, 1698–1713. [CrossRef]
14. Currie, D.J. Energy and large-scale patterns of animal and plant–species richness. *Am. Nat.* **1991**, *137*, 27–49. [CrossRef]
15. Latham, R.E.; Ricklefs, R.E. Global patterns of tree species richness in moist forests: Energy-diversity theory does not account for variation in species richness. *Oikos* **1993**, *67*, 325–333. [CrossRef]
16. Xu, X.; Zhao, H.; Zhang, X.; Hänninen, H.; Korpelainen, H.; Li, C. Different growth sensitivity to enhanced uv-b radiation between male and female populus cathayana. *Tree Physiol.* **2010**, *30*, 1489–1498. [CrossRef] [PubMed]
17. Clarke, A.; Gaston, K.J. Climate, energy and diversity. *Proc. R. Soc.* **2006**, *273*, 2257–2266. [CrossRef] [PubMed]
18. Qin, H.; Dong, G.; Zhang, Y.; Zhang, F.; Wang, M. Patterns of species and phylogenetic diversity of pinus tabuliformis forests in the eastern loess plateau, china. *For. Ecol. Manag.* **2017**, *394*, 42–51. [CrossRef]
19. Thuiller, W.; Lavergne, S.; Roquet, C.; Boulangeat, I.; Lafourcade, B.; Araujo, M.B. Consequences of climate change on the tree of life in europe. *Nature* **2011**, *470*, 531–534. [CrossRef] [PubMed]
20. Pio, D.V.; Engler, R.; Linder, H.P.; Monadjem, A.; Cotterill, F.P.D.; Taylor, P.J.; Schoeman, M.C.; Price, B.W.; Villet, M.H.; Eick, G.; et al. Climate change effects on animal and plant phylogenetic diversity in southern africa. *Glob. Chang. Biol.* **2014**, *20*, 1538–1549. [CrossRef]
21. Hultine, K.R.; Burtch, K.G.; Ehleringer, J.R. Gender specific patterns of carbon uptake and water use in a dominant riparian tree species exposed to a warming climate. *Glob. Chang. Biol.* **2013**, *19*, 3390–3405. [CrossRef] [PubMed]
22. González-Orozco, C.E.; Pollock, L.J.; Thornhill, A.H.; Mishler, B.D.; Knerr, N.; Laffan, S.W.; Miller, J.T.; Rosauer, D.F.; Faith, D.P.; Nipperess, D.A.; et al. Phylogenetic approaches reveal biodiversity threats under climate change. *Nat. Clim. Chang.* **2016**, *6*, 1110–1114. [CrossRef]

23. Bellard, C.; Bertelsmeier, C.; Leadley, P.; Thuiller, W.; Courchamp, F. Impacts of climate change on the future of biodiversity. *Ecol. Lett.* **2012**, *15*, 365–377. [CrossRef] [PubMed]

24. Stocker, T.F.; Qin, D.; Plattner, G.K.; Tignor, M.; Allen, S.K.; Boschung, J.; Nauels, A.; Xia, Y.; Bex, B.; Midgley, B.M. IPCC, 2013: Climate change 2013: The physical science basis. Contribution of working group i to the fifth assessment report of the intergovernmental panel on climate change. *Comput. Geom.* **2013**, *18*, 95–123.

25. Parmesan, C. Ecological and evolutionary responses to recent climate change. *Annu. Rev. Ecol. Evol. Syst.* **2006**, *37*, 637–669. [CrossRef]

26. Myers-Smith, I.H.; Forbes, B.C.; Wilmking, M.; Hallinger, M.; Lantz, T.; Blok, D.; Tape, K.D.; Macias-Fauria, M.; Sass-Klaassen, U.; Lévesque, E.; et al. Shrub expansion in tundra ecosystems: Dynamics, impacts and research priorities. *Environ. Res. Lett.* **2011**, *6*, 045509. [CrossRef]

27. Jones, M.H. Sex- and habitat-specific responses of a high arctic willow, salix arctica, to experimental climate change. *Oikos* **1999**, *87*, 129–138. [CrossRef]

28. Fang, Z.F.; Zhao, S.D.; Skvortsov, A.K. Salicaceae. *Flora China* **1999**, *4*, 139–274.

29. Zhao, S.D. Distribution of *willows (salix)* in china. *Acta Phytotaxon. Sin.* **1987**, *25*, 114–124.

30. Karp, A.; Shield, I. Bioenergy from plants and the sustainable yield challenge. *New Phytol.* **2008**, *179*, 15–32. [CrossRef] [PubMed]

31. Chen, J.H.; Sun, H.; Wen, J.; Yang, Y.P. Molecular phylogeny of *salix l. (salicaceae)* inferred from three chloroplast datasets and its systematic implications. *Taxon* **2010**, *59*, 29–37. [CrossRef]

32. Wang, Q.; Su, X.; Shrestha, N.; Liu, Y.; Wang, S.; Xu, X.; Wang, Z. Historical factors shaped species diversity and composition of salix in eastern asia. *Sci. Rep.* **2017**, *7*, 42038. [CrossRef] [PubMed]

33. Warren, D.L. In defense of 'niche modeling'. *Trends Ecol. Evol.* **2012**, *27*, 497–500. [CrossRef] [PubMed]

34. Peterson, A.T.; Soberón, J.; Pearson, R.G.; Anderson, R.P.; Martínez-Meyer, E.; Nakamura, M.; Araújo, M.B. Ecological niches and geographic distribution. *Monogr. Popul. Biol.* **2011**, *49*, 328.

35. Chen, F.T. *Phylogeography of Rehmannia (Scrophulariaceae)*; Northwest University: Xi'an, China, 2015.

36. Wang, Q.; Wei, Y.K.; Huang, Y.B. Research on distribution pattern of subg. Salvia benth. (lamiaceae), an important group of medicinal plants in east asia. *Acta Ecol. Sin.* **2015**, *5*, 470–479.

37. Beck, J.; Böller, M.; Erhardt, A.; Schwanghart, W. Spatial bias in the gbif database and its effect on modeling species' geographic distributions. *Ecol. Inform.* **2014**, *19*, 10–15. [CrossRef]

38. Fourcade, Y.; Engler, J.O.; Rödder, D.; Secondi, J. Mapping species distributions with maxent using a geographically biased sample of presence data: A performance assessment of methods for correcting sampling bias. *PLoS ONE* **2014**, *9*, e97122. [CrossRef] [PubMed]

39. Zhang, M.G.; Slik, J.W.; Ma, K.P. Using species distribution modeling to delineate the botanical richness patterns and phytogeographical regions of china. *Sci. Rep.* **2016**, *6*, 22400. [CrossRef] [PubMed]

40. Pearson, R.G.; Raxworthy, C.J.; Nakamura, M.; Townsend Peterson, A. Original article: Predicting species distributions from small numbers of occurrence records: A test case using cryptic geckos in madagascar. *J. Biogeogr.* **2007**, *34*, 102–117. [CrossRef]

41. Stevens, G.C. The latitudinal gradient in geographical range: How so many species coexist in the tropics. *Am. Nat.* **1989**, *133*, 240–256. [CrossRef]

42. Yan, Y.; Yang, X.; Tang, Z. Patterns of species diversity and phylogenetic structure of vascular plants on the qinghai-tibetan plateau. *Ecol. Evol.* **2013**, *3*, 4584–4595. [CrossRef] [PubMed]

43. Nybakken, L.; Hörkkä, R.; Julkunen-Tiitto, R. Combined enhancements of temperature and uvb influence growth and phenolics in clones of the sexually dimorphic salix myrsinifolia. *Physiol. Plant.* **2012**, *145*, 551–564. [CrossRef] [PubMed]

44. Randriamanana, T.R.; Nissinen, K.; Moilanen, J.; Nybakken, L.; Julkunen-Tiitto, R. Long-term uv-b and temperature enhancements suggest that females of salix myrsinifolia plants are more tolerant to uv-b than males. *Environ. Exp. Bot.* **2015**, *109*, 296–305. [CrossRef]

45. Feng, L.; Hao, J.; Zhang, Y.; Sheng, Z. Sexual differences in defensive and protective mechanisms of populus cathayana exposed to high uv-b radiation and low soil nutrient status. *Physiol. Plant.* **2014**, *151*, 434–445. [CrossRef]

46. Hageer, Y.; Esperónrodríguez, M.; Baumgartner, J.B.; Beaumont, L.J. Climate, soil or both? Which variables are better predictors of the distributions of australian shrub species? *PeerJ* **2017**, *5*, e3446. [CrossRef] [PubMed]

47. Chen, J.; Dong, T.; Duan, B.; Korpelainen, H.; Niinemets, Ü.; Li, C. Sexual competition and n supply interactively affect the dimorphism and competiveness of opposite sexes in populus cathayana. *Plant Cell Environ.* **2015**, *38*, 1285–1298. [CrossRef] [PubMed]

48. Moor, H.; Hylander, K.; Norberg, J. Predicting climate change effects on wetland ecosystem services using species distribution modeling and plant functional traits. *Ambio* **2015**, *44* (Suppl. 1), S113–S126. [CrossRef]

49. Fitzpatrick, M.C.; Gotelli, N.J.; Ellison, A.M. Maxent versus maxlike: Empirical comparisons with ant species distributions. *Ecosphere* **2013**, *4*, art55. [CrossRef]

50. Guisan, A.; Thuiller, W. Predicting species distribution: Offering more than simple habitat models. *Ecol. Lett.* **2005**, *8*, 993–1009. [CrossRef]

51. Phillips, S.J.; Anderson, R.P.; Schapire, R.E. Maximum entropy modeling of species geographic distributions. *Ecol. Model.* **2006**, *190*, 231–259. [CrossRef]

52. Bertrand, R.; Perez, V.; Gégout, J.C. Disregarding the edaphic dimension in species distribution models leads to the omission of crucial spatial information under climate change: The case of quercus pubescensin france. *Glob. Chang. Biol.* **2012**, *18*, 2648–2660. [CrossRef]

53. Araújo, M.B.; Peterson, A.T. Uses and misuses of bioclimatic envelope modeling. *Ecology* **2012**, *93*, 1527–1539. [CrossRef] [PubMed]

54. Raes, N.; Steege, H.T. A null-model for significance testing of presence-only species distribution models. *Ecography* **2007**, *30*, 727–736. [CrossRef]

55. Radosavljevic, A.; Anderson, R.P.; Araujo, M. Making better maxent models of species distributions: Complexity, overfitting and evaluation. *J. Biogeogr.* **2014**, *41*, 629–643. [CrossRef]

56. Elith, J.; Phillips, S.J.; Hastie, T.; Dudík, M.; Chee, Y.E.; Yates, C.J. A statistical explanation of maxent for ecologists. *Divers. Distrib.* **2010**, *17*, 43–57. [CrossRef]

57. Merow, C.; Smith, M.; Silander, J. A practical guide to maxent for modeling species' distributions: What it does, and why inputs and settings matter. *Ecography* **2013**, *36*, 1058–1069. [CrossRef]

58. Jiménez-Valverde, A.; Lobo, J.M. Threshold criteria for conversion of probability of species presence to either–or presence–absence. *Acta Oecol.* **2007**, *31*, 361–369. [CrossRef]

59. Liu, C.; Berry, P.M.; Dawson, T.P.; Pearson, R.G. Selecting thresholds of occurrence in the prediction of species distributions. *Ecography* **2005**, *28*, 385–393. [CrossRef]

60. Crisp, M.D.; Laffan, S.; Linder, H.P.; Monro, A. Endemism in the australian flora. *J. Biogeogr.* **2001**, *28*, 183–198. [CrossRef]

61. Brown, J.L.; Anderson, B. Sdmtoolbox: A python–based gis toolkit for landscape genetic, biogeographic and species distribution model analyses. *Methods Ecol. Evol.* **2014**, *5*, 694–700. [CrossRef]

62. Brown, J.L.; Bennett, J.R.; French, C.M. Sdmtoolbox 2.0: The next generation python-based gis toolkit for landscape genetic, biogeographic and species distribution model analyses. *PeerJ* **2017**, *5*, e4095. [CrossRef] [PubMed]

63. Diniz-Filho, J.A.F.; Bini, L.M.; Hawkins, B.A. Spatial autocorrelation and red herrings in geographical ecology. *Glob. Ecol. Biogeogr.* **2003**, *12*, 53–64. [CrossRef]

64. Dutilleul, P.; Clifford, P.; Richardson, S.; Hemon, D. Modifying the t test for assessing the correlation between two spatial processes. *Biometrics* **1993**, *49*, 305–314. [CrossRef]

65. Clifford, P.; Richardson, S.; Hémon, D. Assessing the significance of the correlation between two spatial processes. *Biometrics* **1989**, *45*, 123–134. [CrossRef] [PubMed]

66. Karrenberg, S.; Edwards, P.J.; Kollmann, J. The life history of *salicaceae* living in the active zone of floodplains. *Freshw. Biol.* **2002**, *47*, 733–748. [CrossRef]

67. Chao, N.; Liu, J. On the classification and distribution of the family *salicaceae*. *J. Sichuan For. Sci. Technol.* **1998**, *9*, 10–20.

68. Jiang, D.; Wang, H.; Drange, H.; Lang, X. Last glacial maximum over china: Sensitivities of climate to paleovegetation and tibetan ice sheet. *J. Geophys. Res.* **2003**, *108*, 4102. [CrossRef]

69. Fan, L.; Zheng, H.; Milne, R.I.; Zhang, L.; Mao, K. Strong population bottleneck and repeated demographic expansions of populus adenopoda (salicaceae) in subtropical china. *Ann. Bot.* **2018**, *121*, 665–679. [CrossRef] [PubMed]

70. Barnosky, A.D.; Matzke, N.; Tomiya, S.; Wogan, G.O.; Swartz, B.; Quental, T.B.; Marshall, C.; Mcguire, J.L.; Lindsey, E.L.; Maguire, K.C. Has the earth's sixth mass extinction already arrived? *Nature* **2011**, *471*, 51–57. [CrossRef] [PubMed]

71. Wu, J.; Nyman, T.; Wang, D.C.; Argus, G.W.; Yang, Y.P.; Chen, J.H. Phylogeny of salix subgenus salix s.L. (salicaceae): Delimitation, biogeography, and reticulate evolution. *BMC Evol. Biol.* **2015**, *15*, 31. [CrossRef] [PubMed]

72. Qiu, Y.X.; Fu, C.X.; Comes, H.P. Plant molecular phylogeography in china and adjacent regions: Tracing the genetic imprints of quaternary climate and environmental change in the world's most diverse temperate flora. *Mol. Phylogenet. Evol.* **2011**, *59*, 225–244. [CrossRef] [PubMed]

73. Patricola, C.M.; Chang, P.; Saravanan, R. Impact of atlantic sst and high frequency atmospheric variability on the 1993 and 2008 midwest floods: Regional climate model simulations of extreme climate events. *Clim. Chang.* **2013**, *129*, 397–411. [CrossRef]

74. Kodra, E.; Steinhaeuser, K.; Ganguly, A.R. Persisting cold extremes under 21st-century warming scenarios. *Geophys. Res. Lett.* **2011**, *38*, 16. [CrossRef]

75. Planton, S.; Déqué, M.; Chauvin, F.; Terray, L. Expected impacts of climate change on extreme climate events. *C. R. Geosci.* **2008**, *340*, 564–574. [CrossRef]

76. Khanum, R.; Mumtaz, A.S.; Kumar, S. Predicting impacts of climate change on medicinal asclepiads of pakistan using maxent modeling. *Acta Oecol.* **2013**, *49*, 23–31. [CrossRef]

77. Leng, W.; He, H.S.; Liu, H. Response of larch species to climate changes. *Plant Ecol.* **2008**, *1*, 203–205. [CrossRef]

78. Ying, L.; Liu, Y.; Chen, S.; Shen, Z.; Ecology, D.O. Simulation of the potential range of pistacia weinmannifolia in southwest china with climate change based on the maximum-entropy(maxent) model. *Biodivers. Sci.* **2016**, *24*, 453–461. [CrossRef]

79. Guo, Y.L.; Wei, H.Y.; Lu, C.Y.; Zhang, H.L.; Gu, W. Predictions of potential geographical distribution of sinopodophyllum hexandrum under climate change. *Chin. J. Plant Ecol.* **2014**, *38*, 249–261.

80. Cheaib, A.; Badeau, V.; Boe, J.; Chuine, I.; Delire, C.; Dufrene, E.; Francois, C.; Gritti, E.S.; Legay, M.; Page, C.; et al. Climate change impacts on tree ranges: Model intercomparison facilitates understanding and quantification of uncertainty. *Ecol. Lett.* **2012**, *15*, 533–544. [CrossRef] [PubMed]

81. Xu, D.; Yan, H. A study of the impacts of climate change on the geographic distribution of pinus koraiensis in china. *Environ. Int.* **2001**, *27*, 201–205. [CrossRef]

82. Argus, G.W. The genus *salix (salicaceae)* in the southeastern united states. *Syst. Bot. Monogr.* **1986**, *9*, 1–170. [CrossRef]

83. Bertrand, R.; Lenoir, J.; Piedallu, C.; Riofrio-Dillon, G.; de Ruffray, P.; Vidal, C.; Pierrat, J.C.; Gegout, J.C. Changes in plant community composition lag behind climate warming in lowland forests. *Nature* **2011**, *479*, 517–520. [CrossRef] [PubMed]

84. Frei, E.; Bodin, J.; Walther, G.R. Plant species' range shifts in mountainous areas—All uphill from here? *Bot. Helv.* **2010**, *120*, 117–128. [CrossRef]

85. Walther, G.R.; Beißner, S.; Burga, C.A. Trends in the upward shift of alpine plants. *J. Veg. Sci.* **2005**, *16*, 541–548. [CrossRef]

86. Bai, Y.; Wei, X.; Li, X. Distributional dynamics of a vulnerable species in response to past and future climate change: A window for conservation prospects. *PeerJ* **2018**, *6*, e4287. [CrossRef] [PubMed]

87. Wang, C.; Liu, C.; Wan, J.; Zhang, Z. Climate change may threaten habitat suitability of threatened plant species within chinese nature reserves. *PeerJ* **2016**, *4*, e2091. [CrossRef] [PubMed]

88. Puga, N.D.; Corral, J.A.R.; Eguiarte, D.R.G.; Munguia, S.M.; Rosas, G.O.D. Climate change and its impact on environmental aptitude and geographical distribution of salvia hispanica l. In mexico. *Interciencia* **2016**, *41*, 407–413.

89. Hu, X.G.; Jin, Y.; Wang, X.R.; Mao, J.F.; Li, Y. Predicting impacts of future climate change on the distribution of the widespread conifer platycladus orientalis. *PLoS ONE* **2015**, *10*, e0132326. [CrossRef] [PubMed]

90. Garcia, K.; Lasco, R.; Ines, A.; Lyon, B.; Pulhin, F. Predicting geographic distribution and habitat suitability due to climate change of selected threatened forest tree species in the philippines. *Appl. Geogr.* **2013**, *44*, 12–22. [CrossRef]

91. Bomhard, B.; Richardson, D.M.; Donaldson, J.S.; Hughes, G.O.; Midgley, G.F.; Raimondo, D.C.; Rebelo, A.G.; Rouget, M.; Thuiller, W. Potential impacts of future land use and climate change on the red list status of the proteaceae in the cape floristic region, south africa. *Glob. Chang. Biol.* **2005**, *11*, 1452–1468. [CrossRef]

92. Midgley, G.F.; Hannah, L.; Millar, D.; Thuiller, W.; Booth, A. Developing regional and species-level assessments of climate change impacts on biodiversity in the cape floristic region. *Biol. Conserv.* **2003**, *112*, 87–97. [CrossRef]

93. Fischer, J.; Lindenmayer, D.B. Landscape modification and habitat fragmentation: A synthesis. *Glob. Ecol. Biogeogr.* **2007**, *16*, 265–280. [CrossRef]

94. Basile, M.; Valerio, F.; Balestrieri, R.; Posillico, M.; Bucci, R.; Altea, T.; De Cinti, B.; Matteucci, G. Patchiness of forest landscape can predict species distribution better than abundance: The case of a forest-dwelling passerine, the short-toed treecreeper, in central italy. *PeerJ* **2016**, *4*, e2398. [CrossRef] [PubMed]

95. Wang, Z.; Fang, J.; Tang, Z.; Lin, X. Patterns, determinants and models of woody plant diversity in china. *Proc. R. Soc.* **2011**, *278*, 2122–2132. [CrossRef] [PubMed]

96. Collinson, M.E. The early fossil history of *salicaceae*: A brief review. *Proc. R. Soc.* **1992**, *98*, 155–167. [CrossRef]

97. Allen, A.P.; Gillooly, J.F.; Savage, V.M.; Brown, J.H. Kinetic effects of temperature on rates of genetic divergence and speciation. *Proc. Natl. Acad. Sci. USA* **2006**, *103*, 9130–9135. [CrossRef] [PubMed]

98. Xu, X.; Peng, G.Q.; Wu, C.C.; Han, Q.M. Global warming induces female cuttings of populus cathayana to allocate more biomass, c and n to aboveground organs than do male cuttings. *Aust. J. Bot.* **2010**, *58*, 519–526. [CrossRef]

99. Chen, L.H.; Sheng, Z.; Zhao, H.X.; Korpelainen, H.; Li, C.Y. Sex-related adaptive responses to interaction of drought and salinity in populus yunnanensis. *Plant Cell Environ.* **2010**, *33*, 1767–1778. [CrossRef] [PubMed]

100. Xu, X.; Peng, G.; Wu, C.; Korpelainen, H.; Li, C. Drought inhibits photosynthetic capacity more in females than in males of populus cathayana. *Tree Physiol.* **2008**, *28*, 1751–1759. [CrossRef] [PubMed]

101. Taylor, D.W. Paleobiogeographic relationships of angiosperms from the cretaceous and early tertiary of the north american area. *Bot. Rev.* **1990**, *56*, 279–417. [CrossRef]

102. Ding, T.Y. Origin, divergence and geographical distribution of *salicaceae*. *Acta Bot. Yunnanica* **1995**, *17*, 277–290.

103. O'Brien, E.M. Climatic gradients in woody plant species richness: Towards an explanation based on an analysis of southern africa's woody flora. *J. Biogeogr.* **1993**, *20*, 181–198. [CrossRef]

104. O'Brien, E.M. Water–energy dynamics, climate, and prediction of woody plant species richness: An interim general model. *J. Biogeogr.* **1998**, *25*, 379–398. [CrossRef]

105. Evans, K.L.; Warren, P.H.; Gaston, K.J. Species–energy relationships at the macroecological scale: A review of the mechanisms. *Biol. Rev.* **2005**, *80*, 1–25. [CrossRef] [PubMed]

106. Bai, Y.; Wu, J.; Xing, Q.; Pan, Q.; Huang, J.; Yang, D.; Han, X. Primary production and rain use efficiency across a precipitation gradient on the mongolia plateau. *Ecology* **2008**, *89*, 2140–2153. [CrossRef] [PubMed]

107. Whittaker, R.J.; Nogués-Bravo, D.; Araújo, M.B. Geographical gradients of species richness: A test of the water-energy conjecture of hawkins et al. (2003) using european data for five taxa. *Glob. Ecol. Biogeogr.* **2006**, *16*, 76–78. [CrossRef]

108. Zhang, S.; Chen, F.; Peng, S.; Ma, W.; Korpelainen, H.; Li, C. Comparative physiological, ultrastructural and proteomic analyses reveal sexual differences in the responses of populus cathayana under drought stress. *Proteomics* **2010**, *10*, 2661–2677. [CrossRef] [PubMed]

109. Lei, Y.; Chen, K.; Jiang, H.; Yu, L.; Duan, B. Contrasting responses in the growth and energy utilization properties of sympatric populus and salix to different altitudes: Implications for sexual dimorphism in salicaceae. *Physiol. Plant.* **2017**, *159*, 30–41. [CrossRef] [PubMed]

110. Teramura, A.H. Effects of ultraviolet-b radiation on the growth and yield of crop plants. *Physiol. Plant.* **2010**, *58*, 415–427. [CrossRef]

111. Keiller, D.R.; Holmes, M.G. Effects of long-term exposure to elevated uv-b radiation on the photosynthetic performance of five broad-leaved tree species. *Photosynth. Res.* **2001**, *67*, 229–240. [CrossRef] [PubMed]

112. Song, H.; Zhang, S. Sex-related responses to environmental changes in salicaceae. *Mt. Res.* **2017**, *35*, 645–652.

113. Liu, B.; Liu, X.B.; Li, Y.S.; Herbert, S.J. Effects of enhanced uv-b radiation on seed growth characteristics and yield components in soybean. *Field Crops Res.* **2013**, *154*, 158–163. [CrossRef]

114. Svenning, J.C.; Skov, F. The relative roles of environment and history as controls of tree species composition and richness in europe. *J. Biogeogr.* **2005**, *32*, 1019–1033. [CrossRef]

115. Raes, N.; Roos, M.C.; Slik, J.W.F.; Van Loon, E.E.; Ter Steege, H. Botanical richness and endemicity patterns of borneo derived from species distribution models. *Ecography* **2009**, *32*, 180–192. [CrossRef]

116. Stropp, J.; Ter Steege, H.; Malhi, Y. Disentangling regional and local tree diversity in the amazon. *Ecography* **2009**, *32*, 46–54. [CrossRef]
117. Berthel, N.; Schwörer, C.; Tinner, W. Impact of holocene climate changes on alpine and treeline vegetation at sanetsch pass, bernese alps, switzerland. *Rev. Palaeobot. Palynol.* **2012**, *174*, 91–100. [CrossRef]
118. Singh, S.; Kumar, P.; Rai, A. Ultraviolet radiation stress: Molecular and physiological adaptations in trees. In *Abiotic Stress Tolerance in Plants*; Springer: Dordrecht, The Netherlands, 2006.
119. Osprey, S.M.; Butchart, N.; Knight, J.R.; Scaife, A.A.; Hamilton, K.; Anstey, J.A.; Schenzinger, V.; Zhang, C. An unexpected disruption of the atmospheric quasi-biennial oscillation. *Science* **2016**, *353*, 1424. [CrossRef] [PubMed]
120. Dhomse, S.S.; Chipperfield, M.P.; Feng, W.; Hossaini, R.; Mann, G.W.; Santee, M.L. Revisiting the hemispheric asymmetry in midlatitude ozone changes following the mount pinatubo eruption: A 3-d model study. *Geophys. Res. Lett.* **2015**, *42*, 3038–3047. [CrossRef] [PubMed]
121. Dhomse, S.S.; Chipperfield, M.P.; Damadeo, R.P.; Zawodny, J.M.; Haigh, J.D. On the ambiguous nature of the 11-year solar cycle signal in upper stratospheric ozone: Solar signal in upper stratosphere. *Geophys. Res. Lett.* **2016**, *43*, 7241–7249. [CrossRef]
122. Caldwell, M.M. A steep latitudinal gradient of solar ultraviolet-b radiation in the arctic-alpine life zone. *Ecology* **1980**, *61*, 600–611. [CrossRef]

© 2019 by the authors. Licensee MDPI, Basel, Switzerland. This article is an open access article distributed under the terms and conditions of the Creative Commons Attribution (CC BY) license (http://creativecommons.org/licenses/by/4.0/).

forests

MDPI

Article

Species Richness of the Family Ericaceae along an Elevational Gradient in Yunnan, China

Ji-Hua Wang [1,2], Yan-Fei Cai [1,2], Lu Zhang [1,2], Chuan-Kun Xu [3] and Shi-Bao Zhang [4,*]

[1] Flower Research Institute of Yunnan Academy of Agricultural Sciences, Kunming 650051, China; wjh0505@gmail.com (J.-H.W.); caiyanfei1013@126.com (Y.-F.C.); changjiangyulu@163.com (L.Z.)
[2] National Engineering Research Center for Ornamental Horticulture, Kunming 650051, China
[3] Institute of South Asia & Southeast Asia Radiation Center, Yunnan Minzu University, Kunming 650504, China; chkxu@163.com
[4] Key Laboratory for Economic Plants and Biotechnology, Kunming Institute of Botany, Chinese Academy of Sciences, Kunming 650201, China
* Correspondence: sbzhang@mail.kib.ac.cn; Tel.: +86-871-6522-3002

Received: 2 July 2018; Accepted: 22 August 2018; Published: 24 August 2018

Abstract: Knowledge about how species richness varies along spatial and environmental gradients is important for the conservation and use of biodiversity. The Ericaceae is a major component of alpine and subalpine vegetation globally. However, little is known about the spatial pattern of species richness and the factors that drive that richness in Ericaceae. We investigated variation in species richness of Ericaceae along an elevational gradient in Yunnan, China, and used a variation partitioning analysis based on redundancy analysis ordination to examine how those changes might be influenced by the mid-domain effect, the species-area relationship, and climatic variables. Species richness varied significantly with elevation, peaking in the upper third of the elevational gradient. Of the factors examined, climate explained a larger proportion of the variance in species richness along the elevational gradient than either land area or geometric constraints. Species richness showed a unimodal relationship with mean annual temperature and mean annual precipitation. The elevational pattern of species richness for Ericaceae was shaped by the combined effects of climate and competition. Our findings contribute to a better understanding of the potential effects of climate change on species richness for Ericaceae.

Keywords: Ericaceae; variation partitioning; climate; species-area relationship; mid-domain effect

---

1. Introduction

Variation in species richness along geographical and environmental gradients is a central topic in the fields of ecology and biogeography because species diversity can influence community stability and ecosystem functioning [1–3]. Such research is essential for the conservation and sustainable utilization of biodiversity, especially under the circumstances of global climate change [4]. Species richness in relation to elevation has received extensive attention [4,5] because short geographical distances can produce large climatic gradients as the elevation changes [6]. Thus, elevational gradients provide unique opportunities to test ecological theories and explore the effects of climate change on species richness, and may help us interpret deviations from latitude-related patterns [5].

Several elevational patterns for species richness have been described, such as monotonically decreasing patterns with elevation, monotonically increasing patterns, and unimodal/hump-shaped patterns [3,5,7]; unimodal patterns are more prevalent. In contrast, a study of 118 taxa recorded from mountain-temperate forests to a woody shrub community and alpine meadows in Slovakia, no significant elevational trends were found in species richness [8]. Thus, the elevational patterns of species richness may vary because of differences in sampling method and taxonomic group [3,5,9].

Multiple drivers, including environmental variables, mountain topography, and evolutionary history, may interact to affect the elevational pattern of species richness [1,3,10–12]. The mid-domain effect (MDE) and the species-area relationship (SAR) are two hypotheses related to mountain characteristics that have been employed to explain elevational patterns [3,13,14]. The MDE hypothesis states that species richness will peak in the mid-elevation zone because geometric constraints may lead to a greater overlap of species ranges near the midpoint of a mountain [15]. Although the validity of the MDE has been confirmed by many studies [11,13], it has failed to predict plant species richness along a Himalayan elevational gradient [16]. The SAR hypothesis proposes that maximum richness occurs in elevational zones that cover the largest area because more species can be supported there [17]. Its efficacy in explaining the elevational pattern of species richness has been confirmed by an example from the palms of New Guinea [17], but the SAR has little effect on frog species richness in the Hengduan Mountains [13]. This discrepancy among studies may be related to topography because changes in land areas associated with elevation are not uniform among mountain ecosystems [18].

Environmental variables always change with the position and topography of a mountain, and some variables, such as temperature and air pressure, tend to vary predictably along elevational gradients [6]. Because plants species exhibit different degrees of physiological tolerances to environmental stresses, elevational patterns of species richness may be correlated with environmental gradients [4,9,19]. This environment-based hypothesis predicts that species richness will peak at the elevation where the combination of growing conditions is optimal [9,12]. Water and temperature seem to be the main variables that explain elevational gradients of richness [4,19]. Low temperature tolerance may limit the distribution of many plant species [20,21]. For example, a previous study has found that the mean temperature of the coldest quarter of the year is the strongest predictor of species richness for woody plants in China [22]. Higher humidity may contribute to a greater richness of plant species at mid-elevations [4,23]. Species richness at low altitude can be limited by precipitation while at high altitudes it is limited by heat, thus resulting in a high richness at middle altitude [3,24]. However, fern richness shows a unimodal response along a temperature gradient but a linear response to a moisture gradient in the central Himalayas [25]. These differences may be related to the physiological tolerances of species to environmental stresses.

In addition to temperature and water availability, light is important for plant growth and survival. Because light capture by understory species is affected by canopy tree species, competition may be an important factor limiting the number of understory species by affecting seedling recruitment, growth, and survival [26–28]. Although a previous study has suggested that the species richness at large scales is dependent largely upon climatic variables [29], several studies have also confirmed that competition plays an important role in regulating regional species pools and species diversity [3,28]. The outcome of competitive exclusion is also correlated with the environment. Competition between species at low altitudes or in favorable growing conditions may be more intense, and its intensity may decrease with increasing elevation [26,27]. At low elevations in humid areas, the formation of hump-shaped pattern of species richness is due to the shaded canopy, which reduces the richness of understory species [3,24]. The degree to which species extend their ranges downwards is much influenced by competition, while the upper limits are determined by temperature-related factors [30]. Thus, the geographical pattern of species richness may be a result of the combined effects of the environment and competition [12,27].

Species richness in relation to elevational gradients may vary among life forms [5,12,24]. For example, Guo et al. suggested that the elevational peaks of herbaceous species richness are higher than those for trees [5]. Whereas the richness of woody plants decreases monotonically with elevation at high elevations in the Qinling Mountains, herbaceous plants present no significant patterns [31]. However, Wang et al. found that species richness for trees, shrubs, and herbs peak within the same elevational range [32]. These patterns of species richness are probably related to the ecological properties of different life forms. For example, maximum tree height may affect competition for light capture in a vertical gradient [33] and tree hydraulic characteristics [34].

Here, we focus on the species richness of Ericaceae. This family is widely distributed in temperate and subarctic regions, and at high elevations in tropical regions, and has a high rate of speciation and species richness in the Eastern Himalayas and southwestern China [35,36]. *Rhododendron*, the largest genus in Ericaceae, is a major component of alpine and subalpine communities [35]. Yunnan Province in China is recognized as a center of diversification and differentiation for species within Ericaceae, and it harbors more than 50% of all species in this family in China [36,37]. Because spatial patterns of species richness differ among taxonomic groups [38], understanding how the richness of plants in Ericaceae vary with elevation and identifying which factors determine its high species richness in Yunnan would be useful for biogeographical research and biodiversity conservation. However, richness in this family has received little attention so far.

In the current study, we investigated the variation in species richness of Ericaceae along a large elevational gradient in Yunnan Province. We focused on the following questions: (1) How does this richness vary with elevation? And (2) What are the main factors that explain this elevational pattern? We hypothesized that: (1) species richness peaks at the mid-elevation zone due to a combined effect of climatic conditions and mountain geometry, and (2) climatic factors play a more important role in shaping the elevation-related patterns of species richness for Ericaceae than geometric constraints.

2. Materials and Methods

2.1. Study Area

We investigated the species richness of plants within the family Ericaceae over the entire province of Yunnan, China (97°32′–106°15′ E, 21°9′–29°15′ N). Yunnan province in the southwestern part of the country covers approximately 394,000 km^2 at the southeastern edge of the Tibetan Plateau. The elevational gradient ranges from 76.4 m above sea level in the southeast to 6740.0 m in the northwest (Figure 1). More than 60% of this area lies in a zone from 1500 m to 3000 m (Figure S1). Much of the province has mild to warm winters and temperate summers, with obvious wet and dry seasons. Over 80% of the total annual precipitation occurs between May and October. Mean annual temperature (MAT) ranges from 5.3 °C to 23.8 °C at 119 meteorological stations, with an elevational lapse rate of 5.1 °C/1000 m. Mean annual precipitation (MAP) ranges from 573.9 mm to 2305.2 mm. It does not significantly change below 1600 m, but gradually decreases above 1600 m (Figure S2). As an important biodiversity hotspot in the world, more than 18,000 species of higher plants grow there [39]. This province hosts a continuous succession from tropical rain forests, subtropical evergreen broadleaved forests, subalpine conifer forests, temperate deciduous broadleaved forests, mossy evergreen dwarf forest, alpine bushes, and meadows (Table S1) [40]. A combination of high species richness and large elevational gradients over relatively short distances makes this locale an ideal place to evaluate elevational effects on species richness.

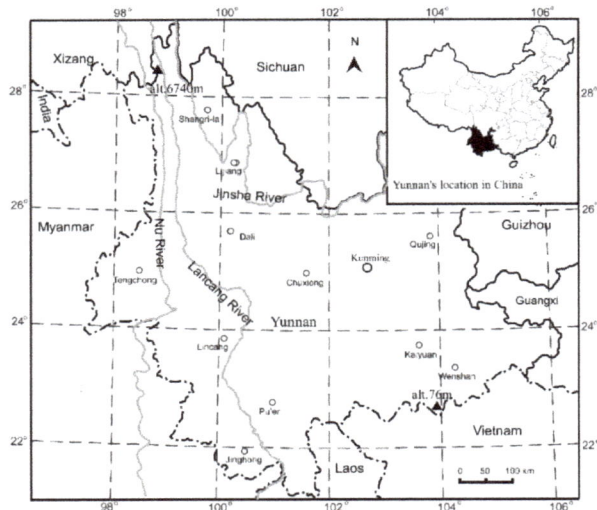

Figure 1. Map of the study area. Insert in top-right corner shows location of Yunnan Province within China. The triangles (▲) show the lowest and highest altitudes (alt.) in the area, respectively.

2.2. Data Sources

All data for species distribution were compiled from two volumes (Tomus 4–5) of the Flora Yunnanica [37,41]. These sources report the lower and upper elevational limits, maximum plant height, and life form for native species, subspecies, and varieties within Ericaceae. Here, we treated "variety" or "subspecies" as individual taxa when diversity is estimated [42]. In all, 425 species belonging to 12 genera were used (Table S2). To explore elevational richness patterns for different life forms, we divided these species into two groups: trees were defined as plants ≥5 m tall while shrubs were defined as plants <5 m high [32]. Thus, species richness was estimated for three species groups: trees, shrubs, and all species. The upper and lower distribution limits were used later for deciding the presence/absence of a species in the elevational band.

The species in our database were distributed from 510 m to 4900 m, and the elevational gradient was divided into 22 elevational bands, at 200-m intervals. Briefly, we firstly decided whether a species was present or not in an elevational band based on its upper and lower limits, and obtained a presence/absence list for each species for each elevational band. Then, we counted the total number of species present in each elevational band. This sum of the species present in each band was considered as the observed species richness [1]. For example, if the elevational range of a species is from 720 m to 1250 m, this species is present at the elevational bands of 701–900 m, 901–1100 m and 1101–1300 m. Then, we summed the number of species present at the elevational band of 901–1100 m to obtain the observed species richness for this elevational band.

We extracted the map of the study area containing elevation information from the global GTOPO 30 map (https://lta.cr.usgs.gov/GTOPO30). This map was then converted into a Lambert-Azimuthal equal area projection map and rasterized on 1 × 1 km grid cells. After counting the number of cells within each elevational band, we calculated the land area from the total number of cells [13] according to a global digital elevation model, with a horizontal grid-spacing of 30 arc-seconds in Envi 4.7 (ITT Exelis, McLean, VA, USA) and ArcGIS 9.3 (ESRI, Redlands, CA, USA).

Climatic data were obtained from the China Meteorological Data Sharing Service System (http://data.cma.cn/). The database includes mean annual temperature (MAT), mean temperature of the coldest month (MTCM), mean temperature of the warmest month (MTWM), mean annual precipitation (MAP), precipitation of the driest month (PDM), and precipitation of the wettest month (PWM). Because each meteorological station has an exact location (longitude, latitude, and elevation) and the latitudinal gradient is relatively small (approximately 8°) in the study area, we were able to derive the values for these climatic variables within each elevational band by fitting regression models from the climate data of 119 meteorological stations from 137 m to 3468 m above sea level recorded between 1961 and 2004 [13] (Figure S2). Temperature seasonality was defined as the difference between MTWM and MTCM, while precipitation seasonality was calculated as 12 × (PWM-PDM)/MAP [43]. Potential evapotranspiration (PET) was calculated following the FAO Penman-Monteith approach [44]. Although these climate variables were significantly correlated (Table S3), they characterize the availability of environmental resources from different perspectives.

2.3. Data Analysis

The mid-domain effect was tested by RangeModel [45], which can generate the mean predicted pattern of richness under "pure" geometric constraints. The richness predicted by RangeModel is affected by the number of elevational bands, total number of species, range size frequency distributions (RSFDs), and frequency distribution of a species midpoint. Using the empirical RSFD Model implemented in RangeModel, we ran 10,000 Monte Carlo simulations to generate the mean predicted pattern of species richness and the 95% confidence interval within each band for all species, trees, and shrubs [10] (Figure 2). The predicted richness was then used as an explanatory variable in variation partitioning analyses to explore the influence of the MDE [17].

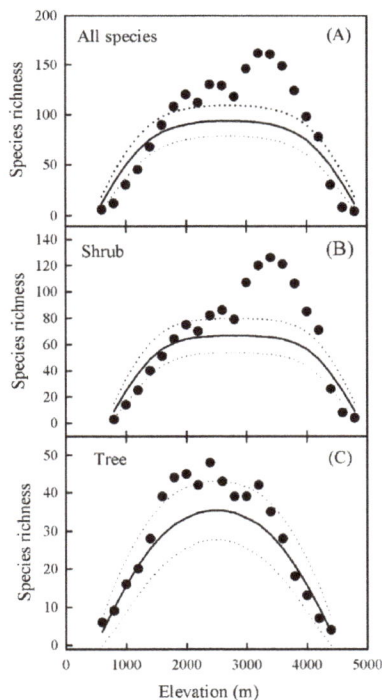

Figure 2. Elevational patterns of richness for all Ericaceae species (**A**), shrubs (**B**), and trees (**C**) in Yunnan, China. Solid dots and solid lines represent observed and predicted species richness, respectively. Dashed lines show 95% confidence intervals predicted by the mid-domain effect (MDE) randomization model.

The species-area relationship was fitted by an Arrhenius's power function. This function used the logarithm of both sides of the equation to obtain the following linear equation: $\log(S) = \log(c) + z \times \log(A)$, where S and A were the number of species and the area covered by each elevational band, respectively. Both c and z were constants.

A variation partitioning analysis, based on redundancy analysis (RDA) ordination, allowed us to determine the independent and shared influences of multiple complementary sets of variables on species richness [46]. We used the "varpart" function of the library "vegan" [47] in the R package 3.4.0 for Windows [48] to determine the contributions of the SAR, the MDE and climate to explain elevational patterns of richness. That function uses an adjusted R^2 to assess the partitions explained by the variables and their combinations. The response variable was the presence/absence of species in 22 elevational bands [1]. The explanatory matrices were species richness predicted by the SAR and the MDE, and the environmental matrix containing seven variables (MAT, MAP, PET, PDM, PWM, MTCM, and MTWM). Prior to the analysis, the presence/absence values of species were Hellinger-transformed because such a species matrix contains many zeros [49]. Meanwhile, we also applied the variation partitioning method to explore the relative importance of temperature (including MAT, MTCM, and MTWM) and water availability (MAP, PET, and PDM) toward species richness. Shared variation can be negative due to either suppressor variables or two closely correlated predictors that have strong influences on response variables with opposite signs [50].

To investigate how species richness might vary along elevational and climatic gradients, we ran generalized linear models (the "glm" function in R) to determine the relationships between richness and elevation, and climatic variables. Species richness was used as the response variable while the elevation and climatic variables were the explanatory variables. Before the "glm" function was used, we tested the data distribution of the response variable by using the "ks.test" function and found that these response variables followed a Normal distribution. The elevational peaks of richness for all species, shrubs, and trees were fitted by generalized linear models. We estimated adjusted R^2 value for each model. A Wilcoxon test was used to determine the elevational difference between shrubs and trees. All statistical analysis were conducted with the R package 3.4.0 for Windows [48].

3. Results

3.1. Elevational Patterns of Species Richness

The elevational patterns of richness were decidedly hump-shaped, and the elevational peaks estimated by polynomial models for all species, shrubs, and trees were 2815 m, 3045 m, and 2454 m, respectively (Figure 3). The mean elevation was significantly lower for trees than for shrubs (Figure S3). Multivariate modeling showed that both elevation and life form had significant impacts on species richness, and it revealed an interaction between these two factors (Table 1). Among these models, the model of MAT combined with PET had the lowest AIC value and the highest R^2 value, and best explained the species richness of Ericaceae.

Table 1. Regression coefficients of the multivariate models for the main variables explaining species richness. *P*-values for all models were less than 0.001. MAT, mean annual temperature; MAP, mean annual precipitation; PET, potential evapotranspiration; AIC, Akaike information criterion.

Intercept	Lifeform	Elevation	MAT	MAP	PET	Interaction	Adjusted R^2 (%)	AIC
−132.83	150.42	0.20	-	-	-	−0.20	77.82	210.5
816.91	−811.15	-	−39.44	-	-	39.93	77.91	210.1
1199.63	−1347.50	-	-	−0.89	-	1.11	72.02	215.3
914.41	−1002.73	-	-	-	−0.79	1.02	77.05	210.9
−93.16	−51.62	-	−21.53	−0.02	0.49	-	83.64	204.2
−116.35	-	-	−18.22	−0.05	0.47	-	83.94	203.1
−316.60	-	-	40.18	0.46	-	−0.04	84.24	202.7
−142.50	-	-	10.72	-	0.39	−0.02	93.01	184.8
−547.20	-	-	-	0.61	0.83	−0.0007	68.37	218.0

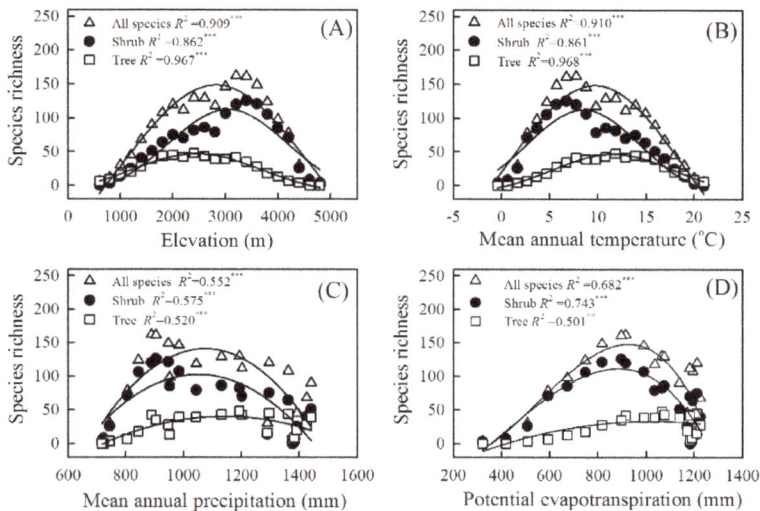

Figure 3. Variations in species richness within Ericaceae for all species, trees, and shrubs as function of elevation (**A**), mean annual temperature (**B**), mean annual precipitation (**C**), and potential evapotranspiration (**D**) in Yunnan Province, China. The explanatory powers (R^2) were evaluated by polynomial models. ** $p < 0.01$; *** $p < 0.001$.

3.2. Explanatory Powers of the MDE, the SAR, and Climate for Elevational Patterns of Richness

Temperature (MAT, MTCM, and MTWM) and water availability (MAP, PDM, and PWM) had significant effects on species richness (Figure 3, Tables 1 and 2), with the explanatory power being higher for the former than for the latter when all species, shrubs, and trees were considered (Figures 3 and 4). In addition, richness was significantly influenced by temperature seasonality but not by precipitation seasonality (Table 2). Although species richness presented unimodal patterns along temperature gradients, it peaked at a higher MAT level for trees than for shrubs (Figure 3).

Table 2. Explanatory powers (R^2) of climatic variables with species richness for all species, and for trees and shrubs in Ericaceae evaluated by polynomial models. ** $p < 0.01$; *** $p < 0.001$.

Environmental Variable	All Species	Shrubs	Trees
Temperature of the coldest month	0.840 ***	0.781 ***	0.838 ***
Temperature of the warmest month	0.905 ***	0.826 ***	0.911 ***
Temperature seasonality	0.662 ***	0.632 ***	0.667 ***
Precipitation of the driest month	0.497 **	0.478 **	0.604 ***
Precipitation of the wettest month	0.758 ***	0.734 ***	0.691 ***
Precipitation seasonality	0.036	0.026	0.065

The partitioning analysis showed that the MDE, the SAR, and climate together explained 54.66%, 52.84%, and 58.44% of the variance for all species, shrubs, and trees, respectively, with climate accounting for the largest proportion (Figure 5). When all species were considered, the individual contributions of the SAR, the MDE, and climate were 5.91%, 3.64%, and 17.93%, respectively. By comparison, the total contributions (including independent and shared effects) of the SAR, the MDE, and climate were 18.73%, 12.22%, and 46.49%, respectively. Finally, the explanatory powers of the SAR, the MDE, and climate were higher for trees than for shrubs.

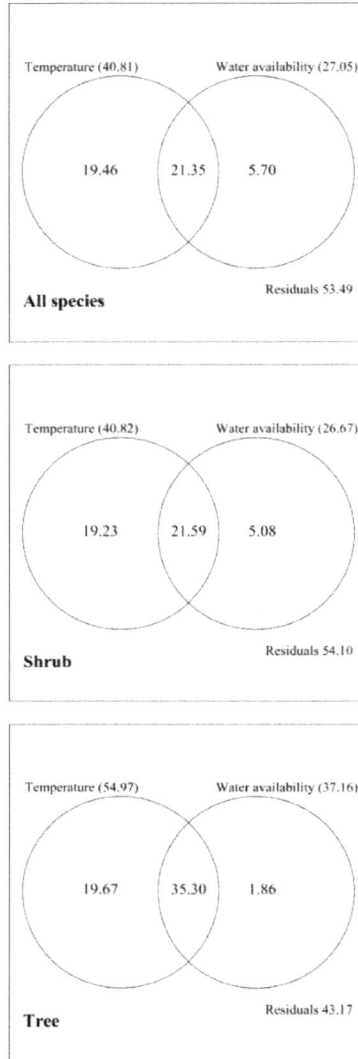

Figure 4. Partitioning of species richness into temperature (including mean annual temperature, temperature of the coldest month, temperature of the warmest month) and water availability (including mean annual precipitation, precipitation of the driest month and precipitation of the wettest month) for all species, and for shrubs and trees in Yunnan. Values represent adjusted R^2 coefficients of independent or shared effects by each variable. Total effect (including independent and shared effects) of each explanatory variable is shown in parentheses.

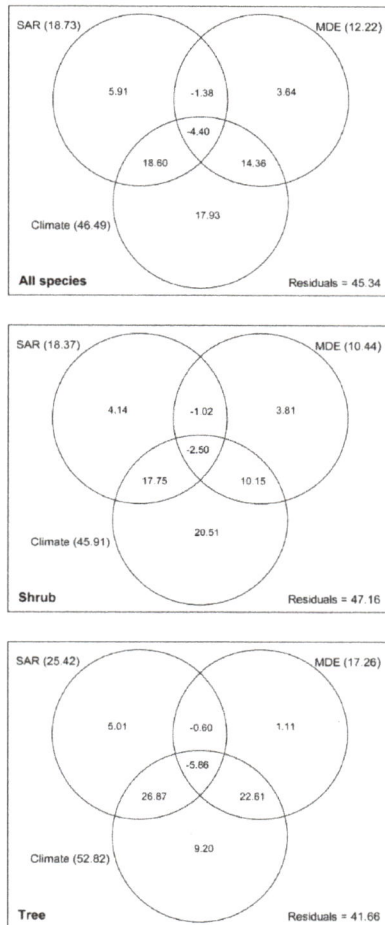

Figure 5. Partitioning of species richness into the species-area relationship (SAR), the mid-domain effect (MDE), and climate components for all species, and for shrubs and trees. Values represent adjusted R^2 coefficients of independent or shared effects by 3 predictors. Total effect (including independent and shared effects) of each explanatory component was shown in parentheses.

4. Discussion

We characterized the elevational patterns of species richness in Ericaceae within Yunnan Province for the first time, and found that richness showed a hump-shaped relationship with elevation, producing the maximum value at 2815 m. Such elevational pattern of species richness is also observed on other woody plants in this area [32,51,52]. The hump-shaped pattern along an elevational gradient has been identified for many other habitats and plant taxa worldwide [5,25], and may result from a variety of factors, such as the combination of suitable temperature with higher humidity at mid-elevations, and the greater overlap of species ranges near the middle of the elevational gradients [7,9,10,38]. However, the main factors that drive these patterns can differ across taxonomical groups or mountains [3,38].

We found that the elevational patterns of species richness in Ericaceae was influenced by climate, the species-area relationship, and the mid-domain effect. However, for trees, shrubs, and all species combined, both the independent contribution and total contribution of climate to species richness were obviously higher than those of the MDE and the SAR. Other researchers have confirmed that climate contributes more than geometric constraints to the hump-shaped pattern of richness along geographic gradients [13,19]. The skew in maximum richness toward higher elevations may indicate the strong influence of climate [10]. Climate potentially affects the metabolism and survival of an organism and, therefore, its geographical distribution [53]. The explanatory power of climate on species richness was higher for trees than for shrubs. This was in line with other data about woody plants in China [22] and reflected functional and physiological differences between trees and shrubs [34,54]. Tree and shrub distributions are associated with environmental gradients [55]. For example, the spatiotemporal variation in soil water availability and species drought resistance may have roles in determining species distribution [56]. Thus, our study indicated that climate might be an important driver for elevational patterns of species richness in Ericaceae, and the distributions of trees were more sensitive to environments than shrubs.

The important roles of temperature and water availability in explaining elevational gradients of species richness have been validated previously [4,42]. Here, we found that the explanatory powers of MAT, MTCM, and MTWM were higher than those of MAP, PDM, and PWM. Previous studies have suggested that the interrelationship between water and energy controls the physiological activities and biomass accumulation of plants, which determines the geographical pattern of species diversity. Species richness of plants can increase with precipitation, but water becomes ice under too low temperature conditions and is easily evaporated under high temperature conditions, which affects the normal physiological processes of plants. Thus, species diversity and energy have a hump-shaped pattern [3,57]. These indicated that the effects of temperature and precipitation on species richness in Ericaceae were coupled.

The freezing-tolerance hypothesis has been validated in a study of species richness for woody plants [22]. We noted that, when compared with temperature seasonality, MTCM and MTWM had relatively larger explanatory powers on species richness. Although many *Rhododendron* species are very tolerant of low temperatures [58], cold climates at high altitudes may slow soil formation, the physiological activities and growth of plants [51]. Embolisms that are induced by freezing conditions can reduce the hydraulic conductivity of species in *Rhododendron*, possibly blocking water transport from the roots to the leaves [58]. Extremely cold temperatures can represent a boundary for the survival of *Rhododendron* species [20]. Here, we found that species richness of plants within Ericaceae peaked at a lower temperature for shrubs than for trees. Wang et al. reported a negative correlation between elevation and plant height in their study of 42 rhododendrons on the Tibetan Plateau [59]. Shorter plants may help compensate for the difficulty in supplying water to leaves [34] and enhance the adaptability of plants to lower temperatures at higher elevations. In addition, the energy-diversity hypothesis states that each population needs a certain amount of energy to maintain its survival and reproduction, and species richness can be affected by the energy per unit area [3,53,57]. In the case of unrestricted water, there is a globally consistent relationship between species richness and temperature for angiosperms [60]. Here, we found a close relationship between MAT and species richness, and the combination of MAT and PET could better explain the elevational patterns of species richness in Ericaceae. These indicated that the energy change along the elevational gradient and temperature filtering may be the main factors affecting species distribution and richness for Ericaceae in Yunnan.

Water availability is another major correlate for the elevational patterns of species richness [3,25], and can regulate the relationship between species diversity and temperature [60]. Under the condition of adequate moisture, the species richness of trees and shrubs are closely related to MAT, while under the condition of insufficient water, the richness of woody plants is positively correlated with precipitation [3]. A previous study has noted that the highest richness for most organisms in the

American Southwest is found at mid-elevations due to the lack of water at lower elevations [23]. Here, we found that species richness for Ericaceae was significantly correlated with MAP, PDM, and PWM, although the relative importance of water availability in explaining the elevational pattern of richness was smaller than temperature. *Rhododendron ferrugineum* L. is thought to be particularly affected by short periods of scarce rainfall, which negatively affects leaf gas exchange [61]. Snow cover during winter provides protection for the plants at high elevation. A reduction in snow cover may increase the frequency and intensity of the freeze-thaw process [62]. Thus, water availability contributed to the humped-pattern of richness for the Ericaceae along an elevational gradient.

A larger land area may contribute to higher species diversity because increased space can accommodate more species and support greater habitat heterogeneity [9,14]. Although some previous studies have found that area effects can account for a substantial portion of the variability in elevational diversity patterns [13,14,17], it is a poor predictor of elevational fern diversity in Costa Rica [11]. This may be because the impact of area on species richness can be counteracted by some environmental factors associated with elevation [6]. Our data showed that over 18% of the variation in species richness for Ericaceae along the elevational gradient was explained by the SAR, but this was shared by climatic factors. Thus, land area had some influence on the elevational pattern of species richness for Ericaceae, even though it contributed more to species richness for trees than for shrubs. Our finding demonstrated that the extent to which SAR explained elevational species richness depended upon the life form of the studied taxa.

The geometric characteristic of mountains has a critical role in shaping elevational patterns [11,17]. The influence of the MDE on species richness has previously been demonstrated [11,15,17]. A previous study has suggested that MDE is an important factor explaining the elevational pattern of species richness for Ericaceae in Gaoligong mountains [52]. However, we found that the explanatory power of the MDE was smaller than that of climate, or of the SAR. Meanwhile, a large portion of the MDE was shared with climate. Similarly, the MDE has not proven to be a main mechanism determining the richness of vascular plants in Crete [14]. In fact, the extent to which the MDE can effectively explain spatial patterns of richness differ among various studies due to the interaction of mountain mass effects with climatic factors [10,11,19]. Thus, for the family Ericaceae in Yunnan, the MDE was not a key mechanism that explained the elevational pattern of species richness.

Last, competition may serve as a filter preventing species from colonizing a site, therefore affecting local species diversity [3,28,63,64]. The environment determines community composition not only via survival, but also by influencing competition. Patterns of species richness across gradients depend on the interaction between individual species' range of tolerance and competition [64]. Here, although we found that climate was the most important driver of elevational pattern of species richness in Ericaceae, the role of competition could not be ignored. Below 1600 m, there was no significant change in precipitation along the elevational gradient, but species richness increased with elevation. This indicated that below 1600 m, precipitation was not responsible for the variation in species richness. Previous studies have confirmed that at lower altitudes, water and heat resources are sufficient and productivity is high, but competition for light resource is intense [26–28], and a dense canopy may buffer climate effects [65]. In fact, the maximum canopy height of main vegetation types below 1600 m in Yunnan are almost above 20 m (Table S1), but only <10 species in Ericaceae have a maximum height of more than 15 m [37,41]. Maximum tree height in part reflects the ability of plants to compete for light resource [33]. Because of their low statures, the growth and diversity of most species in Ericaceae are depressed by competition from tall trees at low elevations, and only a few species (in the genus *Agapetes*) grow as epiphytes in tropical rainforest, tropical monsoon forest, and montane rainforest. With increasing elevation, competition from other tree species is gradually weakened, and many species, such as *Cassiope* and *Rhododendron*, are species-rich, and may dominate mossy evergreen dwarf forests and alpine shrub lands [52]. At higher elevations (>3000 m), the growth and distribution of plants in Ericaceae is limited by lower temperatures. Our study supported the idea that observed elevational gradients in species richness results from a combination of ecological and evolutionary processes, rather than from the independent effect of one overriding force [9,11,12]. Species diversity within Ericaceae was regulated by the competition from other tree species at lower elevations, by the reduced competition and suitable climates at middle elevations, and by the decreased temperature and energy at higher elevations.

5. Conclusions

The species richness within Ericaceae of Yunnan Province in China varied significantly along an elevational gradient, presenting a humped pattern showing that species richness peaked in the upper third of the elevational gradient. We confirmed that the species richness for Ericaceae along the elevational gradient was influenced by multiple interacting factors. Climate explained a larger proportion of the variance in observed richness, but the elevational pattern of species richness for Ericaceae reflected the combined effects of the environment and competition.

We have found that climate had an important role in explaining the elevational pattern of species richness. However, the effects of environmental factors on physiology and productivity in Ericaceae are still unclear, and a deep understanding is lacking about the relationship between richness and climate. Therefore, it will be interesting to conduct field research that explores synthetically the variations in plant physiology, productivity and species richness along elevational gradients.

Supplementary Materials: The following are available online at http://www.mdpi.com/1999-4907/9/9/511/s1, Figure S1: Land area associated with each elevational band, Figure S2: Variations in mean annual temperature and mean annual precipitation with elevation (A,C), and monthly temperature and precipitation at different elevations (B,D) based on the records of 119 meteorological stations between 1961 and 2004 in Yunnan. The value in parenthesis shows the elevations at each meteorological station, Figure S3: Mean elevation of trees and shrubs with Ericaceae in Yunnan. The elevational difference between trees and shrubs was tested by a Wilcoxon test, Table S1: Elevations, dominant species and canopy heights of main vegetations in Yunnan, Table S2: Number of species and life form in each genus of the family Ericaceae in Yunnan, Table S3: Bivariate relationships (r) among explanatory variables.

Author Contributions: J.-H.W. and S.-B.Z. conceived and designed the experiments; Y.-F.C., L.Z. and C.-K.X. collected and analyzed the data; J.-H.W. and S.-B.Z. wrote the manuscript.

Funding: This research was funded by the National Natural Science Foundation of China [grant number 31460217, 31670342], and by the Scientific and Technological Leading Talent Project of Yunnan Province [grant number 2016HA005].

Acknowledgments: We would like to thank Ying-Feng Bi for calculating the land area at each elevational band. We are also very grateful to the editors and two anonymous reviewers for their valuable suggestions and comments to improve the quality of our paper.

Conflicts of Interest: The authors declare no conflict of interest.

References

1. Grytnes, J.A.; Vetaas, O.R. Species richness and altitude: A comparison between null models and interpolated plant species richness along the Himalayan altitudinal gradient, Nepal. *Am. Nat.* **2002**, *159*, 294–304. [CrossRef] [PubMed]

2. Song, C.Y.; Cao, M.C. Relationships between plant species richness and terrain in middle sub-tropical eastern China. *Forests* **2017**, *8*, 344. [CrossRef]

3. Tang, Z.Y.; Fang, J.Y. A review on the elevational patterns of plant species diversity. *Biodivers. Sci.* **2004**, *12*, 20–28.

4. Zhang, S.-B.; Chen, W.-Y.; Huang, J.-L.; Bi, Y.-F.; Yang, X.-F. Orchid species richness along elevational and environmental gradients in Yunnan, China. *PLoS ONE* **2015**, *10*, e0142621. [CrossRef] [PubMed]

5. Guo, Q.F.; Kelt, D.A.; Sun, Z.Y.; Liu, H.X.; Hu, L.J.; Ren, H.; Wen, J. Global variation in elevational diversity patterns. *Sci. Rep.* **2013**, *3*, 3007. [CrossRef] [PubMed]

6. Körner, C. The use of 'altitude' in ecological research. *Trends Ecol. Evol.* **2007**, *22*, 569–574. [CrossRef] [PubMed]

7. Rahbek, C. The elevational gradient of species richness: A uniform pattern? *Ecography* **1995**, *18*, 200–205. [CrossRef]

8. Doležal, J.; Šrůtek, M. Altitudinal changes in composition and structure of mountain-temperate vegetation: A case study from the Western Carpathians. *Plant Ecol.* **2002**, *158*, 201–221. [CrossRef]

9. Lomolino, M.V. Elevation gradients of species-density: Historical and prospective views. *Glob. Ecol. Biogeogr.* **2001**, *10*, 3–13. [CrossRef]

10. McCain, C.M. Elevational gradients in diversity of small mammals. *Ecology* **2005**, *86*, 366–372. [CrossRef]

11. Kluge, J.; Kessler, M.; Robert, R.; Dunn, R.R. What drives elevational patterns of diversity? A test of geometric constraints, climate and species pool effects for pteridophytes on an elevational gradient in Costa Rica. *Glob. Ecol. Biogeogr.* **2006**, *15*, 358–371. [CrossRef]

12. Rai, H.; Khare, R.; Baniya, C.B.; Upreti, D.K.; Gupta, R.K. Elevational gradients of terricolous lichen species richness in the Western Himalaya. *Biodivers. Conserv.* **2015**, *24*, 1155–1174. [CrossRef]

13. Fu, C.Z.; Hua, X.; Li, J.; Chang, Z.; Pu, Z.C.; Chen, J.K. Elevational patterns of frog species richness and endemic richness in the Hengduan Mountains, China: Geometric constraints, area and climate effects. *Ecography* **2006**, *29*, 919–927. [CrossRef]

14. Trigas, P.; Panitsa, M.; Tsiftsis, S. Elevational gradient of vascular plant species richness and endemism in Crete—The effect of post-isolation mountain uplift on a continental island system. *PLoS ONE* **2013**, *8*, e59425. [CrossRef] [PubMed]

15. Colwell, R.K.; Rahbek, C.; Gotelli, N.J. The mid-domain effect and species richness patterns: What have we learned so far? *Am. Nat.* **2004**, *163*, E1–E23. [CrossRef] [PubMed]

16. Carpenter, C. The environmental control of plant species density on a Himalayan elevation gradient. *J. Biogeogr.* **2005**, *32*, 999–1018. [CrossRef]

17. Bachman, S.; Baker, W.J.; Brummitt, N.; Dransfield, J.; Moat, J. Elevational gradients, area and tropical island diversity: An example from the palms of New Guinea. *Ecography* **2004**, *27*, 299–310. [CrossRef]

18. Elsen, P.R.; Tingley, M.W. Global mountain topography and the fate of montane species under climate change. *Nat. Clim. Chang.* **2015**, *5*, 772–776. [CrossRef]

19. Currie, D.J.; Kerr, J.T. Tests of the mid-domain hypothesis: A review of the evidence. *Ecol. Monogr.* **2008**, *78*, 3–18. [CrossRef]

20. Vetaas, O.R. Realized and potential climate niches: A comparison of four *Rhododendron* tree species. *J. Biogeogr.* **2002**, *29*, 545–554. [CrossRef]

21. Körner, C. *Alpine Plant Life: Functional Plant Ecology of High Mountain Ecosystem*, 2nd ed.; Springer: Berlin/Heidelberg, Germany, 2003.

22. Wang, Z.H.; Fang, J.Y.; Tang, Z.Y.; Lin, X. Patterns, determinants and models of woody plant diversity in China. *Proc. R. Soc. B. Biol. Sci.* **2010**. [CrossRef] [PubMed]

23. Brown, J.H.; Lomolino, M.V. *Biogeography*, 2nd ed.; Sinauer Associates: Sunderland, MA, USA, 1998.

24. Wang, G.H.; Zhou, G.S.; Yang, L.M.; Li, Z.Q. Distribution, species diversity and life-form spectra of plant communities along an altitudinal gradient in the northern slopes of Qilianshan Mountains, Gansu, China. *Plant Ecol.* **2003**, *165*, 169–181. [CrossRef]

25. Bhattarai, K.R.; Vetaas, O.R.; Grytnes, J.A. Fern species richness along a central Himalayan elevational gradient, Nepal. *J. Biogeogr.* **2004**, *31*, 389–400. [CrossRef]

26. Grime, J.P. *Plant Strategies and Vegetation Processes*; John Wiley & Sons, Ltd.: Chichester, UK, 1979.

27. Kubota, Y.; Murata, H.; Kikuzawa, K. Effects of topographic heterogeneity on tree species richness and stand dynamics in a subtropical forest in Okinawa Island, southern Japan. *J. Ecol.* **2004**, *92*, 230–240.

28. Michalet, R.; Maalouf, J.-P.; Choler, P.; Clément, B.; Rosebery, D.; Royer, J.-M.; Schöb, C.; Lortie, C.J. Competition, facilitation and environmental severity shape the relationship between local and regional species richness in plant communities. *Ecography* **2015**, *38*, 335–345. [CrossRef]

29. Whittaker, R.J.; Katherine, J.W. Scale and species richness: Towards a general, hierarchical theory of species diversity. *J. Biogeogr.* **2001**, *28*, 453–470. [CrossRef]

30. Ergeton, J.J.G.; Wilson, S.D. Plant competition over winter in alpine shrubland and grassland, Snowy Mountains, Australia. *Arct. Alp. Res.* **1993**, *25*, 124–129.

31. Tang, Z.-Y.; Ke, J.-H. Altitudinal patterns of plant species diversity in Mt. Niubeiliang, Qinling Mountains. *Biodivers. Sci.* **2004**, *12*, 108–114.

32. Wang, Z.; Tang, Z.; Fang, J. Altitudinal patterns of seed plant richness in the Gaoligong Mountains, south-east Tibet, China. *Divers. Distrib.* **2007**, *13*, 845–854. [CrossRef]

33.	Aiba, S.; Kohyama, T. Tree species stratification in relation to allometry and demography in a warm-temperate rain forest. *J. Ecol.* **1996**, *84*, 207–218. [CrossRef]
34.	Ryan, M.G.; Yoder, B.J. Hydraulic limits to tree height and tree growth. *Bioscience* **1997**, *47*, 235–242. [CrossRef]
35.	Singh, K.K.; Kumar, S.; Rai, L.K.; Krishn, A.P. Rhododendrons conservation in the Sikkim Himalaya. *Curr. Sci.* **2003**, *85*, 602–606.
36.	Fang, M.; Fang, R.; He, M.; Hu, L.; Yang, H.; Qin, H.; Min, T.; Chamberlain, D.F.; Stevens, P.F.; Wallace, G.D.; et al. *Flora of China*; Science Press: Beijing, China; Missouri Botanical Garden Press: St. Louis, MO, USA, 2005; Volume 14, pp. 242–517.
37.	Wu, Z.Y. *Flora Yunnanica (Tomus 4)*; Science Press: Beijing, China, 1986.
38.	Kessler, M. Elevational gradients in species richness and endemism of selected plant groups in the central Bolivian Andes. *Plant Ecol.* **2000**, *149*, 181–193. [CrossRef]
39.	Yang, Y.M.; Tian, K.; Hao, J.M.; Pei, S.J.; Yang, Y.X. Biodiversity and biodiversity conservation in Yunnan, China. *Biodivers. Conserv.* **2004**, *13*, 813–826. [CrossRef]
40.	Wu, Z.Y.; Zhu, Y.C. *Vegetation of Yunnan*; Science Press: Beijing, China, 1987.
41.	Wu, Z.Y. *Flora Yunnanica (Tomus 5)*; Science Press: Beijing, China, 1991.
42.	Wolf, J.H.D.; Flamenco-Sandoval, A. Patterns in species richness and distribution of vascular epiphytes in Chiapas, Mexico. *J. Biogeogr.* **2003**, *30*, 1689–1707. [CrossRef]
43.	Qian, H.; Ricklefs, R.E. Global concordance in diversity patterns of vascular plants and terrestrial vertebrates. *Ecol. Lett.* **2008**, *11*, 547–553. [CrossRef] [PubMed]
44.	Allen, R.G.; Pereira, L.S.; Raes, D.; Smith, M. *Crop Evapotranspiration-Guidelines for Computing Crop Water Requirements-FAO Irrigation and Drainage Paper 56*; FAO-Food and Agriculture Organization of the United Nations: Rome, Italy, 1998; Volume 156, p. 178.
45.	Colwell, R.K. RangeModel: Tools for exploring and assessing geometric constraints on species richness (the mid-domain effect) along transects. *Ecography* **2008**, *31*, 4–7. [CrossRef]
46.	Borcard, D.; Legendre, P.; Drapeau, P. Partialling out the spatial component of ecological variation. *Ecology* **1992**, *73*, 1045–1055. [CrossRef]
47.	Oksanen, J.; Blanchet, F.G.; Kindt, R.; Legendre, P.; O'Hara, R.B.; Simpson, G.L.; Solymos, P.; Stevens, M.H.; Wagner, H. Vegan: Community Ecology Package, R Package Version 2.4-3. 2010. Available online: https://cran.r-project.org/web/packages/vegan/index.html (accessed on 16 May 2017).
48.	R Core Team. *R: A Language and Environment for Statistical Computing*; R Foundation for Statistical Computing: Vienna, Austria, 2017; Available online: http://www.R-project.org/ (accessed on 12 May 2017).
49.	Legendre, P.; Gallagher, E.D. Ecologically meaningful transformations for ordination of species data. *Oecologia* **2001**, *129*, 271–280. [CrossRef] [PubMed]
50.	Peres-Neto, P.; Legendre, P.; Dray, S.; Borcard, D. Variation partitioning of species data matrices: Estimation and comparison of fractions. *Ecology* **2006**, *87*, 2614–2625. [CrossRef]
51.	Tao, J.; Zang, R.G.; Yu, C.Y. Altitudinal patterns of plant communities and species diversity in the Habaxueshan Mountains, Yunnan, China. *Sci. Silvae Sin.* **2011**, *47*, 1–6.
52.	Li, X.-H.; Wang, D.-D.; Li, H. Feature of Ericaceae flora in Gaoligong Mountains. *J. West China For. Sci.* **2017**, *46* (Suppl. II), 112–118. Available online: http://159.226.100.28/ZK/detail.aspx?id=674277914 (accessed on 26 July 2018). (In Chinese)
53.	O'Brien, E.M.; Field, R.; Whittaker, R.J. Climatic gradients in woody plant (tree and shrub) diversity: Water-energy dynamics, residual variation, and topography. *Oikos* **2000**, *89*, 588–600. [CrossRef]
54.	Smith, D.D.; Sperry, J.S. Coordination between water transport capacity, biomass growth, metabolic scaling and species stature in co-occurring shrub and tree species. *Plant Cell Environ.* **2014**, *37*, 2679–2690. [CrossRef] [PubMed]
55.	Harms, K.H.; Condit, R.; Hubbell, S.P.; Foster, R.B. Habitat associations of trees and shrubs in a 50-ha neotropical forest plot. *J. Ecol.* **2001**, *89*, 947–959. [CrossRef]
56.	Brodribb, T.J.; Field, T.S. Stem hydraulic supply is linked to leaf photosynthetic capacity: Evidence from New Caledonian and Tasmanian rainforests. *Plant Cell Environ.* **2000**, *23*, 1381–1388. [CrossRef]
57.	O'Brien, E.M. Water-energy dynamics, climate, and prediction of woody plant species richness: An interim general model. *J. Biogeogr.* **1998**, *25*, 379–398. [CrossRef]
58.	Cordero, R.A.; Nilsen, R.T. Effects of summer drought and winter freezing on stem hydraulic conductivity of *Rhododendron* species from contrasting climates. *Tree Physiol.* **2002**, *22*, 919–928. [CrossRef] [PubMed]

59. Wang, Y.J.; Wang, J.J.; Lai, L.M.; Jiang, L.H.; Zhuang, P.; Zhang, L.H.; Zheng, Y.R.; Baskin, J.M.; Baskin, C.C. Geographic variation in seed traits within and among forty-two species of *Rhododendron* (Ericaceae) on the Tibetan plateau: Relationships with altitude, habitat, plant height, and phylogeny. *Ecol. Evol.* **2014**, *4*, 1913–1923. [CrossRef] [PubMed]

60. Francis, A.P.; Currie, D.J. A globally consistent richness-climate relationship for angiosperms. *Am. Nat.* **2003**, *161*, 523–536. [CrossRef] [PubMed]

61. Fernàndez-Martínez, J.; Fransi, M.A.; Fleck, I. Ecophysiological responses of *Betula pendula*, *Pinus uncinata* and *Rhododendron ferrugineum* in the Catalan Pyrenees to low summer rainfall. *Tree Physiol.* **2016**, *36*, 1520–1535. [CrossRef] [PubMed]

62. Caroline, M.W.; Hugh, A.L.H.; Brent, J.S. Cold truths: How winter drives responses of terrestrial organisms to climate change. *Biol. Rev.* **2015**, *90*, 214–235.

63. Rajaniemi, T.K. Why does fertilization reduce plant species diversity? Testing three competition-based hypotheses. *J. Ecol.* **2002**, *90*, 316–324. [CrossRef]

64. Brown, R.L.; Jacobs, L.A.; Peet, R.K. Species richness: Small scale. In *Encyclopedia of Life Sciences*; John Wiley & Sons Ltd.: Hoboken, NJ, USA, 2007.

65. Dobrowski, S.Z.; Swanson, A.K.; Abatzoglou, J.T.; Holden, Z.A.; Safford, H.D.; Schwartz, M.K.; Gavin, D.G. Forest structure and species traits mediate projected recruitment declines in western US tree species. *Glob. Ecol. Biogeogr.* **2015**, *24*, 917–927. [CrossRef]

© 2018 by the authors. Licensee MDPI, Basel, Switzerland. This article is an open access article distributed under the terms and conditions of the Creative Commons Attribution (CC BY) license (http://creativecommons.org/licenses/by/4.0/).

forests

MDPI

Article

Spatial Association and Diversity of Dominant Tree Species in Tropical Rainforest, Vietnam

Hong Hai Nguyen [1], Yousef Erfanifard [2], Van Dien Pham [1], Xuan Truong Le [1], The Doi Bui [1] and Ion Catalin Petritan [3,*]

[1] Faculty of Silviculture, Vietnam National University of Forestry, 02433840 Hanoi, Vietnam; hainh@vfu.edu.vn (H.H.N.); phamvandien100@gmail.com (V.D.P.); Truongfuv@gmail.com (X.T.L.); Doibt@vfu.edu.vn (T.D.B.)
[2] Department of Natural Resources and Environment, Shiraz University, 7144165186 Shiraz, Iran; erfanifard@shirazu.ac.ir
[3] Transilvania University, Sirul Beethoven 1, ROU-500123 Brasov, Romania
[*] Correspondence: petritan@unitbv.ro; Tel.: +40-765-369-782

Received: 5 September 2018; Accepted: 4 October 2018; Published: 7 October 2018

Abstract: Explaining the high diversity of tree species in tropical forests remains a persistent challenge in ecology. The analysis of spatial patterns of different species and their spatial diversity captures the spatial variation of species behaviors from a 'plant's eye view' of a forest community. To measure scale-dependent species-species interactions and species diversity at neighborhood scales, we applied uni- and bivariate pair correlation functions and individual species area relationships (ISARs) to two fully mapped 2-ha plots of tropical evergreen forests in north-central Vietnam. The results showed that (1) positive conspecific interactions dominated at scales smaller than 30 m in both plots, while weak negative interactions were only observed in P2 at scales larger than 30 m; (2) low numbers of non-neutral interactions between tree species were observed in both study plots. The effect of scale separation by habitat variability on heterospecific association was observed at scales up to 30 m; (3) the dominance of diversity accumulators, the species with more diversity in local neighborhoods than expected by the null model, occurred at small scales, while diversity repellers, the species with less diversity in local neighborhoods, were more frequent on larger scales. Overall, the significant heterospecific interactions revealed by our study were common in highly diverse tropical forests. Conspecific distribution patterns were mainly regulated by topographic variation at local neighborhood scales within 30 m. Moreover, ISARs were also affected by habitat segregation and species diversity patterns occurring at small neighborhood scales. Mixed effects of limited dispersal, functional equivalence, and habitat variability could drive spatial patterns of tree species in this study. For further studies, the effects of topographical variables on tree species associations and their spatial autocorrelations with forest stand properties should be considered for a comprehensive assessment.

Keywords: spatial patterns; individual species-area relationship; tropical evergreen mixed forest; competition and facilitation; Vietnam

1. Introduction

One of the fundamental goals of community ecology is to understand the underlying ecological processes that form the patterns of species and species diversity in space. Previous studies have investigated conspecific and heterospecific interactions [1,2], dispersal limitation [3], environmental heterogeneity [2,4,5], neutral theory [6], and the individual species-area relationships [7,8] of plant communities. The authors of these studies proposed the underlying ecological processes that regulate the distribution patterns of species and species diversity in space. The neutral theory assumes demographic equivalence of individuals in terms of their birth, reproduction, and death, regardless of species identity [6]. McGill [9] argued that a stochastic geometry of biodiversity is exposed by intraspecific clustering and independent species placement from other species. Uriarte et al. [10] also found that the majority of species did not respond to the identity of neighbors in Barro Colorado Island, which supports the neutral theory. Recent studies have predicted that stochastic dilution effects may result in species rich communities with independence of species spatial distribution [9,11,12], and even the underlying ecological processes structuring the community are driven by deterministic niche differences. Consequently, the identities of the nearest neighborhood individuals of a given species could be unpredictable because each individual may be neighbored by a different set of competitors (e.g., [13,14]). However, several studies using individual-based analyses of local neighborhoods pointed out that direct plant-plant interactions may operate within local plant neighborhoods on scales smaller than 30 m, and fade away on larger scales [2,10,15]. Under habitat heterogeneity, direct plant interactions on small scales (e.g., facilitation or competition) may be marked by habitat preference on large scales (e.g., shading, nutrients, soil moisture) [2,4,16]. Additionally, habitat conditions (e.g., elevation, soil moisture, aspect) vary typically on large scales along environmental topographical gradients [17,18]. Therefore, using the appropriate null models can account for the effects of habitat heterogeneity and plant interactions [19]. In heterogeneous habitats, using inhomogeneous Poisson processes [20,21] as null models retains the large scales of pattern structure but removes the small-scale correlation structure.

The analysis of scale-dependent patterns of diversity in neighborhoods of individuals may also provide insights into the mechanisms of community assembly. Analysis of individual species-area relationships (ISAR) [7,21] is an efficient technique to quantify changes in species richness and species-specific effects on local diversity. ISAR is used to detect spatial patterns in diversity from the perspective of individual plants, and to relate them to the underlying mechanisms [19]. If positive interactions with other species occur, such as shared responses to abiotic conditions or dispersal by the same frugivores, the target species accumulates and maintains an over-representative proportion of species diversity in its proximity [22]; therefore, it is known as a diversity *accumulator*. In contrast, if negative interactions occur, such as competition for space, it can result in a lower density of heterospecific neighbors. Consequently, there is an under-representative proportion of other species neighboring the focal species, which is therefore named a diversity *repeller*. Hence, the net balance of positive and negative interactions can reduce or elevate the species richness of the neighbors of a focal species, or the richness of neighborhood species may not significantly differ from that of randomly distributed neighbors [22].

In this study, we used point-pattern analyses to examine the species interactions and species diversity in two 2-ha plots of tropical evergreen forest stands in Vietnam, and then investigated the effects of habitat variability on tree species demography. Our general hypothesis is that stochastic effects dilute species associations in species-rich forest communities. Specifically, we hypothesized that: (1) The underlying mechanisms regulating conspecific and heterospecific interactions, such as habitat variability, functional equivalence, limited dispersal or species interaction, are mixed in these species-rich forest communities. (2) Species interactions are commonly observed on local neighborhood scales and blurred on larger scales in species-rich communities. Our aim was to disentangle the effects of environmental variability on species association in space, and the potential ecological processes structuring the spatial patterns of these forests.

2. Materials and Methods

2.1.Study Site and Data Collection

In 2017, two 2-ha study plots (100 m × 200 m) were established at 17°20′11″ N, 106°26′30″ E (P1) and 17°20′15″ N, 106°26′24″ E (P2) in tropical evergreen forest of north-central Vietnam. P1 exhibited an elevation ranging from 119 to 148 m above sea level (a.s.l.) and slope fluctuating from 5 to 40 degrees (Table 1), whereas Plot P2 presented an elevation varying from 137 to 184 m a.s.l. (Table 1), and slope ranging from 5 to 45 degrees, with good drainage. The climate regime characteristic for both plots is tropical monsoon, with an average annual temperature of 23.5 °C and average annual precipitation of 3000 mm. About 60–70% of the total precipitation falls from October to November in the rainy season; the dry season lasts from March to August.

Table 1. Basic description of two sampled plots.

Characteristic	Plot P1	Plot P2
Elevation (m): mean ± standard deviation (min–max)	134 ± 6.3 (119–148)	160 ± 11.1 (137–184)
Slope (degree):mean ± standard deviation (min–max)	20 ± 6.6 (5–40)	26 ± 7.5 (5–45)
Number of individuals	3936	3731
Total basal area (m^2)	48.4	64.6
DBH (cm) (mean, min–max)	8.6 (2.5–79.6)	10.3 (2.5–95.5)
Number of species	61	52
Number of species with one individual	13	7
Number of species with ≥30 individuals	25	28
Number of shared species	47	47
Number of individuals from shared species	3732	3698

All live trees with diameter at breast height (DBH) ≥2.5 cm were mapped, and tree positions and their characteristics (species and DBH) were recorded. The topographic slope and relative coordinates (x,y) of each tree were recorded via a grid system of subplots (10 m × 10 m) by using a laser distance measurer (Leica Disto D2, Leica Geosystems AG, Heerbrugg, Switzerland) and compass. Elevation was calculated for each subplot as the mean of the elevation at its four corners. Both study plots were developed on secondary forest and selectively logged under reduced-impact logging before 2006, and they were highly protected afterward. Seven stumps of *Erythrophloeum fordii* Oliver and *Tarrietia javanica* Blume were recorded in P1, and eighteen logged stumps of *Erythrophloeum fordii*, *Tarrietia javanica*, *Vatica odorata* (Griff.) Symington, and *Garuga pierrei* Guillaumin were found in P2.

2.1. Data Analysis

2.1.1. Uni- and Bivariate Pair Correlation Functions

We used the pair correlation functions [20,23] as summary statistics to quantify the spatial structure of the uni- and bivariate patterns. The pair correlation function $g_{11}(r)$ for the univariate pattern of species 1 can be defined based on the neighborhood density, which is the mean density of trees of species 1 within rings with radius r and width dr centered on the trees of species 1 [23] where λ_1 is the density of species 1 trees in the plot. Therefore, the pair correlation function is the ratio of the observed mean density of trees in the rings to the expected mean density of trees. The univariate pair correlation function $g_{11}(r)$ can be used to find out whether the distribution of a species is random ($g_{11}(r) = 1$), aggregated ($g_{11}(r) > 1$), or segregated ($g_{11}(r) < 1$), and at which distances r these patterns occur. The pair correlation function for bivariate patterns (i.e., species 1 and species 2) follows intuitively: the value of $g_{12}(r)$ is the ratio of the observed mean density of species 2 trees in the rings around species 1 trees to the expected mean density of species 2 trees in these rings [23]. The association of a species pair is that of independence if $g_{12}(r) = 1$, attraction if $g_{12}(r) > 1$, or repulsion if $g_{12}(r) < 1$ at distances r.

The above assessment is applied in cases of environmental homogeneity, where the tree locations are independently and randomly distributed over the entire plot under the null model of complete spatial randomness (CSR). In the case of environmental heterogeneity, where the tree distribution contains areas with low point density, the local neighborhood density is larger than the expected density under CSR; therefore, spurious aggregation appears, which may also obscure an existing small-scale regularity (i.e., virtual aggregation; [4]). Considering this issue, we used the inhomogeneous Poisson process (IPP) as a null model [4,23]. This null model can approximately factor out the effects of heterogeneity by placing the points of the tree distribution only within areas of radius R. It maintains the observed large-scale structure, but removes potential non-random local spatial structures at distances r below R [4]. The constant density of points in the null model of CSR is replaced by an intensity function $\lambda(x,y)$ that varies with location (x,y) in the null model of IPP [4,23]. To estimate the intensity function, a non-parametric kernel estimation of the intensity function based on the Epanechnikov kernel with a bandwidth $R = 50$ m was used. In the estimation, all potential spatial structures in the pattern of the target species on scales up to 50 m were removed, while spatial structures were maintained on larger scales.

To test species association, we used different null models. Under CSR, we applied the null model of independence, in which the locations of tree species 1 are fixed and that of tree species 2 are randomly shifted within the study plot. With environmental heterogeneity, we kept the locations of the first species fixed, and randomized the locations of the second species by using the IPP null model. We assessed both $g_{12}(r)$ and $g_{21}(r)$ for each species association due to the asymmetric interactions of species.

2.1.2. Individual Species—Area Relationship

The ISAR(r) function is the expected number of species within circular areas of radius r, with $a = \pi r^2$, around an arbitrarily chosen individual of a target species t [7]. ISAR is used to analyze the spatial diversity structure in forest ecosystems and combine the species–area relationship with the individual perspective of point pattern analysis [21]. For a species, the ISAR can be estimated as:

$$\text{ISAR}(r) = \sum_{j=1}^{N} \left[1 - P_{tj}(0,r)\right]$$

$P_{tj}(0, r)$ is the emptiness probability that species j is not present in the circle with radius r around individuals of the target species t. If $a = \pi r^2$, the ISAR function can be expressed in terms of circular area a to resemble the common species area relationship [7].

2.1.3. Used Software Package

For both analyses types (pair-correlation function and ISAR), to assess departure from the null models on different scales r, the 5th lowest and 5th highest values of 199 Monte Carlo simulations were used to construct confidence envelopes by using the grid-based software Programita ([23], updated version of February 2014 from http://programita.org/) with spatial resolution of 2 m. Then, the Goodness-of-Fit (GoF) test was applied with $\leq 5\%$ error to reduce type I error inflation [24]. In our study, we selected a distance interval of 0–30 m to assess overall departures from the null model. Significant deviations from the null model were only considered as such if the observed p value of the GoF test was ≤ 0.05.

Thus, the univariate analysis indicated aggregation if the observed $g_{11}(r)$ was above the simulation envelopes and regularity if it was below. Conversely, the bivariate analysis indicated a positive interaction if the observed $g_{12}(r)$ was above the simulation envelopes, and a negative interaction if it was below.

In ISAR analysis, a species is regarded as a diversity accumulator with an approximate α level of 0.05 if the empirical ISAR(r) is above the confidence envelopes on scale r. That means that the target species has more diverse local neighborhoods on scale r than expected by the null model. Conversely, a species is regarded as a diversity repeller if the empirical ISAR(r) is below simulation envelopes on scale r, thereby having a less diverse local neighborhood on scale r than expected by the null model.

3. Results

Plot P2 had fewer individuals and lower species richness, but it had a higher basal area and higher dominance of the common species (having ≥ 30 individuals) than plot P1 (Table 1). In plot

P1, the most common tree species were *Ormosia balansae* Drake, *Garuga pierrei*, *Erythrophfloeum fordii*, *Paviesia annamensis* Pierre, *Castanopsis tonkinensiss* (Rox. Ex Lin.) A., and *Tarrietia javanica*, which contributed 44% of the total basal area and 32% of all individuals. In plot P2, the dominant species were *Cinnamomun obtusifolium* Nees, *Garuga pierrei*, *Ormosia balansae*, *Engelhardtia roxburghiana* Lindl., and *Endosperrmun sinensis* Benth., covering 42% of the total basal area and 22% of all individuals. Figure 1 shows distribution maps of studied species in plot P1 and P2. For characteristics of other common species, see Appendix A Table A1.

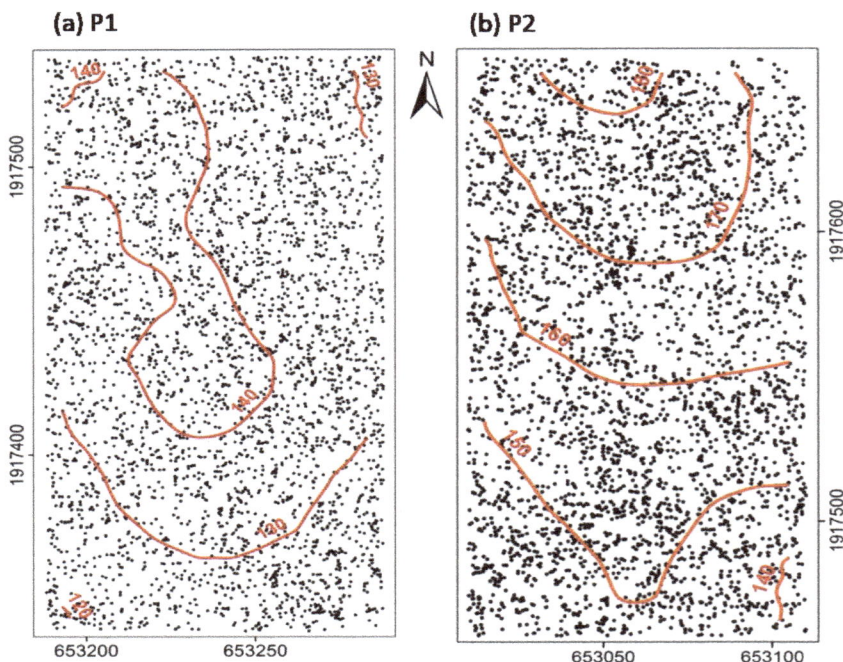

Figure 1. Distribution maps of investigated trees on the two 2-ha plots P1 (**a**) and P2 (**b**) with 10 m contour lines of altitude.

3.1. Species Distribution Patterns

We performed a total of 53 univariate point-pattern analyses for 30 species with at least 30 individuals in both plots (Table A1). The number of species that showed a significant departure from the null models substantially decreased on scales up to 20–30 m and saturated at larger scales (Figure 2).

Of the 25 most abundant species in plot P1, the individual distribution of 15 species (60%) showed a significant departure from the null model of CSR ($p \leq 0.05$) (Figure 2a), the individual distribution of another 12 species (48%) showed aggregated patterns, and three species (12%) had random distribution patterns (not shown) on scales up to 30 m. Strongly aggregated species on different scales were shade intolerant species, including *Bursera tonkinensis* Guillaumin, *T. javanica*, *Polyalthia nemoralis* Aug. DC, *Mallotus kurzii* Hook.f., *C. indica*, *O. balansae*, *E. sinensis*, and *V. odorata*.

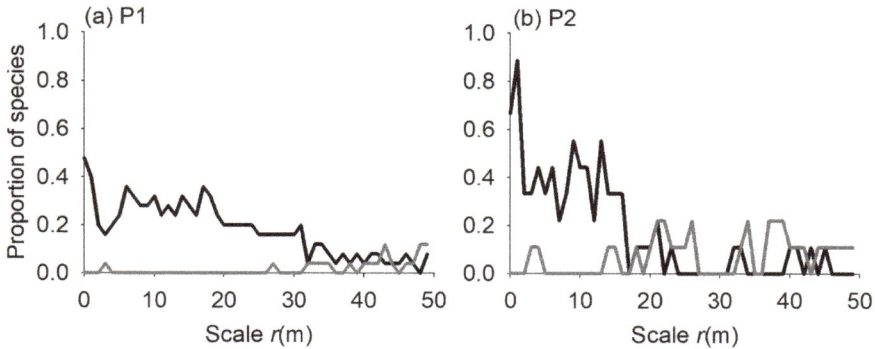

Figure 2. Significant conspecific associations analyzed by the univariate pair-correlation function $g_{11}(r)$ from 199 Monte Carlo simulations under the null models of CSR at P1 (**a**) and IPP at P2 (**b**) at *p* values ≤ 0.05. Black lines indicate aggregation, grey lines indicate regularity.

The heterogeneity of plot P2 was established based on the rejection of the null model of CSR for trees with DBH ≥10 cm of all species on large scales, as described in [23] (results shown in Appendix A Figure A1). Therefore, the null model of IPP was applied for further simulation processes on P2. Among 28 analyzed species in P2, individuals of 9 species (32%) displayed a significant departure under the null model of IPP (*p* ≤ 0.05) (Figure 2b). Of these nine, six species (21%) exhibited the individuals aggregated patterns up to scales of 20 m included *T. javanica, C. obtusifolium, Polyalthia cerasoides* (Roxb.) Bedd, *M. kurzii* and *E. sinensis,* and three species (11%) showed no difference from the null model (data not shown). On scales larger than 20 m, three species (11%), including *C. obtusifolium, M. kurzii* and *E. sinensis,* also showed regular distribution patterns.

In comparison, positive conspecific interactions dominated on small scales up to 30 m, while weak negative interactions were only observed in P2 on large scales.

3.2. Species Associations

For species associations, 600 species pairs were tested for P1 (Figure 3a) and 756 species pairs were tested for P2 (Figure 3b). Most of the species pairs (i.e., 89.7% in P1 and 81.8% in P2) showed no significant deviation from the null models (independence) for species associations at *p* ≤ 0.05. In P1, 12 species pairs (2%) were positively associated (attraction) on scales smaller than 30 m, and 14 species pairs (2.3%) were negatively associated (repulsion) on these scales at *p* ≤ 0.05 (Figure 3a). The number of repulsions were higher than that of attractions for all scales up to 50 m. Positive associations were found in species such as *P. nemoralis, Alangium ridleyi* King, *Vitex trifolia* L., *M. kurzii, Litsea vang* Lecomte, *E. sinensis, C. indica, V. odorata, O. balansae,* and *C. obtusifolium.* Negative associations were found in species pairs of *M. kurzii* vs. (*G. pierrei, P. annamensis,* and *V. trifolia*); *C. indica* vs. (*P. nemoralis, Koilodepas hainanense* (Merr.) Croizat and *C. obtusifolium*).

In P2, the number of positive associations, with 23 species (3%) pairs balanced that of negative associations, with 12 species (1.5%) combinations for all scales from 0–50 m (Figure 3b). Positive associations were found mainly for *C. obtusifolium, B. tonkinensis, P. annamensis, O. balansae,* and *Symplocos laurina* (Retz.) Wall. Ex G. while negative associations were observed in *P. nemoralis, M. kurzii, Canarium album* (Lour.) DC., and *Gironniera Subaequalis* Planch.

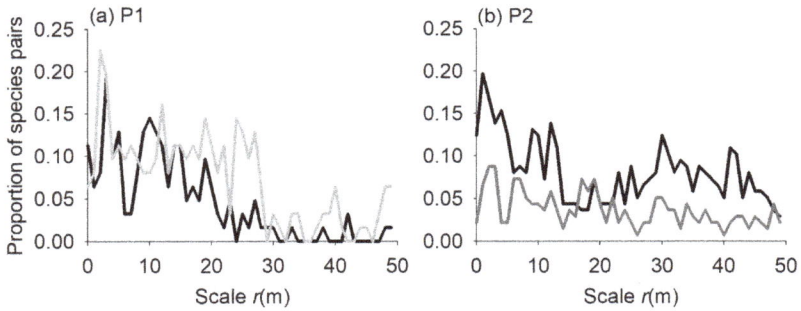

Figure 3. Significant heterospecific associations analyzed by the bivariate pair-correlation function $g_{12}(r)$ from 199 Monte Carlo simulations at p values ≤ 0.05. The null models were independence at P1 (**a**), and pattern 1 fixed and pattern 2 randomized by inhomogeneous Poisson process at P2 (**b**). Black lines indicate attraction, grey lines indicate repulsion.

3.3. Individual Species Area Relationship

ISAR analyses for 25 focal species of P1 were performed and 10 species (40%) were significantly different from the null model of CSR at $p \leq 0.05$ (Figure 4a,b). The results showed a similar trend, where 10 focal species had up to 40 different species within neighborhoods of 40 m; species identity slightly differed on larger scales up to 50 m (Figure 4a). Among these, the number of accumulators (for example, *B. tonkinensis*, *T. javanica*, *P. annamensis*, *Amoora dasyclada* C.Y. Wu, and *O. balansae*) dominated and dropped linearly from 0–10 m and slowly decreased on larger scales, while that of repellers (such as *M. kurzii*, *Syzygium wightianum* Wall. Ex Wig.&Arn., *Garcinia oblongifolia* Chanp. Ex Benth., and *V. odorata*) dominated on spatial scales of 25–50 m (Figure 4b).

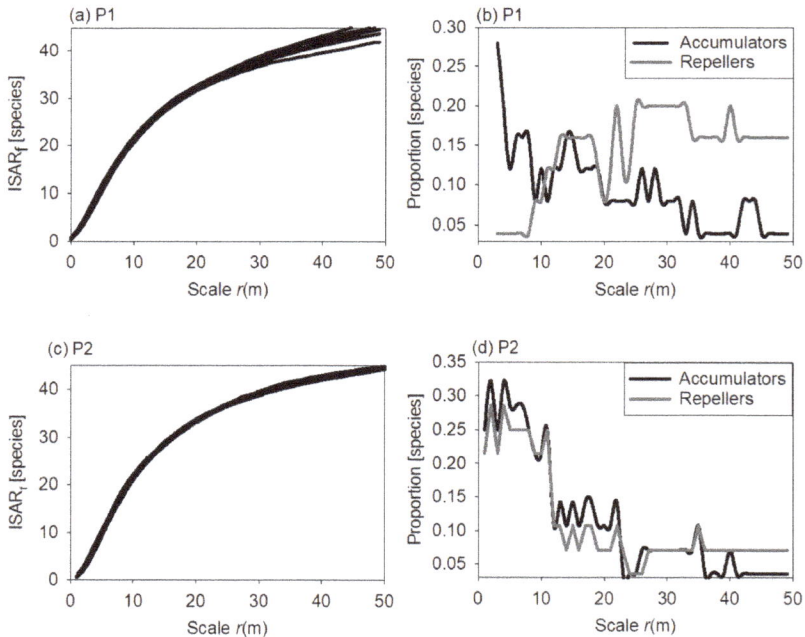

Figure 4. Results of ISAR analyses from 199 Monte Carlo simulations under null models of CSR in P1 (**a–b**) and IPP in P2 (**c–d**); significant at $p \leq 0.05$.

In P2, 28 focal species were analyzed, and 10 species (35.7%) showed a significant departure from the null model of IPP at $p \leq 0.05$ under the GoF test (Figure 4c,d). The ISARs for these 10 focal species were similar to the explored 45 different species neighborhoods (Figure 4c). The number of accumulators (such as *T. javanica*, *C. obtusifolium*, *P. annamensis*, *G. pierrei*, *S. laurina*, *A. dasyclada*, *E. roxburghiana*, *P. cerasoides*, and *Madhuca pasquieri* (Dubard) H.J. Lam) was slightly higher than that of repellers (e.g, *M. kurzii*) on scales up to 10 m (Figure 4d). These numbers decreased linearly up to the scale of 25 m, and became lower than that of repellers on larger scales of 35–50 m.

4. Discussion

We performed spatial pattern analysis for 53 intra- and 1056 interspecific associations respectively, within two fully mapped 2-ha plots of tropical evergreen forest stands. One methodological challenge was to separate first-order effects (e.g., caused by environmental heterogeneity) and second-order effects (e.g., small-scale tree–tree interactions), in order to identify appropriate biological interpretations of the underlying mechanisms or processes regulating spatial association of tree species in space. To deal with this challenge, we applied different null models according to different environmental conditions (less homogeneity in P2). Our findings are consistent with the general hypotheses that (1) stochastic effects dilute species associations in highly diverse communities, making them weak on average, and (2) non-random patterns occur mainly on local neighborhood scales.

Dominance of diversity accumulators occurred on small scales, while diversity repellers were more frequent on larger scales up to 50 m. The assemblage of species richness varied up to scales of 30 m. *M. kurzii* was found to be a diversity repeller in both investigated plots; this finding was also in accordance with results of species association analyses.

4.1. Species Distribution Patterns

Aggregation is a common distribution pattern of tree species in nature, particularly in species-rich tropical forests [2,25,26]. Aggregated distributions may result mainly from limited seed dispersal [27] and/or environmental heterogeneity [2,4]. In this study, the null models of CSR and IPP were applied to P1 and P2, respectively, to control for a variety of habitat conditions in space. The results showed that aggregated patterns were dominant in both study sites with 48% in P1 and 21% in P2. *T. javanica*, *M. kurzii* and *E. sinensis* were aggregated in both plots, while segregated patterns were found mainly in P2 on scales larger than 20 m, e.g., spatial arrangement of *M. kurzii* and *E. sinensis*. These species are shade intolerant (Table A1), having similar habitat requirements [28]; therefore, they showed similar spatial distributions. Debski et al. [29] concluded that species sharing a habitat preference exhibit similar distribution patterns. In our study, the evidence suggests that on small scales, dispersal limitation could be one of the main drivers of species aggregation, while environmental variability could regulate species distribution on larger scales by separating conspecific trees in space [1,2]. Moreover, a wider extent of the clumped patterns of tree species in P1 than P2, and the occurrence of a dispersed pattern in P2, suggest effects of scale separation on scales of 30 m.

4.2. Species Associations

Low numbers of non-neutral interactions between tree species were observed in both plots. In comparison, positive associations were more frequent in P2 than P1, while negative associations exhibited the opposite trend. The effect of scale separation by habitat heterogeneity on interspecies associations occurred on scales of 30 m. It is not surprising that only about 5% of all species–species pairs where significantly non-neutral (i.e., attraction or repulsion) at $p \leq 0.05$ in this study. These results can be interpreted as equilibrated neutral bivariate patterns in the spatial distribution of tree species. This argument is in accordance with the unified neutral theory [15], which hypothesizes that species may be functionally identical and drift randomly in abundance. In addition, this viewpoint is also supported by [11], wherein the authors found only 5.8% of non-independent species associations in a Dipterocarp forest, Sri Lanka, and 5.3% in a tropical forest, Barro Colorado Island. In our study, the percentage of non-neutral species interactions is quite similar, regardless of environmental variability between sampled plots. Moreover, negative interactions decreased and disappeared on scales larger than 30 m, showing that scale separation matched environmental heterogeneity. Hai et al. [2] also found that most of the tree–tree interactions occurred on scales smaller than 30 m in a tropical evergreen forest of Vietnam. Hurtt and Pacala [30] argued that if demographic heterogeneity is strong, it will not allow similar or identical responses at the species level, which would leave neighborhood patterns as predicted by coexistence theory. In this study, under habitat variability in P2, competitive interactions between species were prevented, and that could be a consequence of dispersal limitation, and because the best-adapted species are not always able to colonize newly available gaps. In addition, past logging activities could also affect the low frequency of non-neutral interactions between tree species. These findings also support our hypothesis of scaling on local neighborhood interactions.

4.3. Individual Species—Area Relationship

The presence of diversity-accumulating species suggests that negative intraspecific interactions were stronger than negative interspecific interactions [31], and negative density dependence within species was stronger than that between species. The negative density dependence within species is an ecological process which is considered fundamental to maintaining species diversity in tropical forests [16], while also supporting the segregation hypothesis [32]. Wiegand et al. [4] argued that a species with a highly aggregated distribution will be surrounded by more conspecific individuals and fewer heterospecific individuals than expected on average; therefore, it would appear as a diversity repeller. Conversely, a segregated species may be surrounded by more heterospecific individuals and would appear as a diversity accumulator.

Using data from the bivariate spatial patterns of 30 species in two contrasting environmental habitats, we found that about 40% of the species effected local tree diversity on spatial scales below 30 m. The results from ISAR analysis showed assemblage of species richness on local scales smaller than 25 m in both study plots, which also agreed with uni- and bivariate point pattern analyses of species associations. This confirmed that species–species interactions in the studied forest stands operate only in local plant neighborhoods on scales smaller than 30 m. Our finding is also similar to that of previous studies in tropical forests [4,10,33], where neighborhood effects responded within a distance of 20–30 m.

Environmental conditions were considered as homogeneous in P1; the diversity accumulators decreased while the diversity repellers increased with increasing spatial scales, emphasizing a clear effect of limited dispersal. Under habitat conditions of variability in P2, both diversity accumulators and repellers decreased with increasing spatial scales, showing mixed effects of limited dispersal and habitat variability. Moreover, neutral diversity patterns can be explained by 'balanced' species–species interactions, as are characteristic of species-rich forests [7]; therefore, this finding is in accordance with a low-significance departure from the null models of species–species interactions in our study.

5. Conclusions

The general hypotheses—that stochastic effects dilute species associations in highly diverse communities making them weak on average and that non-random patterns occur mainly on local neighborhood scales—are confirmed by our findings. Intra-species distribution patterns at local scales may be regulated by dispersal limitation and habitat variability. On small scales, dispersal limitation was the main driver of species aggregation. However, environmental variability regulated species distribution on larger scales by separating conspecific trees in space. Overall, low numbers of non-neutral interactions (i.e., attraction or repulsion) between tree species were observed in both plots, mainly due to tree performance that equilibrates and produces neutral bivariate patterns in the spatial distribution of tree species. Individual species-area relationship analysis confirmed that species–species interactions operate only in local plant neighborhoods on scales smaller than 30 m.

Moreover, individual species-area relationships are also affected by the mixed effects of limited dispersal and habitat segregation, and occur on local neighborhood scales. For further studies, the effects of topographical variables on tree species associations and their spatial autocorrelations with forest stand properties should be considered for a comprehensive assessment.

Author Contributions: H.H.N., V.D.P., X.T.L. and T.D.B. conceived and designed the experiment and collected the data; H.H.N., V.D.P., X.T.L. and T.D.B. analyzed the data; H.H.N., Y.E., V.D.P., X.T.L. and T.D.B. and I.C.P. wrote the paper.

Funding: This research was funded by Vietnam National Foundation for Science and Technology Development (NAFOSTED) under grant number 106-NN.06-2016.22.

Acknowledgments: We also would like to thank Thorsten Wiegand that kindly made Programita software available for this work and guided data analysis as well as our team work including Ngoc Bac Nguyen, Van Cuong Kim, Van Thoai Bui, Quoc Khanh Truong, Tran Manh Pham and Ngoc Tra Nguyen for data collection. We are grateful for constructive comments and suggestions of two anonymous reviewers and subject editor, which were really helpful to improve the manuscript.

Conflicts of Interest: The authors declare no conflict of interest.

Appendix A

Table A1. Characteristics of common species in both study plots. N—number of individuals, IVI—Important Value Index, (relative abundance + relative basal area)/2, expressed as percentage proportion. DBH—Diameter at Breast Height (mean ± Standard deviation).

No	Tree Species	P1			P2			Shade Tolerance
		N	DBH (cm)	IVI (%)	N	DBH (cm)	IVI (%)	
1	*Garuga pierrei* Guillaumin	282	10.08 ± 10.89	8.985	232	11.30 ± 13.26	7.72	Tolerant
2	*Tarrietia javanica* Blume	383	5.62 ± 6.39	7.285	330	4.52 ± 3.58	5.14	Intolerant
3	*Ormosia balansae* Drake	138	17.05 ± 12.97	7.26	187	14.75 ± 10.81	6.605	Intolerant
4	*Bursera tonkinensis* Guillaumin	384	6.15 ± 4.16	6.72	253	6.67 ± 4.12	4.41	Medium
5	*Paviesia annamensis* Pierre.	240	9.18 ± 7.64	6.025	239	6.94 ± 4.86	4.325	Intolerant
6	*Litsea glutinosa* (Loureiro) C.B. Rob.	229	8.06 ± 6.21	4.965	264	8.26 ± 6.70	5.495	Intolerant
7	*Castanopsis indica* (Rox. ex Lin.) A.	168	10.21 ± 8.27	4.65	-	-		Intolerant
8	*Polyalthia nemoralis* Aug.DC.	303	5.02 ± 1.77	4.58	244	5.53 ± 1.88	3.78	Intolerant
9	*Syzygium wightianum* Wall. ex Wig. & Arn.	179	9.36 ± 7.04	4.405	81	11.56 ± 8.17	1.545	Intolerant
10	*Erythrophloeum fordii* Oliver	63	18.52 ± 15.35	3.96	36	19.33 ± 21.97	2.475	Medium
11	*Mallotus kurzii* Hook.f.	265	4.01 ± 0.98	3.76	114	3.71 ± 0.73	1.63	Intolerant
12	*Amoora dasyclada* C.Y. Wu	148	7.99 ± 6.73	3.285	96	8.89 ± 6.93	2.08	Medium
13	*Cinnamomun obtusifolium* Nees	100	10.71 ± 9.25	3.005	267	13.01 ± 10.59	8.51	Intolerant
14	*Vatica odorata* (Griff.) Symington	48	17.67 ± 15.90	2.945	48	23.46 ± 16.73	3.24	Intolerant
15	*Gironniera Subaequalis* Planch	92	9.71 ± 6.65	2.27	137	11.19 ± 9.28	3.73	Medium
16	*Endospermun sinensis* Benth.	54	11.77 ± 13.18	2.14	83	21.67 ± 13.33	4.63	Intolerant
17	*Sindora cochinchinensis* auct. non Baill.	41	16.52 ± 13.44	2.125	33	15.77 ± 14.19	2.77	Intolerant
18	*Garcinia oblongifolia* Chanp. ex Benth.	121	6.23 ± 4.08	2.115	67	6.22 ± 3.48	1.115	Tolerant
19	*Canarium album* (Lour.) DC.	46	15.01 ± 8.88	1.79	155	11.03 ± 6.04	3.685	Intolerant
20	*Koilodepas hainanense* (Merr.) Croizat	104	5.83 ± 2.61	1.685	80	8.41 ± 4.52	1.545	Tolerant
21	*Cassine glauca* (Rottb.) Kuntze	74	8.41 ± 5.51	1.59	89	8.69 ± 7.66	1.97	Tolerant
22	*Vitex trifolia* L.	33	14.83 ± 9.63	1.305	-	-		Intolerant
23	*Litsea vang* Lecomte	71	6.54 ± 3.30	1.27	76	8.72 ± 4.67	1.5	Intolerant
24	*Symplocos laurina* (Retz.) Wall. ex G.	55	9.31 ± 5.61	1.255	145	11.81 ± 6.86	3.715	Intolerant
25	*Alangium ridleyi* King	40	7.89 ± 5.19	0.81	49	9.10 ± 6.27	1.045	Tolerant
26	*Engelhardtia roxburghiana* Lindl.	-	-		63	28.78 ± 11.91	4.845	Tolerant
27	*Antheroporum pierrei* Gagnep	-	-		47	19.81 ± 7.39	2	Intolerant
28	*Knema pierrei* Warb.	-	-		46	10.05 ± 4.86	0.99	Tolerant
29	*Polyalthia cerasoides* (Roxb.) Bedd.	-	-		33	25.59 ± 20.80	1.4	Intolerant
30	*Madhuca pasquieri* (Dubard) H.J. Lam	-	-		32	14.69 ± 9.76	1.07	Medium

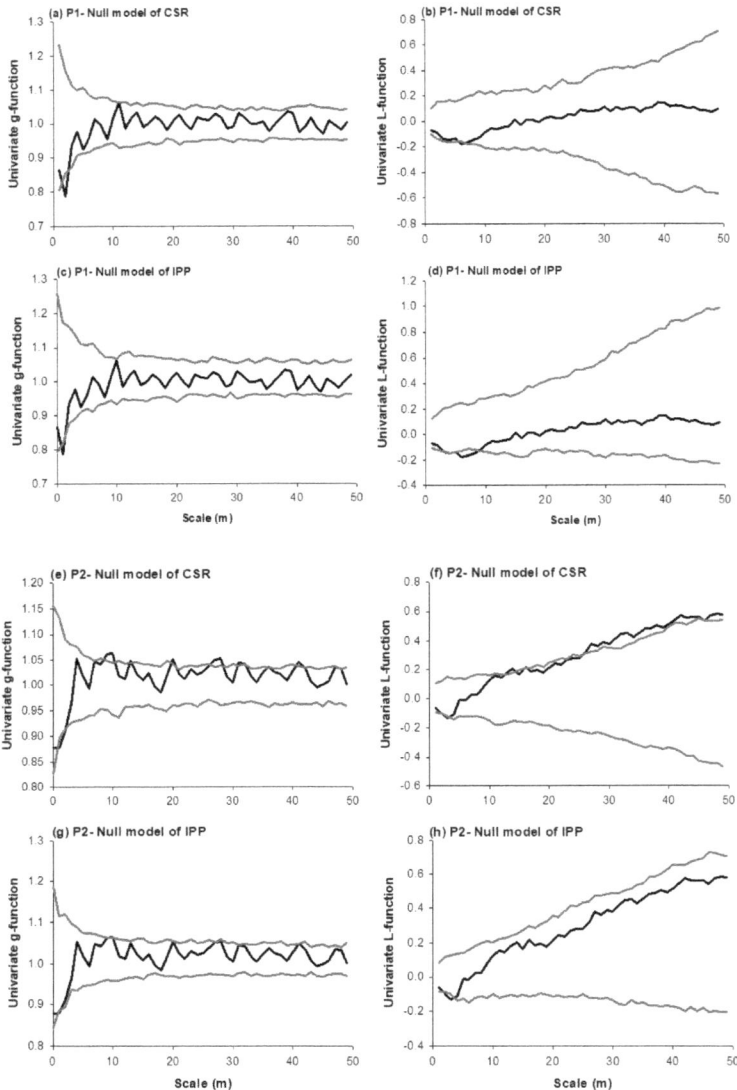

Figure A1. Spatial patterns of all trees with DBH $\geq$10 cm in two plots P1 and P2 using univariate pair-correlation g-function (**a,c,e,g**) and L-function (**b,d,f,h**). Null models were CSR (**a–b,e–f**) and IPP (**c–d,g–h**) with R = 50 m. In the P1, the g-function and L-function showed no large scale departure from the null model of CSR (**a,b**) indicating environmental homogeneity. In the P2, the g-function was significant (>1) for scales larger than 10 m (**e**), the L-function showed a clear departure from scales r > 25 m and did not approach value 0 (**f**) under the null model of CSR. Moreover, the spatial arrangement of trees >10 cm fitted very well under the null model of IPP with R = 50 m (**c–d,g–h**) when analyzed by the g- and L-functions. These evidences significantly exhibited large scale homogeneity at P1 and heterogeneity at P2. Note: L-function is a transformation of Ripley's K-function, $L(r) = (K(r)/\pi)^{0.5} - r$. The g pair-correlation function is the derivative of the K function, $g(r) = K'(r)/(2\pi r)$ and r is radius of the circle from a randomly chosen tree (please see more details about K function in [23]).

References

1. Hai, N.H.; Wiegand, K.; Getzin, S. Spatial distributions of tropical tree species in northern Vietnam under environmentally variable site conditions. *J. For. Res.* **2014**, *25*, 257–268. [CrossRef]
2. Hai, N.H.; Uria-Diez, J.; Wiegand, K. Spatial distribution and association patterns in a tropical evergreen broad-leaved forest of north-central Vietnam. *J. Veg. Sci.* **2016**, *27*, 318–327.
3. Hubbell, S.P. Tree Dispersion, Abundance, and Diversity in a Tropical Dry Forest. *Science* **1979**, *203*, 1299–1309. [CrossRef] [PubMed]
4. Wiegand, T.; Gunatilleke, S.; Gunatilleke, N. Species associations in a heterogeneous Sri Lankan dipterocarp forest. *Am. Nat.* **2007**, *170*. [CrossRef] [PubMed]
5. Getzin, S.; Wiegand, T.; Wiegand, K.; He, F. Heterogeneity influences spatial patterns and demographics in forest stands. *J. Ecol.* **2008**, *96*, 807–820. [CrossRef]
6. Hubbell, S.P. Neutral theory in community ecology and the hypothesis of functional equivalence. *Funct. Ecol.* **2005**, *19*, 166–172. [CrossRef]
7. Wiegand, T.; Savitri Gunatilleke, C.V.; Nimal Gunatilleke, I.A.U.; Huth, A. How individual species structure diversity in tropical forests. *Proc. Natl. Acad. Sci. USA* **2007**, *104*, 19029–19033. [CrossRef] [PubMed]
8. Tsai, C.H.; Lin, Y.C.; Wiegand, T.; Nakazawa, T.; Su, S.H.; Hsieh, C.H.; Ding, T.S. Individual Species-Area Relationship of Woody Plant Communities in a Heterogeneous Subtropical Monsoon Rainforest. *PloS ONE* **2015**, *10*. [CrossRef] [PubMed]
9. McGill, B.J. Towards a unification of unified theories of biodiversity. *Ecol. Lett.* **2010**, *13*, 627–642. [CrossRef] [PubMed]
10. Uriarte, M.; Condit, R.; Canham, C.D.; Hubbell, S.P. A spatially explicit model of sapling growth in a tropical forest: does the identity of neighbours matter? *J. Ecol.* **2004**, *92*, 348–360. [CrossRef]
11. Wiegand, T.; Huth, A.; Getzin, S.; Wang, X.; Hao, Z.; Savitri Gunatilleke, C.V.; Nimal Gunatilleke, I.A.U. Testing the independent species' arrangement assertion made by theories of stochastic geometry of biodiversity. *Proc. R. Soc. B* **2012**, *279*, 3312–3320. [CrossRef] [PubMed]
12. Wang, X.; Wiegand, T.; Kraft, N.J.B.; Swenson, N.G.; Davies, S.J.; Hao, Z.; Howe, R.; Lin, Y.; Ma, K.; Mi, X.; et al. Stochastic dilution effects weaken deterministic effects of niche-based processes in species rich forests. *Ecology* **2016**, *97*, 347–360. [CrossRef] [PubMed]
13. Hubbell, S.P.; Foster, R.B. Biology, Chance, and History and the Structure of Tropical Rain Forest Tree Communities. In *Community Ecology*; Diamond, J., Case, T.J., Eds.; Harper and Row Publishers: New York, NY, USA, 1986; pp. 314–329.
14. Hubbell, S. P. Neutral Theory and the Evolution of Ecological Equivalence. *Ecology* **2006**, *87*, 1387–1398. [CrossRef]
15. Hubbell, S.P. *The Unified Neutral Theory of Biodiversity and Biogeography (MPB-32)*; Princeton University Press: Princeton, NJ, USA, 2001.
16. Wright, J.S. Plant diversity in tropical forests: A review of mechanisms of species coexistence. *Oecologia* **2002**, *130*, 1–14. [CrossRef] [PubMed]
17. Harms, K.E.; Condit, R.; Hubbell, S.P.; Foster, R.B. Habitat associations of trees and shrubs in a 50-ha neotropical forest plot. *J. Ecol.* **2001**, *89*, 947–959. [CrossRef]
18. John, R.; Dalling, J.W.; Harms, K.E.; Yavitt, J.B.; Stallard, R.F.; Mirabello, M.; Hubbell, S.P.; Valencia, R.; Navarrete, H.; Vallejo, M.; et al. Soil nutrients influence spatial distributions of tropical tree species. *Proc. Natl. Acad. Sci. USA.* **2007**, *104*, 864–869. [CrossRef] [PubMed]
19. Rayburn, A.P.; Wiegand, T. Individual species–area relationships and spatial patterns of species diversity in a Great Basin, semi-arid shrubland. *Ecography* **2012**, *35*, 341–347. [CrossRef]
20. Stoyan, D.; Stoyan, H. *Random Shapes and Point Fields: Methods of Geometrical Statistics*; John Wiley & Sons: Chichester, UK, 1994.
21. Wiegand, T.; Moloney, K.A. *Handbook of Spatial Point-Pattern Analysis in Ecology*; Chapman and Hall/CRC: Boca Raton, FL, USA, 2014.
22. Punchi-Manage, R.; Wiegand, T.; Wiegand, K.; Getzin, S.; Huth, A.; Savitri Gunatilleke, C.V.; Nimal Gunatilleke, I.A.U. Neighborhood diversity of large trees shows independent species patterns in a mixed dipterocarp forest in Sri Lanka. *Ecology* **2015**, *96*, 1823–1834. [CrossRef] [PubMed]

23. Wiegand, T.; Moloney, K.A. Rings, circles, and null-models for point pattern analysis in ecology. *Oikos* **2004**, *104*, 209–229. [CrossRef]

24. Loosmore, N.B.; Ford, E.D. Statistical Inference Using the *G* or *K* Point Pattern Spatial Statistics. *Ecology* **2006**, *87*, 1925–1931. [CrossRef]

25. Itoh, A.; Yamakura, T.; Ogino, K.; Lee, H.S.; Ashton, P.S. Spatial distribution patterns of two predominant emergent trees in a tropical rainforest in Sarawak, Malaysia. *Plant Ecol.* **1997**, *132*, 121–136. [CrossRef]

26. Plotkin, J.B.; Potts, M.D.; Leslie, N.; Manokaran, N.; LaFrankie, J.; Ashton, P.S. Species-area Curves, Spatial Aggregation, and Habitat Specialization in Tropical Forests. *J. Theor. Biol.* **2000**, *207*, 81–99. [CrossRef] [PubMed]

27. Dalling, J.W.; Hubbell, S.P.; Silvera, K. Seed dispersal, seedling establishment and gap partitioning among tropical pioneer trees. *J. Ecol.* **1998**, *86*, 674–689. [CrossRef]

28. Chinh, N.N.; Dung, V.V.; Dai, T.D.; Dao, N.K; Hien, N.H; Lien, T.K. *Vietnam Forest Trees*; Agricultural Publishing House: Hanoi, Vietnam, 1996.

29. Debski, I.; Burslem, D.F.R.P.; Palmiotto, P.A.; Lafrankie, J.V.; Lee, H.S.; Manokaran, N. Habitat Preferences of Aporosa in Two Malaysian Forests: Implications for Abundance and Coexistence. *Ecology* **2002**, *83*, 2005–2018. [CrossRef]

30. Hurtt, G.C.; Pacala, S.W. The consequences of recruitment limitation: reconciling chance, history and competitive differences between plants. *J. Theor. Biol.* **1995**, *176*, 1–12. [CrossRef]

31. Comita, L.S.; Hubbell, S.P. Local neighborhood and species' shade tolerance influence survival in a diverse seedling bank. *Ecology* **2009**, *90*, 328–334. [CrossRef] [PubMed]

32. Pacala, S.W.; Levin, S.A. Biologically Generated Spatial Pattern and the Coexistence of Competing Species. In *Spatial Ecology: The Role of Space in Population Dynamics and Interspecific Interactions (MPB-30)*; Tilman, D., Kareiva, P., Eds.; Princeton University Press: Princeton, NJ, USA, 1997; pp. 204–232.

33. Hubbell, S.P.; Ahumada, J.A.; Condit, R.; Foster, R.B. Local neighborhood effects on long-term survival of individual trees in a Neotropical forest. *Ecol. Res.* **2001**, *16*, 859–875. [CrossRef]

© 2018 by the authors. Licensee MDPI, Basel, Switzerland. This article is an open access article distributed under the terms and conditions of the Creative Commons Attribution (CC BY) license (http://creativecommons.org/licenses/by/4.0/).

forests

MDPI

Article

Species Diversity Associated with Foundation Species in Temperate and Tropical Forests

Aaron M. Ellison [1,*]**, Hannah L. Buckley** [2]**, Bradley S. Case** [2]**, Dairon Cardenas** [3]**, Álvaro J. Duque** [4]**, James A. Lutz** [5]**, Jonathan A. Myers** [6]**, David A. Orwig** [1] **and Jess K. Zimmerman** [7]

[1]   Harvard Forest, Harvard University, Petersham, MA 01366, USA; orwig@fas.harvard.edu
[2]   School of Science, Auckland University of Technology, Auckland Central 1010, New Zealand;
     Hannah.Buckley@aut.ac.nz (H.L.B.); Bradley.Case@aut.ac.nz (B.S.C.)
[3]   Instituto Amazónico de Investigaciones Científicas Sinchi, Leticia, Colombia; dcardenas@sinchi.org.co
[4]   Universidad Nacional de Colombia Sede Medellín, Medelliín, Colombia; ajduque@unal.edu.co
[5]   S.J. & Jessie E. Quinney College of Natural Resources and the Ecology Center, Utah State University,
     Logan, UT 84322, USA; james.lutz@usu.edu
[6]   Department of Biology, Washington University in St. Louis, Saint Louis, MO 63130, USA; jamyers@wustl.edu
[7]   Department of Environmental Sciences, University of Puerto Rico–Rio Piedras,
     Rio Piedras 00925, Puerto Rico; jesskz@ites.upr.edu
*   Correspondence: aellison@fas.harvard.edu; Tel.: +1-978-756-6178

Received: 11 January 2019; Accepted: 4 February 2019; Published: 5 February 2019

Abstract: Foundation species define and structure ecological communities but are difficult to identify before they are declining. Yet, their defining role in ecosystems suggests they should be a high priority for protection and management while they are still common and abundant. We used comparative analyses of six large forest dynamics plots spanning a temperate-to-tropical gradient in the Western Hemisphere to identify statistical "fingerprints" of potential foundation species based on their size-frequency and abundance-diameter distributions, and their spatial association with five measures of diversity of associated woody plant species. Potential foundation species are outliers from the common "reverse-J" size-frequency distribution, and have negative effects on alpha diversity and positive effects on beta diversity at most spatial lags and directions. Potential foundation species also are more likely in temperate forests, but foundational species groups may occur in tropical forests. As foundation species (or species groups) decline, associated landscape-scale (beta) diversity is likely to decline along with them. Preservation of this component of biodiversity may be the most important consequence of protecting foundation species while they are still common.

Keywords: abundance; Bray-Curtis; codispersion analysis; Smithsonian ForestGEO; Shannon diversity; Simpson diversity; spatial analysis; species richness

---

1. Introduction

Foundation species (*sensu* [1,2]) define and structure ecological communities and entire ecosystems through bottom-up control of species diversity and non-trophic modulation of energy and nutrient cycles [3]. Foundation species tend to be common and abundant and generally receive less attention from ecologists, regulatory agencies, or conservation biologists who are otherwise focused on the study, management or protection of rare, threatened, or endangered species [4,5]. However, because foundation species are likely to control the distribution and abundance of such rare species, it has been argued that foundation species should be protected before their populations decline to non-functional levels or disappear entirely [6].

Identification of foundation species, however, is challenging. It can take many years—often decades—to acquire sufficient data to distinguish foundation species from species that also are common or abundant but lack "foundational" characteristics. In part, this is because the non-trophic effects of foundation species [3] usually are more subtle and harder to detect than trophic ("who-eats-whom") interactions characteristic of keystone species (*sensu* [7,8]) or competitive interactions characteristic of "dominant" (*sensu* [9]) species. If indeed we are in the midst of Earth's sixth mass extinction [10], it is imperative to prioritize the identification—and subsequent protection—of foundation species because of the unique role they play in structuring diversity of associated species at a range of spatial scales.

Data from networks of observations and experiments focused on species distributions and interactions (e.g., [11–14]) provide an opportunity to use specific statistical criteria to sieve large numbers of species for candidate foundation species. Because one characteristic of foundation species is that they control the diversity of associated species, we would predict that either the number of species (species richness) or measures of diversity weighted by the abundance of individual species would differ between plots that differ in size or abundance of foundation species [15]. We note that there is no explicit or implicit magnitude or directionality of the effect of a foundation species on associated diversity, but that foundational tree species tend to have lower species richness of associated plants whereas foundational herbaceous species tend to have higher species richness of associated plants [15]. Concomitantly, between-plot (beta) diversity is generally enhanced by the size or abundance of foundation species [15].

Here, we leverage our knowledge that *Tsuga canadensis* (eastern hemlock) is a foundation species in eastern North American forests [16] to explore whether statistical patterns of its abundance, size, and effects on associated diversity of woody plant species in the large (35-ha) forest dynamics plot at Harvard Forest have analogs among species in other large forest dynamics plots elsewhere in the Western Hemisphere. The myriad foundational roles of *Tsuga canadensis* have been documented by over a century of research [17], and it was the species that was first used to characterize foundation tree species [2]. The effects of *T. canadensis* on associated species and the forest ecosystems they create [15,16,18] also are similar to those of other dominant forest species since revealed as foundation species [6].

For this study, we first identified candidate foundation species (henceforth "focal species") in five other forests across a temperate-to-tropical latitudinal gradient by comparing their size-frequency and diameter-abundance distributions with those observed for *T. canadensis* at Harvard Forest. We then used codispersion analysis [19] to explore positive and negative spatial associations between foundation species and the diversity of co-occurring woody taxa. Codispersion analysis has been used previously in forest ecology to explore patterns and processes in pairwise species co-occurrences [20], relationships between species occurrences and underlying environmental gradients [21], and temporal changes in spatial patterns of species abundances [22,23].

2. Materials and Methods

2.1. Forest Dynamics Plots in the Americas

We sought to identify potential foundation species and their relationships to diversity of associated woody plant species in six of the large, permanent Smithsonian ForestGEO (Forest Global Earth Observatory: https://forestgeo.si.edu/) Forest Dynamics Plots. These six plots (three in the north temperate region of the United States and three in the American tropics) span a temperate-to-tropical latitudinal and species-richness gradient in the Western Hemisphere (Figure 1). They are part of the 67-plot global network of ForestGEO research sites that have been established continually since 1980 and re-surveyed at ≈5-year intervals to monitor forest dynamics, effects of climate, and more recently, carbon fluxes and associated microbial and macrobial fauna [24]. Importantly, data among these plots are directly comparable because tree-census protocols are common to all ForestGEO plots. All individuals of woody plant species (canopy and understory trees, understory shrubs, woody vines and lianas) ≥1 cm in diameter at breast height ("DBH;" 1.3 m) are tagged, identified, measured, and mapped at ≈5-year intervals using standardized protocols [25].

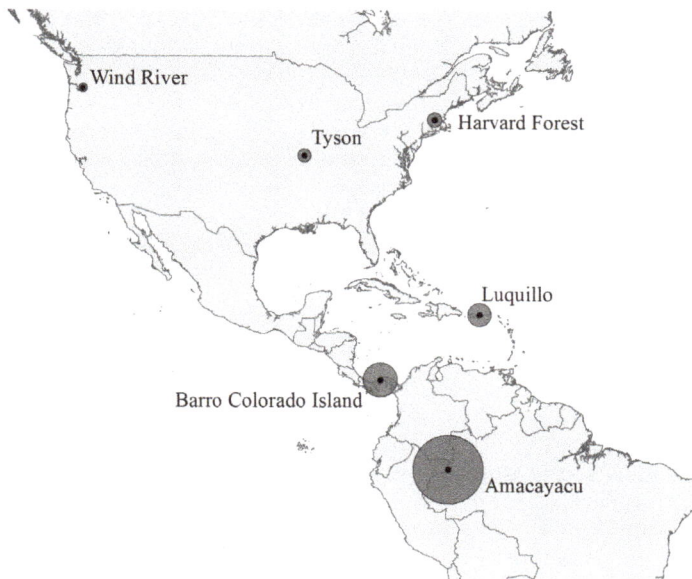

Figure 1. The geographic distribution of the six ForestGEO study sites we assessed for foundation species and their effects on diversity of associated woody plants. The size of the symbols are proportional to species richness of living woody species at each site: Wind River (20 species), Tyson (42), Harvard Forest (51), Luquillo (135), Barro Colorado Island (297), and Amacayacu (1232).

The three temperate plots are all in the USA, and include Wind River in southwest Washington State, Harvard Forest in Massachusetts, and Tyson Research Center in Missouri (Figure 1). The 25.6-ha Wind River plot ("WR"; 45.82 °N) is located in a late-seral *Pseudotsuga menziesii*/*Tsuga heterophylla* forest of the T.T. Munger Research Natural Area of the Gifford Pinchot National Forest. WR is at ≈365 m a.s.l. and receives ≈2500 mm of precipitation yr^{-1}, much of which falls as snow, leading to a 2–3 month summer dry season [26]. Half of the aboveground biomass at WFDP is concentrated in trees ≥ 93 cm DBH [27]. Regeneration of most species occurs preferentially in gaps created by windthrow [28], and large-diameter trees strongly inhibit recruitment [29].

The 35-ha Harvard Forest plot ("HF"; 42.54 °N) is a mix of deciduous and evergreen trees located at the Harvard Forest Long Term Ecological Research Site within the Worcester/Monadnock Plateau ecoregion [30] of transition hardwood/*Pinus strobus*/*Tsuga canadensis* forests [31]. HF is at ≈350 m a.s.l. and receives ≈1150 mm of precipitation (as rain and snow) yr^{-1} [32]. The HF plot is part of the Prospect Hill tract at Harvard Forest and has a mixed history of land-use and disturbance that includes selective logging and partial clearing for pasturage through the late 1800s, various outbreaks of nonnative pathogens and insects in the 1900s, and periodic hurricanes (in 1815 and 1938) and ice storms (most recently in 2008) [33–35].

The 20-ha Tyson plot ("TY"; 38.52 °N) is a late-successional temperate deciduous oak-hickory forest located at Washington University in St. Louis' Tyson Research Center on the northeast edge of the Ozark plateau. TY is at ≈200 m a.s.l. (elevational range 172–233 m a.s.l.), receives ≈ 1000 mm precipitation (as rain and snow) yr^{-1}, and contains strong edaphic and topographic gradients typical of the Missouri Ozarks [36,37]. Disturbance history at TY includes moderate grazing and selective logging in the early 1900s [38] and extreme droughts in 1988 and 2012. TY has remained relatively undisturbed by humans for ≈80 years [38].

The three tropical plots are in the Caribbean, Central America, and South America. The 16-ha Luquillo plot ("LFDP": 18.20 °N) is located within the University of Puerto Rico's El Verde Research

Area of the Luquillo Experimental Forest on the Caribbean island of Puerto Rico. This wet tropical forest ($\approx$3500 mm rainfall yr^{-1}) includes a mix of well-conserved and secondary stands (logged and used for subsistence agriculture approximately a century ago) located at $\approx$380 m a.s.l. [39]. Hurricanes strike the region on average $\approx$ every 50 years. The most recent large hurricanes (Category 2 or higher) that affected the plot before the census data analyzed here were Hurricanes Hugo (1989) and Georges (1998) [40]. These caused defoliation and branch breakage but relatively little damage to stems, resulting in widespread recruitment of shrubs and sapling trees followed by subsequent thinning as the canopy re-establishes [41].

The 50-ha plot on Barro Colorado island, Panama ("BCI"; 9.15 °N) is on a hilltop that became an island when the Panama Canal was filled and Gatun Lake was created in 1914; the associated Barro Colorado Nature Monument has been managed since 1923 by the Smithsonian Institution. BCI is at $\approx$140 m a.s.l. and receives $\approx$2500 mm rainfall yr^{-1}; five–six months of the year make up a pronounced dry season that receives little rain [42,43]. Hurricanes have never been recorded at BCI, and gaps created by single-to-multiple treefalls are the primary type of disturbance observed there [44]. At scales from individual gaps to the entire BCI plot, species diversity is more strongly controlled by recruitment limitation than gap dynamics [44].

The 25-ha Amacayacu plot ("AM": 3.8 °S) is located within the Amacayacu National Natural Park in Colombia, near the joint border of Colombia, Peru, and Brazil. AM is at $\approx$93 m a.s.l. within a wet tropical ($\approx$3200 mm rainfall^{-1}) *terra-firme* forest in the Colombian Amazon [45]. Although topographic relief at Amacayacu varies only over 20 m, it is more tightly associated with woody plant distributions than soil chemistry alone or in combination with topography [45].

2.2. Identifying Candidate Foundation Species

At HF, four trees numerically dominate the assemblage: *Acer rubrum, Quercus rubra, Pinus strobus* and *Tsuga canadensis*. Decades of observational work throughout New England and experimental work at Harvard Forest and elsewhere have consistently supported the hypothesis that *T. canadensis* is a foundation species whereas the other three are not [16,17]. We started, therefore, by asking whether commonly used graphical assessments of forest structure could distinguish *T. canadensis* from the other three dominant species. Specifically, we first plotted a basic size-frequency plot for the entire 35-ha Harvard Forest plot; we used DBH of the living woody stems as the measure of tree size and plotted the species-specific mean DBH against the total number of living stems of each of the 51 living woody species in the plot (Figure 2). This size-frequency plot showed a typical "reverse-J" shape [46], but the aforementioned four dominant species were "outliers", well to the right of the "reverse-J". To explore the size-frequency relationship of these four species in a more fine-grained way—at the scale at which interactions among individual trees is stronger—we rasterized the plot with a 20-m grid [25]. We then plotted the total basal area of each of the four dominant tree species in each of the 20 × 20-m contiguous subplots within HF (Figure 3). This diameter-abundance plot showed that the foundation species *T. canadensis* dominated each subplot both numerically and in total basal area, whereas in any given subplot, the other three dominant species either had many individuals or large total basal area, but not both.

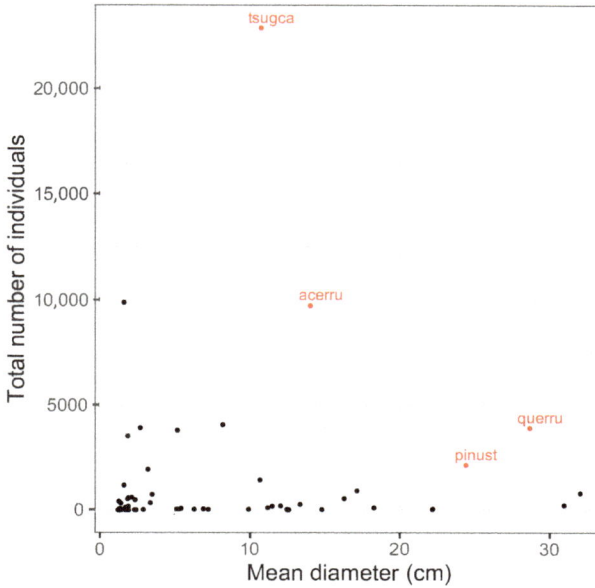

Figure 2. Size-frequency distribution of the 51 living woody species in the Harvard Forest 35-ha plot. Each point is the average diameter of all living stems ≥1-cm diameter for a single species. The four dominant species (labeled in red; key to abbreviations in Table 1) fall well away from the expected "reverse-J" distribution.

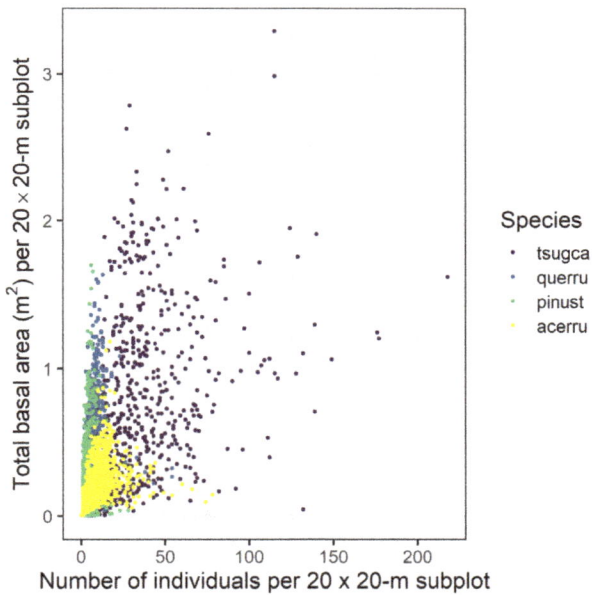

Figure 3. Total basal area and abundance (per 20 × 20-m subplot) of the four dominant species in the Harvard Forest 35-ha plot. Species abbreviations are given in Table 1.

We then used these two graphical "fingerprints" to identify dominant species (in terms of abundance and size) in the other five plots (Figures 4 and 5). This analysis identified distinctive species in terms of departure from the "reverse-J" size-frequency distribution at the three temperate plots (*Tsuga heterophylla* and possibly *Pseudotusga menziesii* and the subcanopy tree *Acer circinatum* at WR; three *Quercus* spp. at TY), and at two of the tropical plots (the canopy tree *Dacryodes excelsa* and the understory palm *Prestoea acuminata* at LFDP; and *Eschweilera coriacea* at Amacayacu) (Figure 4). Other than *P. menziesii*, *A. circinatum*, and *E coriacea*, these species also had basal area-abundance distributions similar to that seen for *Tsuga canadensis* at HF (Figure 5). For the tropical sites, we also included in subsequent analyses species identified by the site PIs, who established these forest dynamics plots and are co-authors of this paper, as being of known importance for forest dynamics (Table 1). Among these, *Oenocarpus mapora* at BCI had a basal area-abundance distribution similar to that seen for *T. canadensis* at HF (Figure 5), but it fell within the "reverse-J" cloud of points in the diameter-abundance distribution (Figure 4). We also included in our analysis several additional species at WR and TY that were identified by the site PIs as ecologically important. These included the canopy species *Abies amabilis* at WR and three *Carya* species at TY; and the understory or lower canopy species *Taxus brevifolia* at WR; and *Asimina triloba*, *Cornus florida*, *Lindera benzoin*, and *Frangula caroliniana* at TY.

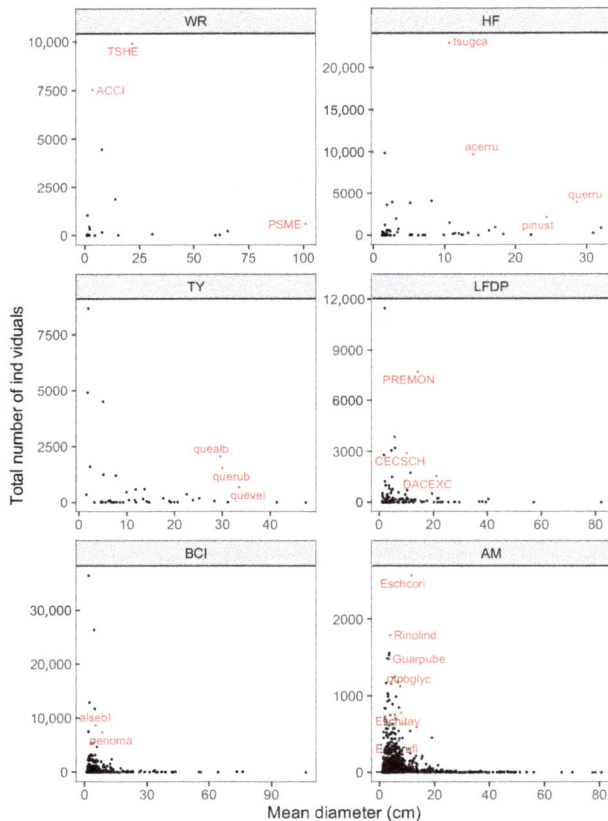

Figure 4. Size-frequency distributions of the species in each of the six studied ForestGeo plots. Panels are ordered from the northernmost temperate (top left) to equatorial (bottom right). Species that do not lie on the "reverse-J" line or that were otherwise are thought to be "important" species (in the tropical plots) are identified in red. Species abbreviations are given in Table 1.

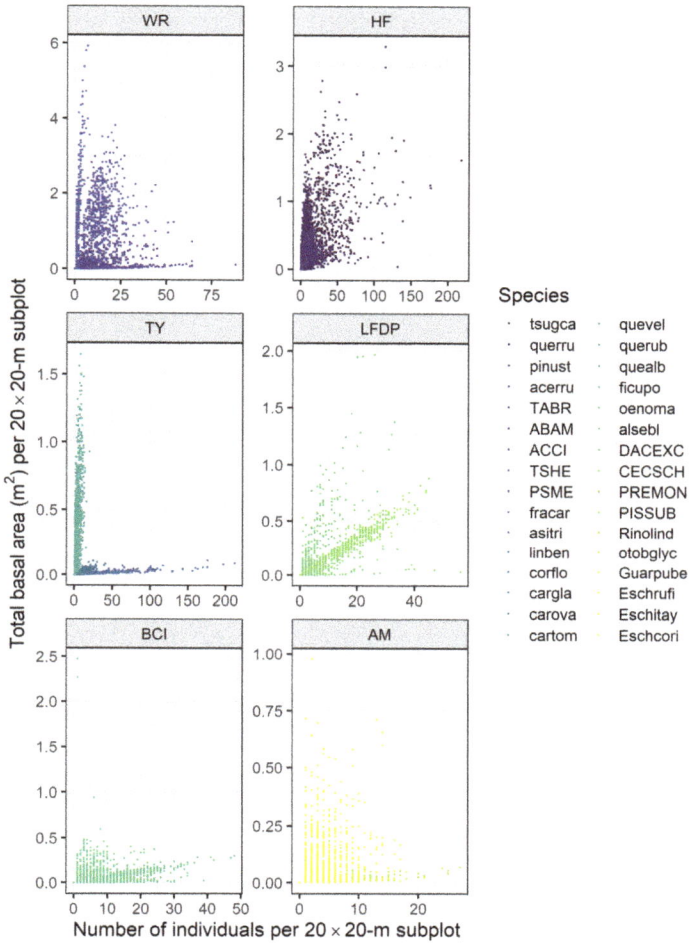

Figure 5. Total basal area and abundance (per 20 × 20-m subplot) of canopy tree species and two tropical understory species identified in Figure 4 as "important" in the six studied ForestGeo plots. Plots are ordered from the northernmost temperate (top left) to equatorial (bottom right). Species abbreviations are given in Table 1.

Table 1. List of focal canopy tree and understory species identified at the six sites using graphic visualizations (Figures 4 and 5), and additional species included following discussions with site PIs. Sites are ordered top-to-bottom from the northernmost temperate to equatorial, and species within sites are ordered by basal area within canopy and understory species. Abbreviations are those used at each site as species codes, and which are used in the legends and labeling of Figures 2–5. *N* is the total number of live individuals of each species in each forest dynamics plot; DBH is the average diameter (in cm) measured at 1.3 m aboveground (standard deviation in parentheses) for all individuals of a given species in each forest dynamics plot; and BA is the total basal area (m^2) of the species in each ForestGEO plot.

Site	Species	Abbreviation	*N*	DBH (SD)	BA
Wind River	*Tsuga heterophylla*	TSHE	9903	21.7 (25.35)	865.90
	Pseudotsuga menzeisii	PSME	564	101.1 (26.38)	483.32
	+*Abies amabilis*	ABAM	4481	7.9 (9.73)	55.08
	Taxus brevifolia	TABR	1848	14.1 (8.15)	38.30
	Acer circinatum	ACCI	7514	3.6 (1.99)	9.93
Harvard Forest	*Tsuga canadensis*	tsugca	22,880	10.8 (12.22)	476.03
	Quercus rubra	querru	3896	28.6 (13.38)	305.79
	Acer rubrum	acerru	9723	14.1 (8.94)	212.01
	Pinus strobus	pinust	2126	24.4 (18.30)	155.30
Tyson Research Center	*Quercus alba*	quealb	2061	29.6 (16.24)	184.06
	Quercus rubra	querub	1547	30.1 (17.63)	147.49
	Quercus velutina	quevel	690	33.5 (13.91)	71.25
	+*Carya tomentosa*	cartom	373	22.6 (12.76)	19.72
	+*Carya glabra*	cargla	188	25.2 (14.96)	12.66
	+*Carya ovata*	carova	190	17.6 (11.16)	6.48
	Cornus florida	corflo	4500	5.1 (2.43)	11.23
	Frangula caroliniana	fracar	8674	2.0 (0.85)	3.32
	Lindera benzoin	linben	4907	1.8 (0.66)	1.47
	Asimina triloba	asitri	1610	2.4 (1.32)	0.94
Luquillo	*Dacryodes excelsa*	DACEXC	1544	21.2 (15.71)	84.28
	+*Cecropia schreberiana*	CECSCH	2902	10.0 (6.65)	32.95
	Prestoea acuminata	PREMON	7707	14.3 (2.96)	128.82
Barro Colorado Island	*Alseis blackiana*	alsebl	8680	5.4 (8.96)	74.79
	Oenocarpus mapora	oenoma	7387	8.6 (1.84)	44.57
Amacayacu	*Eschweilera coriacea*	Eschcori	2574	11.7 (12.53)	59.43
	Eschweilera itayensis	Eschitay	776	7.9 (13.74)	15.34
	Otoba glycyarpa	otobglyc	1124	7.5 (10.00)	13.78
	Eschweilera rufifolia	Eschrufi	414	7.1 (8.72)	4.10
	Guarea pubescens	Guarpube	1483	3.4 (3.95)	3.17
	Rinorea lindeniana	Rinolind	1793	4.0 (2.93)	3.44

+ Additional tree species suggested by site PI; *Understory species

2.3. Metrics of Forest Structure and Species Diversity

For each of the focal species (Table 1), we calculated the total number of stems, the total basal area and the mean basal area at both the plot scale (Table 1) and within cells of the 20-m raster (i.e., each of the contiguous 20 × 20-m subplots within each plot). In each subplot, for the remainder of the woody plant community (i.e., all species other than the focal species), we calculated the total abundance of stems, the total number of species, the Hill-number equivalents of Shannon's diversity and the inverse Simpson's indices [47], and the mean Bray-Curtis dissimilarity (of each subplot relative to each of the other subplots in the raster [48]; the latter is a measure of how "modal" or "outlying" each subplot is in its species composition relative to all other grid cells. For the calculation of Bray-Curtis dissimilarity, empty grid cells (i.e., those with no species other than the focal species) were assigned a dummy-value ("empty") to allow the calculation of all pairwise dissimilarities; this means that all empty cells appear to be maximally different from occupied cells, but maximally similar to each other. For all datasets, we used only the records for the main stem of each species, thus removing any additional stems; this eliminated the problem in spatial analyses of having multiple stems at the same point location. Initial exploration of the data showed that the effect of removing additional stems (primarily of small shrubs) on relative patterns in stem abundance and basal area was negligible for most species across all datasets. The diversity() and vegdist() functions in the R vegan package [49] were used for calculating each diversity metric.

2.4. Codispersion Analysis

Codispersion analysis quantifies the spatial association between two variables measured at individual spatial locations or on a grid [19]. Here, we examined codispersion between the total basal area of each focal species in each subplot of the 20-m raster versus each of the community metrics summarizing the diversity and composition of the other species in same subplot [20]. The spatial association quantified by the codispersion coefficient ranges from −1 to 1, where positive values represent a positive spatial association and negative values represent a negative spatial association. The results are presented as a hemispherical codispersion graph, which shows how the coefficient changes with distance (i.e., spatial lags) and direction (angles around a hemisphere). Thus, differences observed at different spatial lags indicate the scale(s) over which spatial processes are operating within the forest plots, whereas differences in the strength or sign of the correlation in different directions are diagnostic of anisotropic spatial processes [20]. To ensure sufficient sample size when computing codispersion coefficients between measurements made in 20-m grid cells (the size of each subplot), we used spatial lags in 20-m intervals from 20 to one-quarter the size of the minimum dimension of each plot [50]. We calculated codispersion coefficients using code custom-written and compiled in C (to reduce computation time), but with a link to R that allows for easy manipulation of input and output datasets [51].

2.5. Significance Testing with Null Model Analysis

For each dataset, focal species, and community metric combination, we recalculated the codispersion 199 times using the observed focal species total basal area raster and both (1) community metric rasters calculated from the point-pattern of the other species under a random shift around a torus ("toroidal-shift" null model) and (2) a spatially randomized raster of the community metric for the other species in the community ("CSR" null model) [21]. For each null model, the observed codispersion value for each cell was then compared to the distribution of the 199 null values and deemed significant if it fell outside 95 percent of the values (i.e., a two-tailed test). The toroidal-shift null model breaks the spatial association between the focal species and the other species in the community but retains any larger-scale spatial patterns, making this test slightly more conservative because it accounts for spatial patterns caused by the environment. In contrast, the CSR null model breaks the spatial association between the focal species and the other species in the community and simultaneously randomizes any larger-scale spatial patterns caused by environmental patchiness in the plots.

2.6. Data and Code Availability

Each of the ForestGeo plots were established at different times and have been censused every five years. To maximize comparability among datasets, we used data collected within a fifteen year period: the first censuses at Wind River (2010–2011), Harvard Forest (2010–2012), Tyson (2011–2013), and Amacayacu (2008–2009), the third census at Luquillo (2000), and the eighth census at BCI (2015). All datasets are available from the ForestGEO data portal https://ctfs.si.edu/datarequest). R code for all analyses is available on GitHub (https://github.com/buckleyhannah/FS_diversity).

3. Results

3.1. Forest Structure and Species Diversity

Across the entire plots and within the 20 × 20-m subplots, abundance (number of main stems) of the hypothesized foundation species and other focal species was highest in the three temperate plots (Table 1, Figure 6). The average and total basal area of these same species were highest at Wind River (*Tsuga heterophylla*) and Harvard Forest (*Tsuga canadensis*) in the temperate region, at least 2–4-fold greater than the most abundant important species in the tropical plots (*Dacryodes excelsa* at Luquillo and *Alseis blackiana* at BCI) (Table 1, Figure 6). Note that focal species were absent in some of 20 × 20-m subplots at both BCI and AM (white cells in the bottom two rows of Figure 6). In contrast, abundance, richness, Shannon diversity and Simpson's diversity of associated species were substantially higher in the three tropical plots than in any of the temperate plots (Figure 7). Average subplot-wise Bray-Curtis dissimilarity of associated species was higher in the three temperate plots and at Amacayacu than at Luquillo or BCI (Figure 7). Note that only focal species were present in many of the 20 × 20-m subplots at WR and in a handful of 20 × 20-m subplots at HF and TY (white cells in Figure 7).

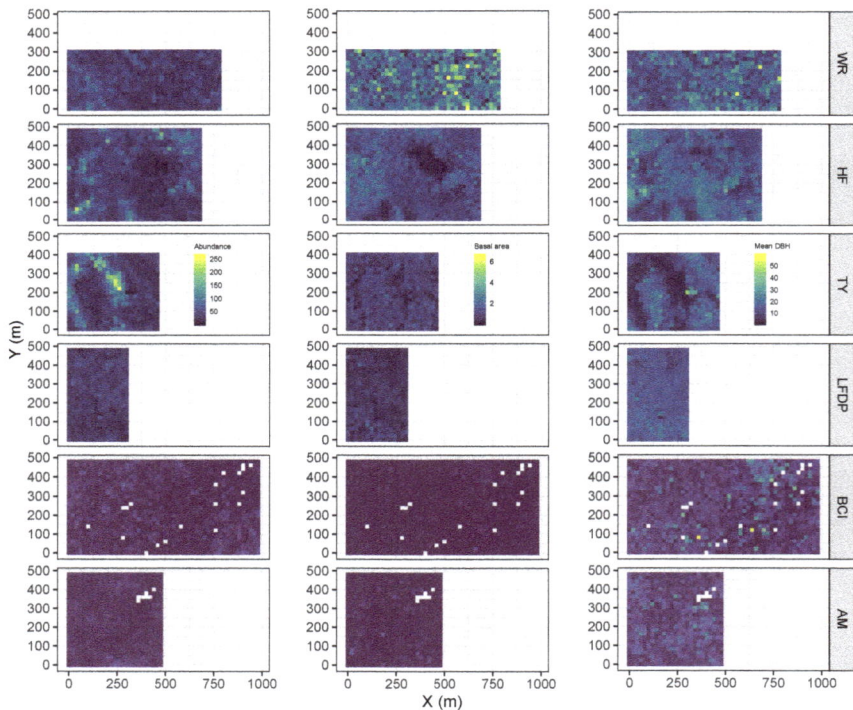

Figure 6. Abundance, total basal area, and mean basal area of all hypothesized foundation species and other focal species (Table 1) in the six 20-m rasterized forest dynamics plots. Sites are ordered top-to-bottom by decreasing latitude, and each plot is scaled relative to the 50-ha BCI plot. White squares in the BCI and AM panels indicate none of the focal species were present in that subplot.

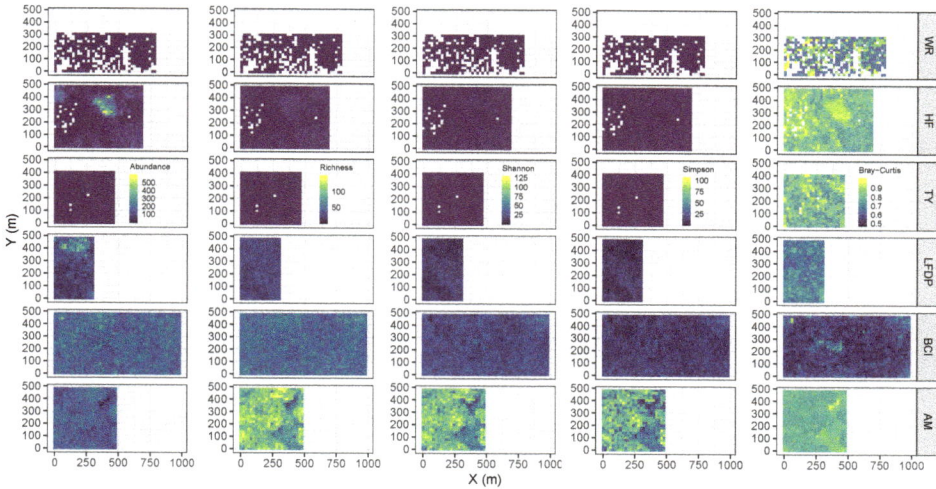

Figure 7. Abundance, species richness, two measures of species diversity, and one measure of beta diversity for all species other than the focal species (Table 1) in the six 20-m rasterized forest dynamics plots. Sites are ordered top-to-bottom by decreasing latitude, and each plot is scaled relative to the 50-ha BCI plot. White squares in the WR, HF, and TY panels indicate that only the focal species were present in that subplot.

3.2. Codispersion Between Focal Species and Associated-species Diversity at Harvard Forest

As with our initial screen for candidate foundation species, in which we used a graphical "fingerprint" observed for *Tsuga canadensis* at HF to screen for important species at other sites (Figures 2–5), we first examined the codispersion statistics and graphs illustrating spatial associations between metrics of woody plant diversity and total basal area of *Tsuga canadensis* relative to the other three dominant tree species at HF within 20 × 20-m subplots.

At all spatial lags (computed from 20–125 m) and directions, the strongest and most consistently significant measures of codispersion between total basal area and metrics of associated-species diversity at Harvard Forest were observed for the foundation species, *T. canadensis* (Figure 8, Table A1). Total basal area of *T. canadensis* was negatively spatially associated at all spatial lags and directions with the total abundance, species richness, Shannon diversity, and inverse Simpson's diversity of associated-woody species, but was positively spatially associated at all spatial lags and directions with average Bray-Curtis dissimilarity (Figure 8, Table A1). All of these codispersion coefficients—at virtually all spatial lags and directions—were significant relative to those generated with the more conservative toroidal-shift null models (Figure 9). The few codispersion coefficients that were not significant when tested with the more conservative toroidal-shift null model were significant when tested with the CSR null model (Figure 10). In contrast, codispersion between total basal area of the other three focal species at HF and all metrics of associated-species diversity were weaker and inconsistently significant (Figures 8–10, Table A1). Only the consistently positive codispersion between *Q. rubra* and either abundance of associated woody species or Bray-Curtis dissimilarity were always significant relative to null expectation (Figures 9 and 10). Overall, the absolute value of the mean or median codispersion between total basal area of *T. canadensis* and metrics of associated-species diversity ranged from 2–10-times greater than the mean or median codispersion between the total basal area of any of the other three dominant tree species and measures of associated-species diversity (Table A1).

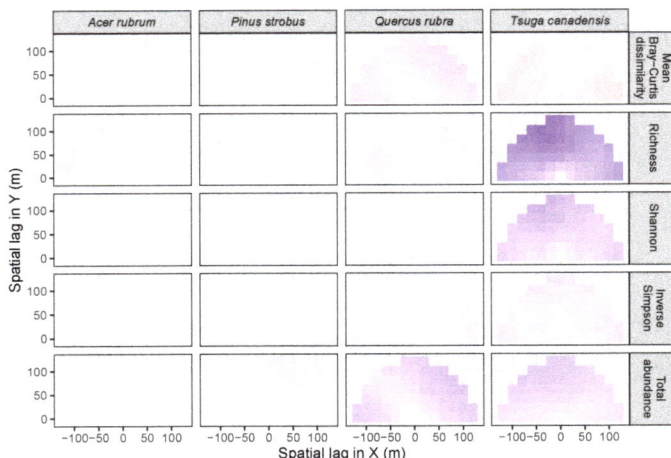

Figure 8. Codispersion between measures of diversity and total basal area of the four focal species in 20 × 20-m subplots in the 35-ha Harvard Forest plot. Codispersion coefficients were calculated for spatial lags ranging from 0–125 m at 20-m intervals. The value of the codispersion coefficient can range from −1 (dark blue) through 0 (white) to 1 (dark red). Statistical significance of each codispersion coefficient is shown in Figure 10. Summaries of codispersion values are given in Table A1).

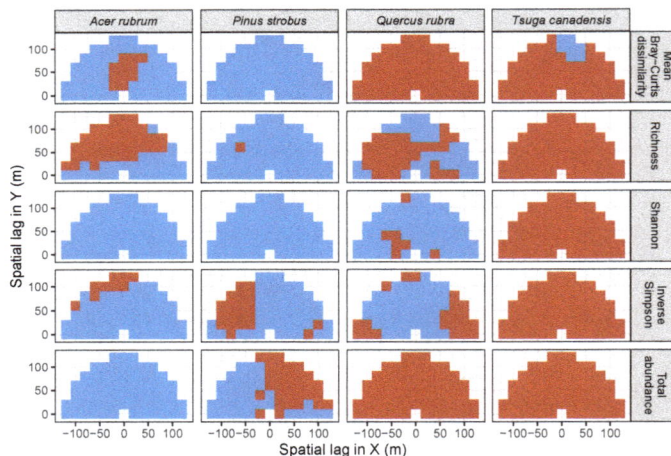

Figure 9. Statistical significance (red: $P \leq 0.05$; blue: $P > 0.05$) of the codispersion coefficients calculated between measures of diversity and total basal area of the four dominant species in 20 × 20-m subplots in the 35-ha Harvard Forest plot (Figure 8). Statistical significance was determined by comparing observed codispersion at each spatial lag with a distribution of 199 spatial randomizations of a toroidal-shift null model.

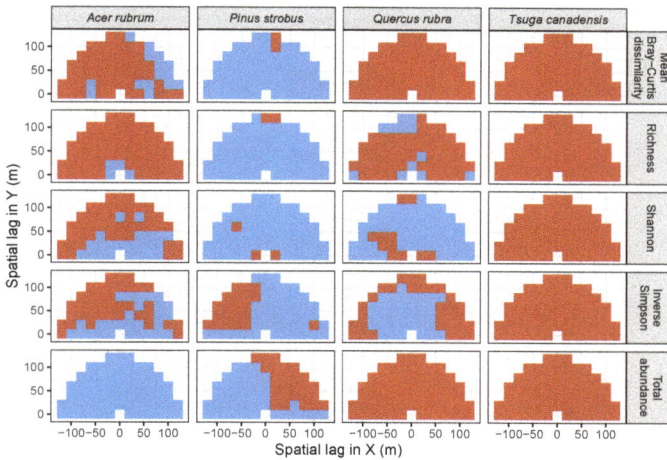

Figure 10. Statistical significance (red: $P \leq 0.05$; blue: $P > 0.05$) of the codispersion coefficients calculated between measures of diversity and total basal area of the four dominant species in 20×20-m subplots in the 35-ha Harvard Forest plot (Figure 8). Statistical significance was determined by comparing observed codispersion at each spatial lag with a distribution of 199 spatial randomizations of a CSR null model.

3.3. Codispersion Between Focal Tree Species and Associated-Species Diversity at the Other Five Sites

3.3.1. Trees at temperate sites

Observed patterns of codispersion between total basal area of focal species and metrics of associated-species diversity at WR and TY were qualitatively similar to those observed for the non-foundation species at HF. At WR, codispersion between total basal area of *Tsuga heterophylla* and all measures of diversity was consistently negative (Figure 11) but only significant at all lags and directions for abundance and richness of associated species (Figure A1). Values of codispersion at WR between *T. heterophylla* and associated-species diversity were much lower than those observed for *T. canadensis* but similar to those observed for *Q. rubra* at HF (Table A1). Codispersion between total basal area of *Pseudotsuga menziesii* and Bray-Curtis similarity was positive, and was significant for about two-thirds of the lags and directions when assessed with the toroidal-shift null model (Figure 12) but not the CSR null model (Figure A1).

At TY, codispersion between associated-species abundance and richness and basal area of two of the oaks—*Quercus alba* and *Quercus rubra*—was negative and consistently significant (Figures 13 and 14 and Figure A2). Codispersion between Shannon diversity of associated-species and basal area of *Q. rubra* was also consistently and significantly negative (Figures 13 and 14 and Figure A2). Codispersion between Bray-Curtis dissimilarity and basal area of *Q. alba* and *Carya ovata* were significantly positive and negative, respectively (Figures 13 and 14 and Figure A2). As at WR, the range of observed, significant, codispersion values measured at TY paralleled those measured for *Q. rubra* at HF (Table A1). A notable exception were the values of codispersion between basal area of *Q. rubra* at TY and associated-species richness, which were nearly identical to that observed for *T. canadensis* at HF (Table A1).

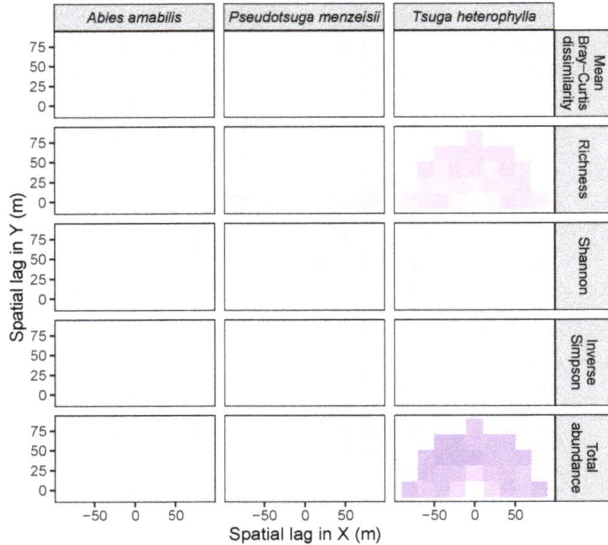

Figure 11. Codispersion between measures of diversity and total basal area of the three focal tree species in 20 × 20-m subplots in the 25.6-ha Wind River plot. Codispersion coefficients were calculated for spatial lags ranging from 0–105 m at 20-m intervals. The value of the codispersion coefficient can range from −1 (dark blue) through 0 (white) to 1 (dark red). Statistical significance of each codispersion coefficient is shown in Figures 12 and A1. Summaries of codispersion values are given in Table A1).

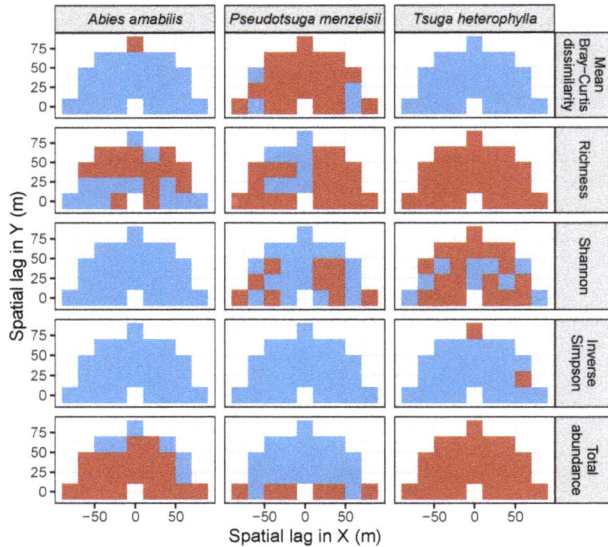

Figure 12. Statistical significance (red: $P \leq 0.05$; blue: $P > 0.05$) of the codispersion coefficients calculated between measures of diversity and total basal area of the three focal tree species in 20 × 20-m subplots in the 25.6-ha Wind River plot (Figure 11). Statistical significance was determined by comparing observed codispersion at each spatial lag with a distribution of 199 spatial randomizations of a toroidal-shift null model. See Figure A1 for significance testing using a CSR null model.

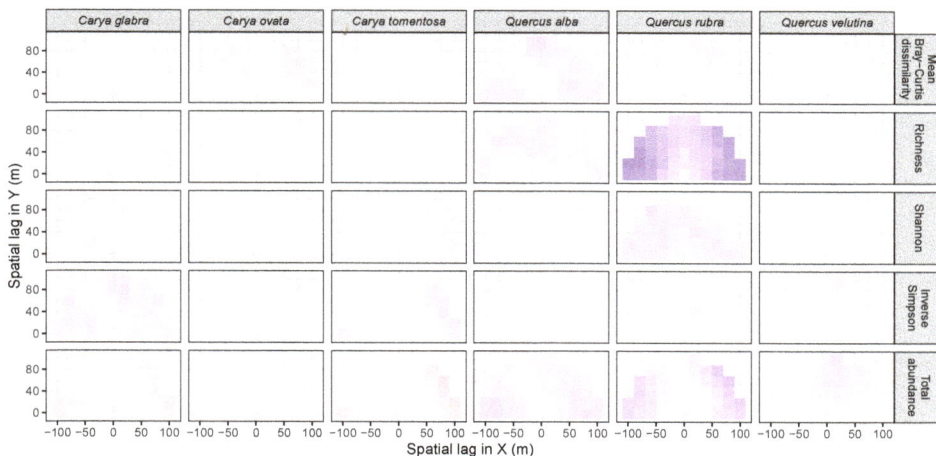

Figure 13. Codispersion between measures of diversity and total basal area of the six focal tree species in 20 × 20-m subplots in the 20-ha Tyson Research Center plot. Codispersion coefficients were calculated for spatial lags ranging from 0–150 m at 20-m intervals. The value of the codispersion coefficient can range from −1 (dark blue) through 0 (white) to 1 (dark red). Statistical significance of each codispersion coefficient is shown in Figures 14 and A2. Summaries of codispersion values are given in Table A1).

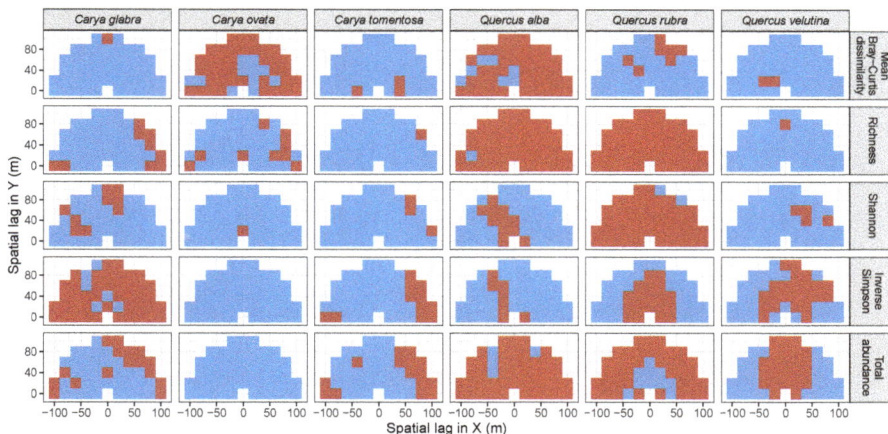

Figure 14. Statistical significance (red: $P \leq 0.05$; blue: $P > 0.05$) of the codispersion coefficients calculated between measures of diversity and total basal area of the six focal tree species in 20 × 20-m subplots in the 20-ha Tyson Research Center plot (Figure 13). Statistical significance was determined by comparing observed codispersion at each spatial lag with a distribution of 199 spatial randomizations of a toroidal-shift null model. See Figure A2 for significance testing using a CSR null model.

3.3.2. Trees at Tropical Sites

Codispersion between tree size or abundance and measures of associated-species diversity at the three tropical sites were much weaker than at the temperate sites (Figures 15–17), and were less frequently statistically significant for most spatial lags or directions (Figures 18–20 and Figures A3–A5). Notable exceptions included: positive codispersion between basal area of *Dacryodes excelsa* and Bray-Curtis dissimilarity at LFDP (Figure 15); negative codispersion between basal area of *Oenocarpus mapora* and associated-species richness and abundance at BCI (Figure 16); and positive codispersion between basal area of *O. mapora* and associated-species Simpson's diversity at BCI (Figure 16). In all cases, however, values of codispersion were much less than that observed for *Q. rubra* at HF (except for codispersion between basal area of *O. mapora* and associated-species abundance, which was similar to that of *Q. rubra* at HF) (Table A2). At Amacayacu, the most species-rich site, total basal area of the dominant tree species had small and rarely significant spatial associations with metrics of diversity of associated woody species (Figures 17 and 20 and Figure A5).

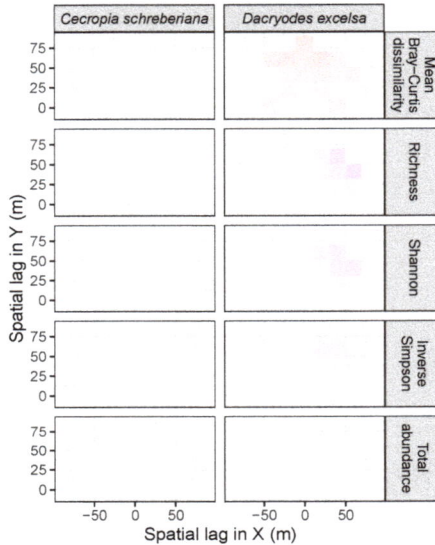

Figure 15. Codispersion between measures of diversity and total basal area of the two focal tree species in 20 × 20-m subplots in the 16-ha Luquillo plot. Codispersion coefficients were calculated for spatial lags ranging from 0–80 m at 20-m intervals. The value of the codispersion coefficient can range from −1 (dark blue) through 0 (white) to 1 (dark red). Statistical significance of each codispersion coefficient is shown in Figures 18 and A3.

3.4. Codispersion Between Focal Understory Species and Associated-species Diversity

Understory species were included in our focal species list at two temperate sites (WR, TY) and two tropical sites (LFDP, AM) (Table 1; Figures A6–A11; Tables A3 and A4). Of all of these, consistently significant codispersion between basal area and any metrics of associated-species diversity was observed only for *Acer circinatum* at WR, *Cornus florida* at TY, and *Rinorea lindeniana* at AM (Figures A10 and A11). At WR, codispersion between basal area of *A. circinatum* and both Shannon and Simpson diversity was consistently negative, but was consistently positive with Bray-Curtis dissimilarity (Figures A6–A8). At AM, the pattern was reversed. There, codispersion between basal area of *R. lindeniana* and both Shannon and Simpson's diversity were consistently positive, but was consistently negative with Bray-Curtis dissimilarity (Figures A9–A11), and were of the same magnitude or lower than that observed for *Q. rubra* at HF (Table A4). At TY, codispersion between basal area of *C. florida* and both associated-species richness and Bray-Curtis similarity were consistently negative (Figure A6 and of low magnitude (Table A3).

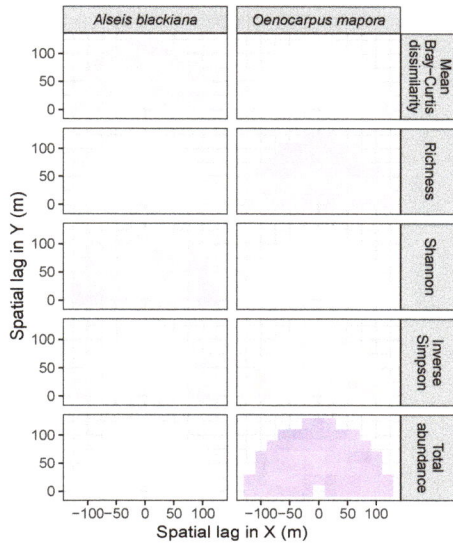

Figure 16. Codispersion between measures of diversity and total basal area of the two focal tree species in 20 × 20-m subplots in the 50-ha BCI plot. Codispersion coefficients were calculated for spatial lags ranging from 0–125 m at 20-m intervals. The value of the codispersion coefficient can range from −1 (dark blue) through 0 (white) to 1 (dark red). Statistical significance of each codispersion coefficient is shown in Figures 19 and A4.

Figure 17. Codispersion between measures of diversity and total basal area of the five focal tree species in 20 × 20-m subplots in the 25-ha Amacayacu plot. Codispersion coefficients were calculated for spatial lags ranging from 0–125 m at 20-m intervals. The value of the codispersion coefficient can range from −1 (dark blue) through 0 (white) to 1 (dark red). Statistical significance of each codispersion coefficient is shown in Figures 20 and A5.

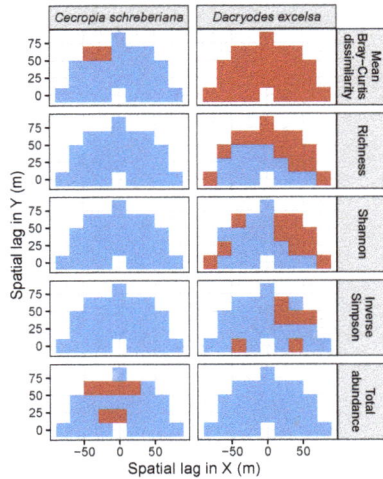

Figure 18. Statistical significance (red: $P \leq 0.05$; blue: $P > 0.05$) of the codispersion coefficients calculated between measures of diversity and total basal area of the two focal tree species in 20×20-m subplots in the 16-ha Luquillo plot (Figure 15). Statistical significance was determined by comparing observed codispersion at each spatial lag with a distribution of 199 spatial randomizations of a toroidal-shift null model. See Figure A3 for significance testing using a CSR null model.

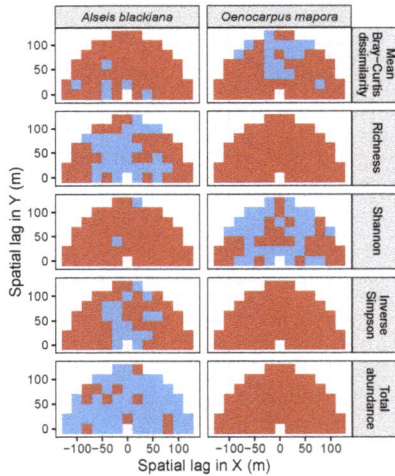

Figure 19. Statistical significance (red: $P \leq 0.05$; blue: $P > 0.05$) of the codispersion coefficients calculated between measures of diversity and total basal area of the two focal tree species in 20×20-m subplots in the 50-ha BCI plot (Figure 16). Statistical significance was determined by comparing observed codispersion at each spatial lag with a distribution of 199 spatial randomizations of a toroidal-shift null model. See Figure A4 for significance testing using a CSR null model.

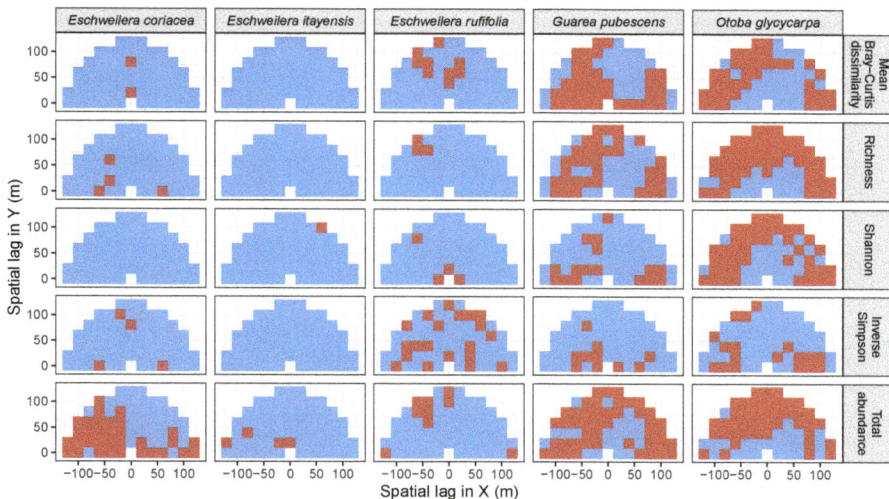

Figure 20. Statistical significance (red: $P \leq 0.05$; blue: $P > 0.05$) of the codispersion coefficients calculated between measures of diversity and total basal area of the five focal tree species in 20 × 20-m subplots in the 25-ha Amacayacu plot (Figure 17). Statistical significance was determined by comparing observed codispersion at each spatial lag with a distribution of 199 spatial randomizations of a toroidal-shift null model. See Figure A5 for significance testing using a CSR null model.

4. Discussion

Foundation species in forests define and structure entire ecological communities primarily through non-trophic effects such as providing physical points of attachment for associated species (e.g., epiphytes), creating habitat for associated fauna (e.g., tree-holes), modulating hydrological flow of adjacent streams, or altering the chemical composition of soils [2,3,52]. Forest foundation species most frequently are common, abundant, and large, and have been identified more frequently in temperate forests [2,53] than in tropical ones [2]. Ellison et al. [2] suggested that foundation tree species would be more likely in comparatively species-poor temperate forests because functional redundancy among trees there would be much less common than in species-rich tropical forests (e.g., [54,55]). In partial support of this idea, Lamanna et al. presented data showing a peak in the size of functional-trait space for tree assemblages growing at mid-latitudes (where our temperate plots are located) but more narrow functional-trait space for tropical assemblages [56]; see also [57]. These analyses imply that foundation tree species should be less common in very high latitude ecosystems (examined in more detail by [56] than by [57]). In those ecosystems, low-growing perennial, cushion- and tussock-forming plants have been found to be foundation species (e.g., [58]).

Indeed, the goal of testing community ecology's neutral theory (which posits that differences among trophically similar species, such as trees, are irrelevant to their demographic success and subsequent spatial distribution), deriving from Hubbell's work in a 13-ha plot in Costa Rica [59,60], in part motivated the establishment of the 50-ha plot at BCI in 1980 and subsequently other tropical forest dynamics plots. Although neutral theory has been supported for some distributional patterns in many tropical forests, it fails in others [61]. Niche theory (which posits functional non-equivalence of co-occurring, trophically similar species) more strongly fits data observed in temperate forests [62] and many tropical ones, too [61]. Thus, there is no *a priori* reason to suspect on functional grounds that foundation species should be less likely to occur in tropical forests than in temperate ones. We note also that the definition of foundation species [1,2] implicitly assumes that they always have non-neutral effects on the systems they define. However, it is plausible that the high abundance of these species causes them to have neutral influences on co-occurring species via ecological drift (i.e., by keeping local populations of associated species at low abundance, contributing to their demographic stochasticity). This might help explain why the basal area of the potential foundation species we examined was often negatively correlated with local species richness but positively correlated with beta diversity of associated species.

Despite the ecological importance of foundation species, it has proven challenging to identify plausible foundation species from among groups of locally dominant species or those that are perceived by various criteria to have importance in particular systems. It has taken decades of observational and experimental work to support the hypothesis that *Tsuga canadensis* is a foundation species in eastern North American forests [16] and that *Pinus albicaulis* is a foundation species in high-elevation western North American forests [53]. In both cases, their foundational characteristics were identified only as these species were declining [2]. If foundation species really are of critical importance for associated biodiversity, it is crucial that they be identified and protected while they are still common, locally abundant, and present in their full size distribution [6,27]. Because experiments on forest trees can take decades to centuries (and are generally impractical with longer-lived species), alternative approaches to identifying foundation forest species are needed.

Our statistical screening of abundance and size data from six large forest dynamics plots suggested particular characteristics of potential foundation species from two different size-frequency and diameter-abundance distributions (Figures 2–5). The foundation species *T. canadensis* was a notable outlier from the "reverse-J" distribution of abundance versus mean diameter in the entire Harvard Forest 35-ha forest dynamics plot (Figure 2) *and* had a much greater range and spread of values of total basal area versus abundance in the 20-m rasterized plot (Figure 3). A similarly small set of species that had similar characteristics were identified in the other forest dynamics plots we analyzed (Figures 4 and 5), with more candidates identified in the temperate plots than the tropical ones. This result is not especially surprising, as the more even abundance distributions of most tropical forests suggest that any species with both of these screening characteristics would be less common in the tropics. Alternatively, groups of species with similar traits (e.g., the *Eschweilera* species [Lecythidaceae] at AM, Dipterocarpaceae in Southeast Asia, or *Quercus* species at TY) or particular characteristics of individuals themselves (e.g., very large trees [26,27]) could have disproportionate influence akin to foundational characteristics in the forests in which they occur. This hypothesis is in need of additional exploration.

We also note that departures from the "reverse-J" size-frequency distribution may depend on when in succession a stand is sampled [63]. For example, abundance of early seral species (such as *Pseudotsuga menziesii* at WR) may initially be very abundant but relatively small. Because *P. menziesii* has been decreasing in abundance for at least the last 200 years at WR [64], it might be identified as a candidate foundation species in a 300-year-old stand, but not in the older forest dynamics plot at WR. That both the subcanopy *A. circinatum* and the late-seral *T. heterophylla* basically mirror the position of *P. menziesii* in the size-frequency plot (Figure 4) adds some support for this hypothesis. Conversely, the foundational characteristics of *T. heterophylla* suggested by Figures 4 and 5 might not yet be manifest this early in succession (*cf.* [63]). Related to this, some of the species that fall off the "reverse-J" have intermediate shade-tolerance and their role is likely to change during succession.

Subsequent analysis of spatial relationships between these candidate foundation species, along with others of known ecological importance (Table 1), and metrics of diversity of associated species pointed to attributes of *T. canadensis* shared by only a few other species. These included codispersion values $> |0.25|$ at virtually all spatial lags and directions for two or more metrics of diversity that were often negative for measures of abundance or alpha-diversity (within-subplot species richness, Shannon diversity, or inverse-Simpson's diversity) but positive for a measure of beta diversity (average subplot Bray-Curtis dissimilarity). That is, foundation species are negatively correlated with diversity of associated species at local (within-subplot) spatial scales where they dominate the vegetation but are positively associated with beta diversity at the larger (between subplot) spatial scale of the whole-plot. In virtually all cases (a notable exception being *Pseudotsuga menziesii* at WR), little effect of underlying environmental patchiness was observed: the toroidal-shift null model and the CSR null model gave nearly identical results, with the toroidal-shift null model being predictably more conservative.

By these criteria, candidates for additional, more detailed study for their foundational roles include *Acer circinatum* at WR, *Dacryodes excelsa* at Luquillo, and *Oenocarpus mapora* at BCI (Figures 15 and 16 and Figure A6). In the western Cascades of Oregon, *A. circinatum* is a major component of the subcanopy, where it can grow rapidly, have high biomass, and form broad canopies that suppress other species [65,66]. This suppression of recruiting species could lead to negative association between *A. circinatum* basal area and local diversity of associated species (as is also seen with *T. canadensis* in eastern North American forests). At Luquillo, adult *D. excelsa* dominates ridges where trees form root grafts that may, in part, explain their relatively low hurricane mortality (<1%) [67,68]. *Dacryodes excelsa* saplings also grow more rapidly than saplings of other late successional species [68], which may explain the lack of pole-sized individuals in their populations [67]. Finally, *O. mapora* alters litter quality and quantity and the local microenvironment, and limits establishment of small-seeded and shade-intolerant species at BCI [69].

Disturbance frequency also may play a role in both statistical "fingerprints" of foundation species. For example, intermediate disturbance regimes (moderately frequent disturbances of moderate size and severity, such as small understory fires, multi-tree blowdowns, or selective harvesting) could push mid-successional, shade-intermediate species such as *Q. rubra* or *A. rubrum* at Harvard Forest back into the "reverse-J" distribution. At the same time, successional dynamics of these species (especially recruitment and prolonged growth) could result in transient conditions whereby they have both relatively high basal area and large numbers of small individuals in any given subplot. Where these species fall within the fingerprints, therefore, may indicate how their life-history traits are aligned with the frequency and type of local disturbances. How foundation species exert control on disturbance regimes in forests has not been previously explored. It would be useful to explore how positive feedbacks between a forest foundation species and local disturbances (especially fire, small windstorms or microbursts, and treefalls) could affect size-frequency and diameter-abundance distributions of associated species. These processes also are likely to differ in temperate and tropical forests (e.g., [70–72]).

Although local diversity of associated woody species is lower in the presence of foundation species, local diversity of other taxa associated with foundation species is often higher or composed of unique assemblages. Non-woody vascular and especially nonvascular plant taxa, and fauna are rarely sampled at the individual tree scale or in every 20 × 20-m subplot in large forest dynamics plots, but a handful of other studies have examined associations between these taxa and either *T. canadensis* or our candidate foundation species. For example, *T. canadensis* hosts unique assemblages of arthropods [73–76]. *Acer circinatum* is associated with a diverse and abundant assemblage of epiphytes [77]. At Luquillo, higher numbers of invertebrate species are associated with *Dacryodes excelsa* than with other dominant canopy species (including *Cecropia schreberiana*) [78], but abundances and herbivory rates of invertebrates on *D. excelsa* are lower than on other co-occurring species [79]. Several understory palms, including *O. mapora*, recently have been described as "terrestrial litter trappers" [80] that collect leaf litter at the base of their foliage. Two other litter-trapping palms, *Asterogyne sociale* and *A. spectabilis*, have very large numbers of arthropods in the trapped litter [81]; comparable studies with *O. mapora* would provide a useful comparison with the known and potential foundation species discussed here.

Finally, species like *T. canadensis*, *P. menziesii* and *D. excelsa* "name" their communities: hemlock, Douglas-fir, and *tabonuco* forests, respectively. The forest stands they dominate are easily identifiable as distinctive patches on the landscape. When a foundation species like *T. canadensis* declines, the associated regional or landscape-scale beta diversity is likely to decline along with it. Such regional homogenization—as indicated by lower Bray-Curtis values—is more characteristic of the majority of our focal species and would be a likely consequence of the loss of our candidate foundation species. It is precisely this larger-scale heterogeneity that may be the greatest value gained by protecting foundation species while they are still common.

5. Conclusions

Foundation species define and structure ecological communities but are difficult to identify before they are declining. Comparative analyses of six large forest dynamics plots spanning a temperate-to-tropical gradient in the Western Hemisphere identified statistical "fingerprints" of potential foundation species. Specifically, known and potential foundation species are outliers from the common "reverse-J" size-frequency distribution, and have negative effects on alpha diversity and positive effects on beta diversity at most spatial lags and directions. Potential foundation species appear to be more likely in temperate forests, whereas in tropical forests, foundational characteristics may be more likely to occur in species groups within genera or families. As foundation species (or species groups) decline, local (alpha) diversity may increase but landscape-scale (beta) diversity is likely to decline along with them. Preservation of beta biodiversity may be the most important consequence of protecting foundation species while they are still common.

Author Contributions: Conceptualization, A.M.E., H.L.B. and B.S.C.; statistical methodology, A.M.E., H.L.B. and B.S.C.; coding: H.L.B., B.S.C. and A.M.E.; forest dynamics plot establishment, data collection, and curation: D.C.; Á.J.D., J.A.L., J.A.M., D.A.O. and J.K.Z.; writing and editing, all authors.

Funding: This research was funded by the Smithsonian Institution, Utah State University, and the Utah Agricultural Experiment Station (projects 1153 and 1398) (WR plot); NSF grant 123749, the Smithsonian Institution, and the Harvard Forest (HF plot); Tyson Research Center, the International Center for Energy, Environment and Sustainability (InCEES) at Washington University in St. Louis, Smithsonian ForestGEO, and NSF grant DEB 1557094 (TY plot); the NSF-supported Luquillo Experimental Forest Long-Term Ecological Research Program, the University of Puerto Rico, and the International Institute of Tropical Forestry (LFDP plot); NSF grants 021104, 021115, 0212284, 0212818 and 0314581, and the Center for Tropical Forest Science (CTFS) (BCI plot); and Parques Nacionales Naturales de Colombia, Eliana Martínez and staff members of the Amacayacu Natural National Park, and the Center for Tropical Forest Science (CTFS) (AM plot).

Acknowledgments: An early version of this paper was presented at the 2014 International LTER meeting in Valdivia, Chile. We especially thank Ronny Vallejos for development of software to measure and plot codispersion, and for his continuing collegial collaboration with A.M.E., H.L.B. and B.S.C. Two anonymous reviewers provided helpful comments that substantially improved the final version of the paper. We acknowledge logistical support from the Gifford Pinchot National Forest, the USDA Forest Service Pacific Northwest Research Station, and the volunteers individually listed at http://wfdp.org for the WR plot. We also thank the many field technicians who helped census the HF plot; Jason Aylward for field supervision, data screening and database management at HF; and John Wisnewski and the Harvard Forest Woods Crew for providing materials, supplies, and invaluable field assistance with plot logistics. We thank the Tyson Research Center staff and more than 100 students and researchers that have contributed to tree censuses at TY. We sincerely thank the many volunteers who collected tree census data at LFDP. Vegetation data from BCI are part of the BCI forest dynamics research project founded by S.P. Hubbell and R.B. Foster and now managed by R. Condit, S. Lao, and R. Pérez through the Center for Tropical Forest Science (CTFS) and the Smithsonian Tropical Research Institute (STRI) in Panamá. We are very grateful for the assistance of coworkers from the Comunidad de Palmeras and the students of forest engineering from the Universidad Nacional de Colombia who collected tree census data at the AM plot. This paper is a contribution of the Harvard Forest and Luquillo LTER programs.

Conflicts of Interest: The authors declare no conflict of interest. The funders had no role in the design of the study; in the collection, analyses, or interpretation of data; in the writing of the manuscript, or in the decision to publish the results.

Appendix A. Supplemental Tables and Figures

Table A1. Codispersion statistics for focal tree species in temperate forests (Figures 8–20 and Figures A1 and A2).

Site	Species	Diversity Metric	Min	Median	Mean (SD)	Max
WR	*Pseudotsuga menzeisii*	Bray-Curtis	0.05	0.11	0.11 (0.02)	0.15
		Richness	−0.21	−0.14	−0.13 (0.05)	−0.06
		Shannon	−0.15	−0.07	−0.07 (0.04)	0.02
		Simpson	−0.08	−0.01	0 (0.04)	0.08
		Abundance	−0.11	−0.03	−0.04 (0.04)	0.03
	Tsuga heterophylla	Bray−Curtis	−0.04	0.01	0.01 (0.03)	0.06
		Richness	−0.31	−0.27	−0.27 (0.03)	−0.21
		Shannon	−0.13	−0.09	−0.09 (0.03)	−0.03
		Simpson	−0.09	−0.04	−0.04 (0.03)	0.03
		Abundance	−0.47	−0.44	−0.43 (0.03)	−0.35
	Abies amabilis	Bray−Curtis	−0.14	0.03	0.02 (0.04)	0.07
		Richness	0.04	0.09	0.1 (0.04)	0.17
		Shannon	−0.02	0.01	0.02 (0.03)	0.08
		Simpson	−0.06	0.01	0 (0.03)	0.06
		Abundance	0.03	0.13	0.13 (0.04)	0.19
HF	*Acer rubrum*	Bray−Curtis	0.01	0.06	0.07 (0.03)	0.13
		Richness	−0.01	0.11	0.11 (0.04)	0.19
		Shannon	−0.05	0.05	0.05 (0.03)	0.11
		Simpson	−0.04	0.07	0.07 (0.04)	0.14
		Abundance	−0.06	−0.01	0 (0.02)	0.05
	Pinus strobus	Bray−Curtis	−0.1	−0.04	−0.04 (0.04)	0.07
		Richness	−0.09	−0.01	0 (0.05)	0.09
		Shannon	−0.06	0.03	0.02 (0.04)	0.09
		Simpson	−0.03	0.07	0.07 (0.05)	0.17
		Abundance	−0.21	−0.09	−0.11 (0.06)	−0.02
	Quercus rubra	Bray−Curtis	−0.28	−0.21	−0.2 (0.05)	−0.07
		Richness	−0.2	−0.11	−0.12 (0.05)	0
		Shannon	−0.11	0	0 (0.06)	0.12
		Simpson	−0.05	0.09	0.09 (0.08)	0.24
		Abundance	−0.4	−0.28	−0.27 (0.07)	−0.1
	Tsuga canadensis	Bray−Curtis	0.12	0.21	0.21 (0.05)	0.28
		Richness	−0.64	−0.49	−0.49 (0.08)	−0.29
		Shannon	−0.46	−0.34	−0.34 (0.07)	−0.17
		Simpson	−0.26	−0.2	−0.2 (0.04)	−0.07
		Abundance	−0.45	−0.33	−0.34 (0.06)	−0.24
TY	*Quercus alba*	Bray−Curtis	−0.27	−0.18	−0.18 (0.05)	−0.04
		Richness	−0.24	−0.18	−0.18 (0.03)	−0.12
		Shannon	−0.16	−0.08	−0.08 (0.04)	0.01
		Simpson	−0.13	−0.03	−0.03 (0.05)	0.05
		Abundance	−0.26	−0.18	−0.19 (0.05)	−0.09
	Quercus rubra	Bray−Curtis	−0.18	−0.08	−0.09 (0.05)	0
		Richness	−0.55	−0.39	−0.39 (0.1)	−0.21
		Shannon	−0.29	−0.23	−0.22 (0.05)	−0.09
		Simpson	−0.2	−0.1	−0.09 (0.06)	0.07
		Abundance	−0.42	−0.21	−0.22 (0.1)	−0.05
	Quercus velutina	Bray−Curtis	−0.16	−0.08	−0.07 (0.04)	0.06
		Richness	−0.14	0	−0.01 (0.06)	0.1
		Shannon	0.02	0.1	0.1 (0.04)	0.17
		Simpson	−0.01	0.13	0.13 (0.05)	0.22
		Abundance	−0.28	−0.14	−0.15 (0.07)	−0.01
	Carya glabra	Bray−Curtis	−0.06	0.06	0.05 (0.04)	0.12
		Richness	−0.05	0.05	0.06 (0.07)	0.22
		Shannon	−0.16	−0.06	−0.07 (0.04)	0.04
		Simpson	−0.26	−0.18	−0.16 (0.05)	0
		Abundance	−0.09	0.11	0.09 (0.1)	0.25

Table A1. *Cont.*

Site	Species	Diversity Metric	Min	Median	Mean (SD)	Max
	Carya ovata	Bray−Curtis	0.04	0.15	0.14 (0.04)	0.22
		Richness	−0.1	0.08	0.07 (0.06)	0.17
		Shannon	−0.13	−0.02	−0.02 (0.04)	0.07
		Simpson	−0.13	−0.05	−0.06 (0.03)	0.03
		Abundance	−0.07	0.03	0.03 (0.05)	0.13
	Carya tomentosa	Bray−Curtis	−0.12	−0.04	−0.04 (0.05)	0.08
		Richness	−0.05	0.05	0.05 (0.05)	0.18
		Shannon	−0.19	−0.04	−0.06 (0.05)	0.04
		Simpson	−0.26	−0.07	−0.09 (0.08)	0.02
		Abundance	−0.03	0.12	0.12 (0.1)	0.35

Table A2. Codispersion statistics for focal tree species in tropical forests (Figures 15–20 and Figures A1–A5).

Site	Species	Diversity Metric	Min	Median	Mean (SD)	Max
		Richness	−0.04	0.04	0.03 (0.03)	0.09
		Shannon	−0.03	0.05	0.04 (0.03)	0.07
		Simpson	0.01	0.07	0.06 (0.04)	0.12
		Abundance	0.01	0.08	0.09 (0.05)	0.2
	Dacryodes excelsa	Bray−Curtis	0.12	0.19	0.19 (0.05)	0.28
		Richness	−0.28	−0.11	−0.11 (0.07)	0.03
		Shannon	−0.24	−0.1	−0.12 (0.06)	−0.02
		Simpson	−0.18	−0.09	−0.09 (0.05)	−0.01
		Abundance	−0.12	−0.03	−0.03 (0.04)	0.04
BCI	*Alseis blackiana*	Bray−Curtis	−0.15	−0.11	−0.1 (0.02)	−0.05
		Richness	−0.12	−0.07	−0.07 (0.03)	0.01
		Shannon	−0.19	−0.12	−0.12 (0.04)	−0.03
		Simpson	−0.16	−0.09	−0.09 (0.03)	−0.02
		Abundance	0.01	0.05	0.05 (0.02)	0.1
	Oenocarpus mapora	Bray−Curtis	0.01	0.08	0.08 (0.03)	0.17
		Richness	−0.2	−0.14	−0.14 (0.02)	−0.08
		Shannon	−0.01	0.06	0.06 (0.03)	0.11
		Simpson	0.08	0.13	0.14 (0.03)	0.2
		Abundance	−0.39	−0.34	−0.34 (0.02)	−0.3
AM	*Eschweilera coriacea*	Bray−Curtis	−0.09	−0.05	−0.04 (0.03)	0.06
		Richness	−0.07	0.05	0.04 (0.03)	0.09
		Shannon	−0.06	0.01	0.01 (0.03)	0.08
		Simpson	−0.02	0.02	0.03 (0.03)	0.11
		Abundance	−0.04	0.09	0.09 (0.04)	0.18
	Eschweilera itayensis	Bray−Curtis	−0.06	0.02	0.02 (0.03)	0.09
		Richness	−0.09	−0.02	−0.01 (0.03)	0.06
		Shannon	−0.11	−0.03	−0.03 (0.03)	0.03
		Simpson	−0.1	−0.05	−0.04 (0.03)	0.04
		Abundance	−0.02	0.04	0.04 (0.03)	0.12
	Eschweilera rufifolia	Bray−Curtis	−0.12	−0.04	−0.05 (0.03)	0.01
		Richness	−0.06	0.03	0.03 (0.04)	0.13
		Shannon	−0.09	0.01	0 (0.04)	0.13
		Simpson	−0.12	−0.05	−0.05 (0.02)	0.02
		Abundance	−0.02	0.05	0.05 (0.03)	0.12
	Guarea pubescens	Bray−Curtis	−0.14	−0.09	−0.09 (0.03)	−0.04
		Richness	−0.01	0.09	0.09 (0.03)	0.14
		Shannon	0	0.08	0.08 (0.03)	0.15
		Simpson	−0.01	0.03	0.04 (0.04)	0.18
		Abundance	0.01	0.1	0.1 (0.03)	0.17
	Otoba glycarpa	Bray−Curtis	−0.16	−0.11	−0.1 (0.04)	0.04
		Richness	−0.02	0.13	0.11 (0.05)	0.19
		Shannon	0	0.11	0.11 (0.04)	0.17
		Simpson	0	0.07	0.07 (0.03)	0.13
		Abundance	−0.05	0.1	0.09 (0.05)	0.17

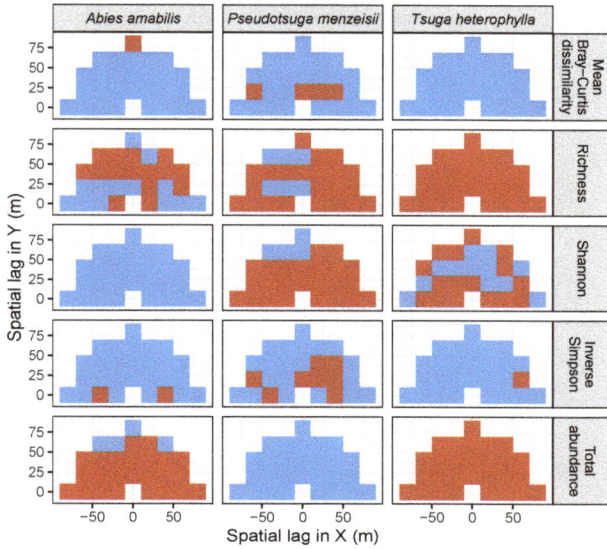

Figure A1. Statistical significance (red: $P \leq 0.05$; blue: $P > 0.05$) of the codispersion coefficients calculated between measures of diversity and total basal area of the four dominant species in 20 × 20-m subplots in the 25.6-ha Wind River plot (Figure 11). Statistical significance was determined by comparing observed codispersion at each spatial lag with a distribution of 199 spatial randomizations of a CSR null model.

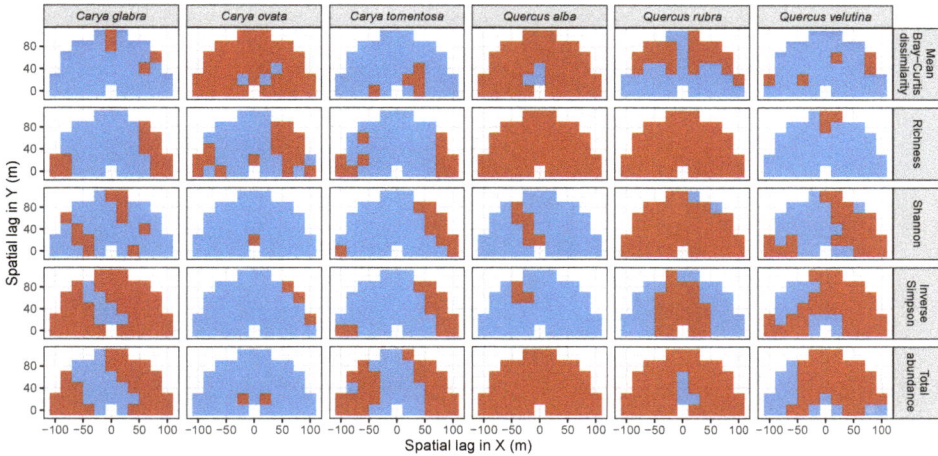

Figure A2. Statistical significance (red: $P \leq 0.05$; blue: $P > 0.05$) of the codispersion coefficients calculated between measures of diversity and total basal area of the six focal tree species in 20 × 20-m subplots in the 20-ha Tyson Research Center plot (Figure 13). Statistical significance was determined by comparing observed codispersion at each spatial lag with a distribution of 199 spatial randomizations of a CSR null model.

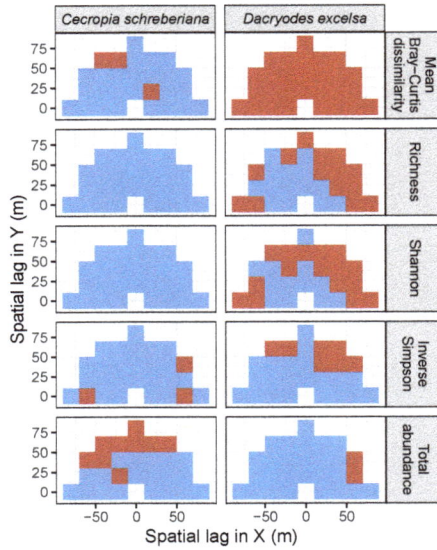

Figure A3. Statistical significance (red: $P \leq 0.05$; blue: $P > 0.05$) of the codispersion coefficients calculated between measures of diversity and total basal area of the two focal tree species in 20 × 20-m subplots in the 16-ha Luquillo plot (Figure 15). Statistical significance was determined by comparing observed codispersion at each spatial lag with a distribution of 199 spatial randomizations of a CSR null model.

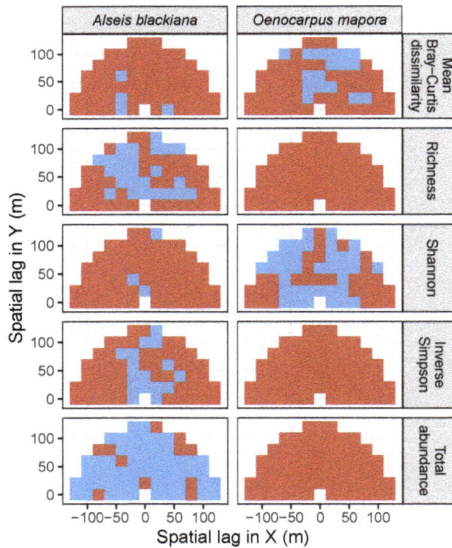

Figure A4. Statistical significance (red: $P \leq 0.05$; blue: $P > 0.05$) of the codispersion coefficients calculated between measures of diversity and total basal area of the two focal tree species in 20 × 20-m subplots in the 50-ha BCI plot (Figure 16). Statistical significance was determined by comparing observed codispersion at each spatial lag with a distribution of 199 spatial randomizations of a CSR null model.

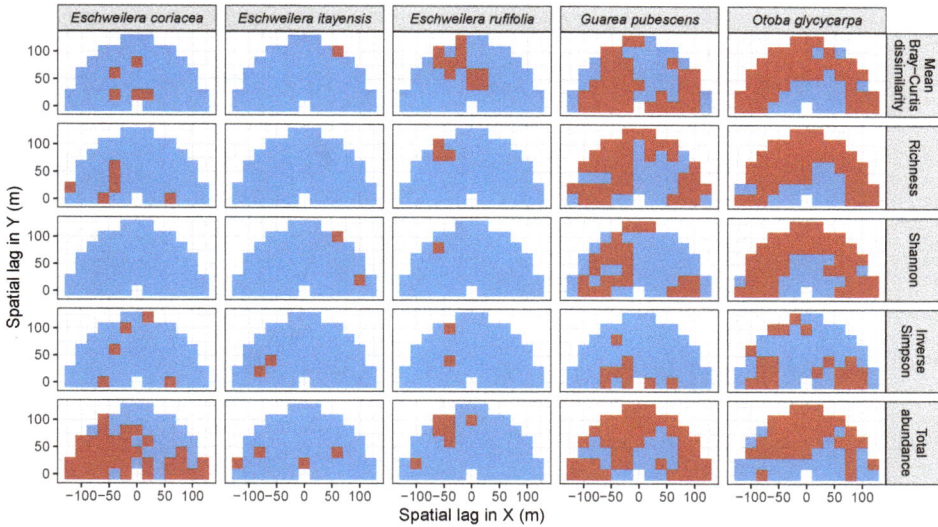

Figure A5. Statistical significance (red: $P \le 0.05$; blue: $P > 0.05$) of the codispersion coefficients calculated between measures of diversity and total basal area of the five focal tree species in 20×20-m subplots in the 25-ha Amacayacu plot (Figure 17). Statistical significance was determined by comparing observed codispersion at each spatial lag with a distribution of 199 spatial randomizations of a CSR null model.

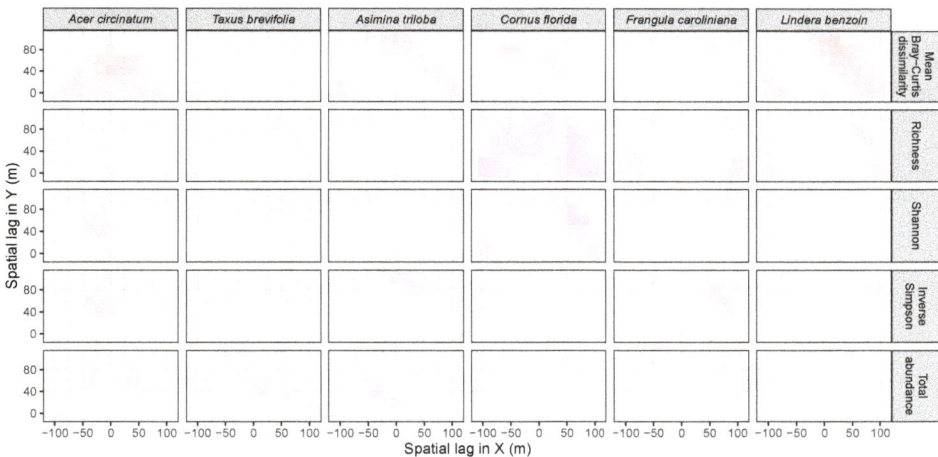

Figure A6. Codispersion between measures of diversity and total basal area of the focal understory species in 20×20-m subplots in the 25.6-ha Wind River (*Acer circinatum*, *Taxus brevifolia*) and 20-ha Tyson Research Center (*Asimina triloba*, *Cornus florida*, *Frangula caroliniana*, *Lindera benzoin*) plots. Codispersion coefficients were calculated for spatial lags ranging from 0–80 m (WR) and 0–105 (TY) at 20-m intervals. The value of the codispersion coefficient can range from -1 (dark blue) through 0 (white) to 1 (dark red). Statistical significance of each codispersion coefficient is shown in Figures A7 and A8.

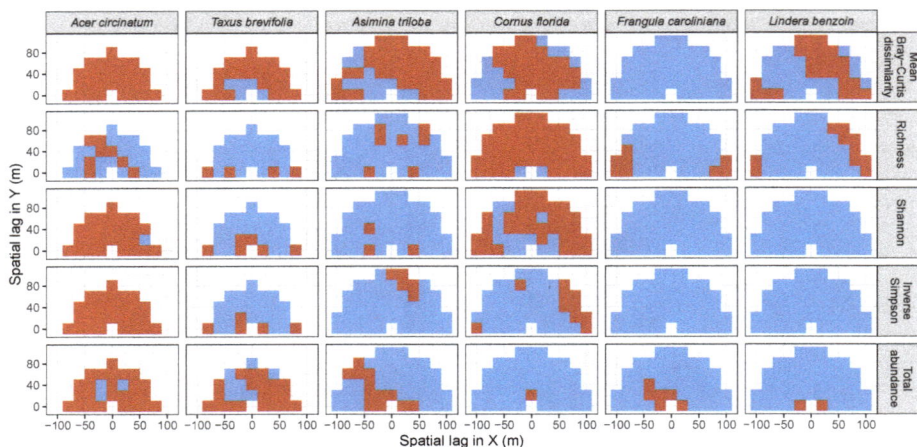

Figure A7. Statistical significance (red: $P \leq 0.05$; blue: $P > 0.05$) of the codispersion coefficients calculated between measures of diversity and total basal area of the focal understory species in 20×20-m subplots in the 25.6-ha Wind River (*Acer circinatum, Taxus brevifolia*) and 20-ha Tyson Research Center (*Asimina triloba, Cornus florida, Frangula caroliniana, Lindera benzoin*) plots (Figure A6). Statistical significance was determined by comparing observed codispersion at each spatial lag with a distribution of 199 spatial randomizations of a toroidal shift model.

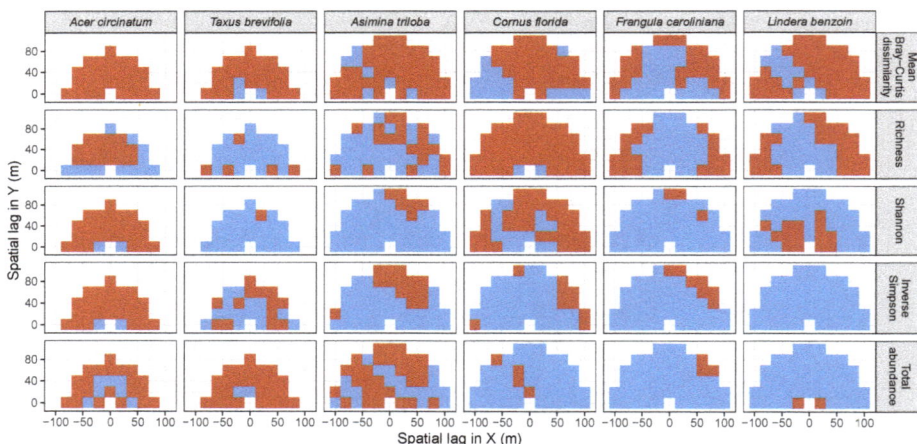

Figure A8. Statistical significance (red: $P \leq 0.05$; blue: $P > 0.05$) of the codispersion coefficients calculated between measures of diversity and total basal area of the focal understory species in 20×20-m subplots in the 25.6-ha Wind River (*Acer circinatum, Taxus brevifolia*) and 20-ha Tyson Research Center (*Asimina triloba, Cornus florida, Frangula caroliniana, Lindera benzoin*) plots (Figure A6). Statistical significance was determined by comparing observed codispersion at each spatial lag with a distribution of 199 spatial randomizations of a CSR null model.

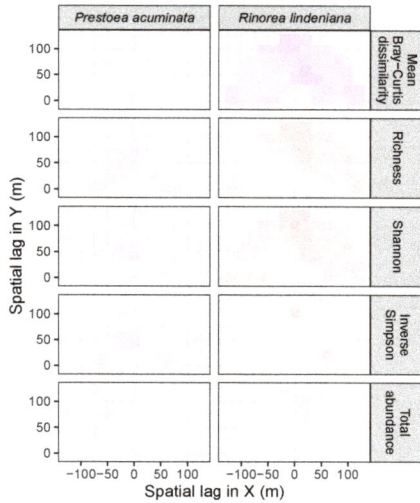

Figure A9. Codispersion between measures of diversity and total basal area of the focal understory species in 20 × 20-m subplots in the 16-ha Luquillo (*Prestoea acuminata*) and 25-ha Amacayacu (*Rinorea lindeniana*) plots. Codispersion coefficients were calculated for spatial lags ranging from 0–80 m (LFDP) and 0–125 (AM) at 20-m intervals. The value of the codispersion coefficient can range from −1 (dark blue) through 0 (white) to 1 (dark red). Statistical significance of each codispersion coefficient is shown in Figures A10 and A11.

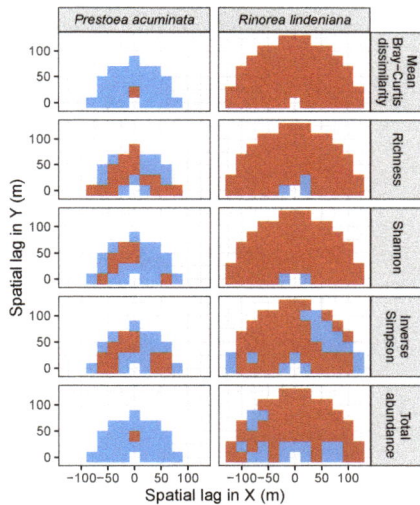

Figure A10. Statistical significance (red: $P \leq 0.05$; blue: $P > 0.05$) of the codispersion coefficients calculated between measures of diversity and total basal area of the focal understory species in 20 × 20-m subplots in the 16-ha Luquillo (*Prestoea acuminata*) and 25-ha Amacayacu (*Rinorea lindeniana*) plots (Figure A9). Statistical significance was determined by comparing observed codispersion at each spatial lag with a distribution of 199 spatial randomizations of a toroidal-shift null model.

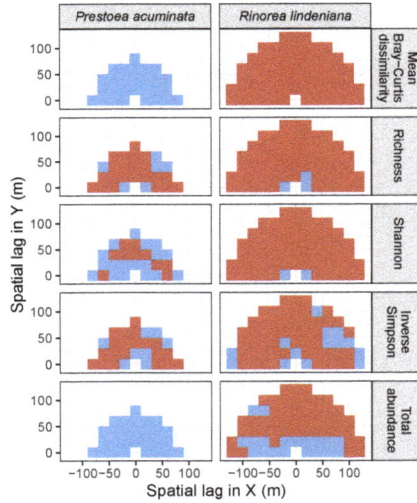

Figure A11. Statistical significance (red: $P \leq 0.05$; blue: $P > 0.05$) of the codispersion coefficients calculated between measures of diversity and total basal area of the focal understory species in 20×20-m subplots in the 16-ha Luquillo (*Prestoea acuminata*) and 25-ha Amacayacu (*Rinorea lindeniana*) plots (Figure A9). Statistical significance was determined by comparing observed codispersion at each spatial lag with a distribution of 199 spatial randomizations of a CSR null model.

Table A3. Codispersion statistics for focal understory species in temperate forests (Figures A6–A8).

Site	Species	Diversity Metric	Min	Median	Mean (SD)	Max
WR	*Acer circinatum*	Bray-Curtis	0.17	0.22	0.22 (0.03)	0.29
		Richness	0.03	0.09	0.08 (0.03)	0.14
		Shannon	0.1	0.14	0.15 (0.04)	0.25
		Simpson	0.1	0.14	0.16 (0.05)	0.26
		Abundance	−0.27	−0.12	−0.14 (0.05)	−0.06
	Taxus brevifolia	Bray−Curtis	0.04	0.14	0.13 (0.05)	0.22
		Richness	0.01	0.07	0.07 (0.03)	0.11
		Shannon	0.02	0.05	0.06 (0.03)	0.1
		Simpson	0	0.05	0.05 (0.03)	0.12
		Abundance	0.03	0.12	0.11 (0.04)	0.17
TY	*Asimina triloba*	Bray−Curtis	0.05	0.16	0.15 (0.05)	0.25
		Richness	0.02	0.08	0.08 (0.04)	0.16
		Shannon	−0.12	0.01	0.02 (0.06)	0.11
		Simpson	−0.18	−0.05	−0.04 (0.07)	0.09
		Abundance	−0.17	0.03	0.01 (0.11)	0.18
	Cornus florida	Bray−Curtis	−0.27	−0.13	−0.14 (0.04)	−0.05
		Richness	−0.26	−0.22	−0.21 (0.04)	−0.07
		Shannon	−0.31	−0.15	−0.16 (0.07)	−0.02
		Simpson	−0.2	−0.07	−0.08 (0.06)	0.03
		Abundance	−0.11	−0.03	−0.03 (0.04)	0.04
	Frangula caroliniana	Bray−Curtis	−0.04	0.08	0.07 (0.05)	0.16
		Richness	−0.07	0.05	0.07 (0.1)	0.24
		Shannon	−0.1	−0.01	0 (0.06)	0.11
		Simpson	−0.2	−0.01	−0.04 (0.06)	0.07
		Abundance	−0.13	−0.02	−0.02 (0.07)	0.16
	Lindera benzoin	Bray−Curtis	0.02	0.21	0.18 (0.09)	0.33
		Richness	−0.04	0.1	0.11 (0.08)	0.25
		Shannon	0.02	0.08	0.08 (0.03)	0.13
		Simpson	−0.1	0	0 (0.05)	0.07
		Abundance	−0.1	0	−0.01 (0.05)	0.08

Table A4. Codispersion statistics for focal understory species in tropical forests (Figures A9–A11).

Site	Species	Diversity Metric	Min	Median	Mean (SD)	Max
LFDP	*Prestoea acuminata*	Bray-Curtis	−0.04	0.05	0.04 (0.05)	0.11
		Richness	−0.18	−0.13	−0.14 (0.03)	−0.08
		Shannon	−0.2	−0.1	−0.11 (0.04)	−0.05
		Simpson	−0.22	−0.13	−0.12 (0.06)	0.02
		Abundance	−0.1	0.07	0.05 (0.06)	0.13
AM	*Rinorea lindeniana*	Bray−Curtis	−0.29	−0.22	−0.22 (0.04)	−0.08
		Richness	−0.02	0.19	0.19 (0.06)	0.3
		Shannon	0.01	0.21	0.21 (0.05)	0.33
		Simpson	0	0.1	0.11 (0.05)	0.3
		Abundance	−0.07	0.12	0.12 (0.06)	0.21

References

1. Dayton, P.K. Toward an understanding of community resilience and the potential effects of enrichments to the benthos at McMurdo Sound, Antarctica. In Proceedings of the Colloquium on Conservation Problems in Antarctica, Blacksburg, VA, USA, 10–12 September 1971; Parker, B.C., Ed.; Allen Press: Lawrence, KS, USA, 1972; pp. 81–95.
2. Ellison, A.M.; Bank, M.S.; Clinton, B.D.; Colburn, E.A.; Elliott, K.; Ford, C.R.; Foster, D.R.; Kloeppel, B.D.; Knoepp, J.D.; Lovett, G.M.; et al. Loss of foundation species: Consequences for the structure and dynamics of forested ecosystems. *Front. Ecol. Environ.* **2005**, *9*, 479–486. [CrossRef]
3. Baiser, B.; Whitaker, N.; Ellison, A.M. Modeling foundation species in food webs. *Ecosphere* **2013**, *4*, 146. [CrossRef]
4. Gaston, K.J.; Fuller, R.A. Biodiversity and extinction: losing the common and the widespread. *Prog. Phys. Geog.* **2007**, *31*, 213–225. [CrossRef]
5. Gaston, K.J.; Fuller, R.A. Commonness, population depletion and conservation biology. *Trends Ecol. Evol.* **2008**, *23*, 14–19. [CrossRef]
6. Ellison, A.M.; Degrassi, A.L. All species are important, but some species are more important than others (Commentary). *J. Veg. Sci.* **2017**, *28*, 669–671. [CrossRef]
7. Paine, R.T. Food web complexity and species diversity. *Am. Nat.* **1966**, *100*, 65–75. [CrossRef]
8. Valls, A.; Coll, M.; Christensen, V. Keystone species: Toward an operational concept for marine biodiversity conservation. *Ecol. Monogr.* **2015**, *85*, 29–47. [CrossRef]
9. Grime, J.P. Dominant and subordinate components of plant communities: implications for succession, stability and diversity. In *Colonization, Succession and Stability*; Gray, A.J., Crawley, M.J., Edwards, P.J., Eds.; Blackwell Scientific Publications: Oxford, UK, 1987; pp. 413–428.
10. Leakey, R.; Lewin, R. *The Sixth Extinction: Patterns of Life and the Future of Humankind*; Anchor Books: New York, NY, USA, 1996.
11. Short, F.; Carruthers, T.; Dennison, W.; Waycott, M. Global seagrass distribution and diversity: A bioregional model. *J. Exp. Mar. Biol. Ecol.* **2007**, *350*, 3–20. [CrossRef]
12. Fischer, M.; Bossdorf, O.; Gockel, S.; Hänsel, F.; Hemp, A.; Hessenmöller, D.; Korte, G.; Nieschulze, J.; Pfeiffer, S.; Prati, D.; et al. Implementing large-scale and long-term functional biodiversity research: The Biodiversity Exploratories. *Basic Appl. Ecol.* **2010**, *11*, 473–485. [CrossRef]
13. LaManna, J.A.; Mangan, S.A.; Alonso, A.; Bourg, N.A.; Brockelman, W.Y.; Bunyavejchewin, S.; Chang, L.-W.; Chiang, J.-M.; Chuyong, G.B.; Clay, K.; et al. Plant diversity increases with the strength of negative density dependence at the global scale. *Science* **2017**, *356*, 1389–1392. [CrossRef]
14. ReefBase. ReefBase—A Global Information System on Coral Reefs. 2018. Available online: http://www.reefbase.org (accessed on 5 October 2018).
15. Ellison, A.M.; Barker Plotkin, A.A.; Khalid, S. Foundation species loss and biodiversity of the herbaceous layer in New England forests. *Forests* **2016**, *7*, 9. [CrossRef]
16. Ellison, A.M. Experiments are revealing a foundation species: A case study of eastern hemlock (*Tsuga canadensis*). *Adv. Ecol.* **2014**, *2014*, 456904. [CrossRef]
17. Foster, D.R. (Ed.) *Hemlock: A Forest Giant on the Edge*; Yale University Press: New Haven, CT, USA, 2014.

18. Orwig, D.A.; Barker Plotkin, A.A.; Davidson, E.A.; Lux, H.; Savage, K.E.; Ellison, A.M. Foundation species loss affects vegetation structure more than ecosystem function in a northeastern USA forest. *PeerJ* **2013**, *1*, e41. [CrossRef] [PubMed]

19. Cuevas, F.; Porcu, E.; Vallejos, R. Study of spatial relationships between two sets off variables: A nonparametric approach. *J. Nonparametr. Statist.* **2013**, *25*, 695–714. [CrossRef]

20. Buckley, H.L.; Case, B.S.; Ellison, A.M. Using codispersion analysis to characterize spatial patterns in species co-occurrences. *Ecology* **2016**, *97*, 32–39. [CrossRef] [PubMed]

21. Buckley, H.L.; Case, B.S.; Zimmermann, J.; Thompson, J.; Myers, J.A.; Ellison, A.M. Using codispersion analysis to quantify and understand spatial patterns in species-environment relationships. *New Phytol.* **2016**, *211*, 735–749. [CrossRef] [PubMed]

22. Case, B.S.; Buckley, H.L.; Barker Plotkin, A.; Ellison, A.M. Using codispersion analysis to quantify temporal changes in the spatial pattern of forest stand structure. *Chil. J. Stat.* **2016**, *7*, 3–15.

23. Case, B.S.; Buckley, H.L.; Barker Plotkin, A.A.; Orwig, D.A.; Ellison, A.M. When a foundation crumbles: Forecasting forest community dynamics associated with the decline of the foundation species. *Ecosphere* **2017**, *8*, e01893. [CrossRef]

24. Anderson-Teixeira, K.J.; Davies, S.J.; Bennett, A.C.; Gonzalez-Akre, E.B.; Muller-Landau, H.C.; Wright, S.J.; Abu Salim, K.; Almeyda Zambrano, A.M.; Alonso, A.; Baltzer, J.L.; et al. CTFS—ForestGEO: A worldwide network monitoring forests in an era of global change. *Glob. Change Biol.* **2015**, *21*, 528–549. [CrossRef]

25. Condit, R. *Tropical Forest Census Plots: Methods and Results from Barro Colorado Island, Panama and a Comparison with Other Plots*; Springer: Berlin, Germany, 1998.

26. Lutz, J.A.; Larson, A.J.; Freund, J.A.; Swanson, M.E.; Bible, K.J. The importance of large-diameter trees to forest structural heterogeneity. *PLoS ONE* **2013**, *8*, e82784. [CrossRef]

27. Lutz, J.A.; Furniss, T.J.; Johnson, D.J.; Davies, S.J.; Allen, D.; Alonso, A.; Anderson-Teixeira, K.; Andrade, A.; Baltzer, J.; Becker, K.M.L.; et al. Global importance of large-diameter trees. *Global Ecol. Biogeogr.* **2018**, *27*, 849–864. [CrossRef]

28. Das, A.J.; Larson, A.J.; Lutz, J.A. Individual species-area relationships in temperate coniferous forests. *J. Veg. Sci.* **2018**, *29*, 317–324. [CrossRef]

29. Lutz, J.A.; Larson, A.J.; Furniss, T.J.; Freund, J.A.; Swanson, M.E.; Donato, D.C.; Bible, K.J.; Chen, J.; Franklin, J.F. Spatially non-random tree mortality and ingrowth maintain equilibrium pattern in an old-growth *Pseudotsuga-Tsuga* forest. *Ecology* **2014**, *95*, 2047–2054. [CrossRef] [PubMed]

30. Griffith, G.E.; Omernik, J.M.; Pierson, S.M.; Kiilsgaard, C.W. *The Massachusetts Ecological Regions Project*; U.S. Environmental Protection Agency: Corvallis, OR, USA, 1994.

31. Westveld, M.; Ashman, R.I.; Baldwin, H.I.; Holdsworth, R.P.; Johnson, R.S.; Lambert, J.H.; Lutz, H.J.; Swain, L.; Standish, M. Natural forest vegetation zones of New England. *J. For.* **1956**, *54*, 332–338.

32. Wang, X.; Wiegand, T.; Anderson-Teixeira, K.J.; Bourg, N.A.; Hao, Z.; Howe, R.; Jin, G.; Orwig, D.A.; Spasojevic, M.J.; Wang, S.; et al. Ecological drivers of spatial community dissimilarity, species replacement and species nestedness across temperate forests. *Global Ecol. Biogeogr.* **2018**, *27*, 581–592. [CrossRef]

33. Raup, H.M.; Carlson, R.E. The history of land use in the Harvard Forest. *Harv. For. Bull.* **1941**, *20*, 4–62.

34. Foster, D.R. Land-use history (1730–1990) and vegetation dynamics in central New England, USA. *J. Ecol.* **1992**, *80*, 753–772. [CrossRef]

35. Foster, D.R.; Zebryk, T.M.; Schoonmaker, P.K.; Lezberg, A.L. Post-settlement history of human land-use and vegetation dynamics of a hemlock woodlot in central New England. *J. Ecol.* **1992**, *80*, 773–786. [CrossRef]

36. Spasojevic, M.J.; Yablon, E.A.; Oberle, B.; Myers, J.A. Ontogenetic trait variation influences tree community assembly across environmental gradients. *Ecosphere* **2014**, *5*, 129. [CrossRef]

37. LaManna, J.A.; Walton, M.L.; Turner, B.L.; Myers, J.A. Negative density dependence is stronger in resource-rich environments and diversifies communities when stronger for common but not rare species. *Ecol. Lett.* **2016**, *19*, 657–667. [CrossRef]

38. Zimmerman, M.; Wagner, W.L. A description of the woody vegetation of oak-hickory forest in the Northern Ozark Highlands. *Bull. Torrey Bot. Club* **1979**, *106*, 117–122. [CrossRef]

39. Thompson, J.; Brokaw, N.; Zimmerman, J.K.; Waide, R.B.; Everham, E.M., III; Lodge, J.; Taylor, C.M.; Garca-Montiel, D.; Fluet, M. Land use history, environment, and tree composition in a tropical forest. *Ecol. Appl.* **2002**, *12*, 1344–1363. [CrossRef]

40. Zimmerman, J.K.; Comita, L.S.; Thompson, J.; Uriarte, M.; Brokaw, N. Patch dynamics and community metastability of a tropical forest: Compound effects of natural disturbance and human land use. *Landsc. Ecol.* **2010**, *25*, 1099–1111. [CrossRef]

41. Hogan, J.A.; Zimmerman, J.K.; Thompson, J.; Nytch, C.J.; Uriarte, M. The interaction of land-use legacies and hurricane disturbance in subtropical wet forest: Twenty-one years of change. *Ecosphere* **2016**, *7*, e01405. [CrossRef]

42. Leigh, E.G., Jr.; Loo de Lao, S.; Condit, R.; Hubbell, S.P.; Foster, R.B.; Pérez, R. Barro Colorado Island Forest Dynamics Plot, Panama. In *Tropical Forest Diversity and Dynamism: Findings from a Large-Scale Plot Network*; Losos, E.C., Leigh, E.G., Jr., Eds.; University of Chicago Press: Chicago, IL, USA, 2004; pp. 451–463.

43. Condit, R.; Pérez, R.; Lao, S.; Aguilar, S.; Hubbell, S.P. Demographic trends and climate over 35 years in the Barro Colorado 50 ha plot. *For. Ecosyst.* **2017**, *4*, 17. [CrossRef]

44. Hubbell, S.P.; Foster, R.B.; O'Brien, S.T.; Harms, K.E.; Condit, R.; Wechsler, B.; Wright, S.J.; Loo de Lao, S. Light-gap disturbances, recruitment limitation, and tree diversity in a Neotropical forest. *Science* **1999**, *283*, 554–557. [CrossRef] [PubMed]

45. Zuleta, D.; Russo, S.E.; Barona, A.; Barreto-Silva, J.S.; Carenas, D.; Castaño, N.; Davies, S.J.; Detto, M.; Sua, S.; Turner, B.L.; Duque, A. Importance of topography for tree species habitat distributions in a terra firme forest in the Colombian Amazon. *Plant Soil* **2018**. [CrossRef]

46. Loehle, C. Species abundance distributions result from body size-energetics relationships. *Ecology* **2006**, *87*, 2221–2226. [CrossRef]

47. Chao, A.; Gotelli, N.J.; Hsieh, T.C.; Snader, E.L.; Ma, K.H.; Colwell, R.K.; Ellison, A.M. Rarefaction and extrapolation with Hill numbers: a framework for sampling and estimation in species diversity studies. *Ecol. Monogr.* **2014**, *84*, 45–67. [CrossRef]

48. Bray, J.R.; Curtis, J.T. An ordination of upland forest communities of southern Wisconsin. *Ecol. Monogr.* **1957**, *27*, 325–349. [CrossRef]

49. Oksanen, J.; Blanchet, F.G.; Friendly, M.; Kindt, R.; Legendre, P.; McGlinn, D.; Minchin, P.R.; O'Hara, R.B.; Simpson, G.L.; Solymos, P.; et al. Vegan: Community Ecology Package. R package version 2.5-3. Available online: https://CRAN.R-project.org/package=vegan (accessed on 3 February 2018).

50. Vallejos, R.; Buckley, H.; Case, B.; Acosta, J.; Ellison, A.M. Sensitivity of codispersion to noise and error in ecological and environmental data. *Forests* **2018**, *9*, 679. [CrossRef]

51. Osorio, F.; Vallejos, R.; Cuevas, F.; Mancilla, D. SpatialPack Package. R package version 0.3. Available online: https://cran.r-project.org/package=SpatialPack (accessed on 3 February 2018).

52. Brantley, S.T.; Ford, C.R.; Vose, J.M. Future species composition will affect forest water use after loss of eastern hemlock from southern Appalachian forests. *Ecol. Appl.* **2013**, *23*, 777–790. [CrossRef] [PubMed]

53. Tomback, D.F.; Resler, L.M.; Keane, R.E.; Pansing, E.R.; Andrade, A.J.; Wagner, A.C. Community structure, biodiversity, and ecosystem services in treeline whitebark pine communities: potential impacts from a non-native pathogen. *Forests* **2016**, *7*, 21. [CrossRef]

54. Whitmore, T.C. *Tropical Rain Forests of the Far East*; Clarendon: Oxford, UK, 1984.

55. Iwasa, Y.; Kubo, T.; Sato, K. Maintenance of forest species diversity and latitudinal gradient. *Vegetatio* **1995**, *121*, 127–134. [CrossRef]

56. Lamanna, C.; Blonder, B.; Violle, C.; Kraft, N.J.B.; Sandel, B.; Šímová, I.; Donoghue, J.C., II; Svenning, J.-C.; McGill, B.J.; Boyle, B.; et al. Functional trait space and the latitudinal diversity gradient. *Proc. Nat. Acad. Sci. USA* **2014**, *111*, 13745–13750. [CrossRef] [PubMed]

57. Wieczynski, D.J.; Boyle, B.; Buzzard, V.; Duran, S.M.; Henderson, A.N.; Hulshof, D.M.; Kerkhoff, A.J.; McCarthy, M.C.; Michaletz, S.T.; Swenson, N.G.; et al. Climate shapes and shifts functional biodiversity in forests worldwide. *Proc. Nat. Acad. Sci. USA* **2019**, *116*, 587–592. [CrossRef] [PubMed]

58. Elumeeva, T.G.; Onipchenko, V.G.; Weger, M.J.A. No other species can replace them: evidence for the key role of dominants in an alpine Festuca varia grassland. *J. Veg. Sci.* **2017**, *28*, 674–683. [CrossRef]

59. Hubbell, S.P. Tree dispersion, abundance, and diversity in a tropical dry forest. *Science* **1979**, *203*, 1299–1309. [CrossRef]

60. Hubbell, S.P. *The Unified Neutral Theory of Biodiversity and Biogeography*; Princeton University Press: Princeton, NJ, USA, 2001.

61. Leigh, E.G., Jr. Neutral theory: A historical perspective. *J. Evol. Biol.* **2007**, *20*, 2075–2091. [CrossRef]

62. Furniss, T.J.; Larson, A.J.; Lutz, J.A. Reconciling niches and neutrality in a subalpine temperate forest. *Ecosphere* **2017**, *8*, e01847. [CrossRef]

63. Ellison, A.M.; Lavine, M.; Kerson, P.B.; Barker Plotkin, A.A.; Orwig, D.A. Building a foundation: land-use history and dendrochronology reveal temporal dynamics of a *Tsuga canadensis* (Pinaceae) forest. *Rhodora* **2014**, *116*, 377–427. [CrossRef]

64. Freund, J.A.; Franklin, J.F.; Lutz, J.A. Structure of early old-growth Douglas-fir forests in the Pacific Northwest. *Forest Ecol. Manag.* **2015**, *335*, 11–25. [CrossRef]

65. Lutz, J.A.; Halpern, C.B. Tree mortality during early forest development: A long-term study of rates, causes, and consequences. *Ecol. Monogr.* **2006**, *76*, 257–275. [CrossRef]

66. Halpern, C.B.; Lutz, J.A. Canopy closure exerts weak controls on understory dynamics: A 30-year study of overstory-understory interactions. *Ecol. Monogr.* **2013**, *83*, 221–237. [CrossRef]

67. Basnet, K.; Scatena, F.N.; Likens, G.E.; Lugo, A.E. Ecological consequences of root grafting in tabonuco (*Dacryodes excelsa*) trees in the Luquillo Experimental Forest, Puerto Rico. *Biotropica* **1993**, *25*, 28–35. [CrossRef]

68. Uriarte, M.; Canham, C.D.; Thompson, J.; Zimmerman, J.K. A neighborhood analysis of tree growth and survival in a hurricane-driven tropical forest. *Ecol. Monogr.* **2004**, *74*, 591–614. [CrossRef]

69. Farris-Lopez, K.; Desnlow, J.S.; Moser, B.; Passmore, H. Influence of a common palm, *Oenocarpus mapora* on seedling establishment in a tropical moist forest in Panama. *J. Trop. Ecol.* **2004**, *20*, 429–438. [CrossRef]

70. Baker, T.R.; Vela Díaz, D.M.; Chama Moscoso, V.; Navarro, G.; Monteagudo, A.; Pinto, R.; Cangani, K.; Fyllas, N.F.; Lopez Gonzalez, G.; Laurance, W.F.; et al. Consistent, small effects of treefall disturbances on the composition and diversity of four Amazonian forests. *J. Ecol.* **2016**, *104*, 497–506. [CrossRef]

71. Battles, J.J.; Cleavitt, N.L.; Saah, D.S.; Pling, B.T.; Fahey, T.J. Ecological impact of a microburst windstorm in a northern hardwood forest. *Can. J. For. Res.* **2017**, *47*, 1695–1701. [CrossRef]

72. Després, T.; Asselin, H.; Doyon, F.; Drobyshev, I.; Bergeron, Y. Gap dynamics of late successional sugar maple-yellow birch forests at their northern range limit. *J. Veg. Sci.* **2017**, *28*, 368–378. [CrossRef]

73. Dilling, C.; Lambdin, P.; Grant, J.; Buck, L. Insect guild structure associated with eastern hemlock in the southern Appalachians. *Env. Entomol.* **2007**, *36*, 1408–1414. [CrossRef]

74. Rohr, J.; Mahan, C.G.; Kim, K.C. Response of arthropod biodiversity to foundation species declines: the case of the eastern hemlock. *Forest Ecol. Manag.* **2009**, *258*, 1503–1510. [CrossRef]

75. Sackett, T.E.; Record, S.; Bewick, S.; Baiser, B.; Sanders, N.J.; Ellison, A.M. Response of macroarthropod assemblages to the loss of hemlock (*Tsuga canadensis*), a foundation species. *Ecosphere* **2011**, *2*, e74. [CrossRef]

76. Record, S.; McCabe, T.; Baiser, B.; Ellison, A.M. Identifying foundation species in North American forests using long-term data on ant assemblage structure. *Ecosphere* **2018**, *9*, e02139. [CrossRef]

77. Ruchty, A.; Rosso, A.L.; McCune, B. Changes in epiphyte communities as the shrub, *Acer circinatum*, develops and ages. *The Bryologist* **2001**, *104*, 274–281. [CrossRef]

78. Schowalter, T.D. Canopy invertebrate community response to disturbance and consequences of herbivory in temperate and tropical forests. *Selbyana* **1995**, *16*, 41–48.

79. Schowalter, T.D. Invertebrate community structure and herbivory in a tropical rain forest canopy in Puerto Rico following Hurricane Hugo. *Biotropica* **1994**, *26*, 312–319. [CrossRef]

80. Weissenhofer, A.; Huber, W.; Wanek, W.; Weber, A. Terrestrial litter trappers in the Golfo Dulce region: Diversity, architecture and ecology of a poorly known group of plant specialists. *Stapfia 88, zugleich Kataloge der oberösterreichischen Landesmuseen (Neue Serie)* **2008**, *80*, 143–154.

81. De Vasconcelos, H.L. Effects of litter collection by understory palms on the associated macroinvertebrate fauna in Central Amazonia. *Pedobiologia* **1990**, *34*, 157–160.

© 2019 by the authors. Licensee MDPI, Basel, Switzerland. This article is an open access article distributed under the terms and conditions of the Creative Commons Attribution (CC BY) license (http://creativecommons.org/licenses/by/4.0/).

![forests logo] *forests*

MDPI

Article

Climate-Related Distribution Shifts of Migratory Songbirds and Sciurids in the White Mountain National Forest

Aimee Van Tatenhove [1,*], Emily Filiberti [1], T. Scott Sillett [1], Nicholas Rodenhouse [2] and Michael Hallworth [1]

[1] Migratory Bird Center, Smithsonian Conservation Biology Institute, Washington, DC 20008, USA; efiliberti@une.edu (E.F.); silletts@si.edu (T.S.S.); mhallworth@gmail.com (M.H.)

[2] Department of Biological Sciences, Wellesley College, 106 Central Street, Wellesley, MA 02481, USA; nrodenho@wellesley.edu

* Correspondence: aimee.van.tatenhove@aggiemail.usu.edu; Tel.: +1-715-529-0159

Received: 31 December 2018; Accepted: 21 January 2019; Published: 23 January 2019

Abstract: Climate change has been linked to distribution shifts and population declines of numerous animal and plant species, particularly in montane ecosystems. The majority of studies suggest both that low-elevation avian and small mammal species are shifting up in elevation and that high-elevation avian communities are either shifting further upslope or relocating completely with an increase in average local temperatures. However, recent research suggests numerous high elevation montane species are either not shifting or are shifting down in elevation despite the local increasing temperature trends, perhaps as a result of the increased precipitation at high elevations. In this study, we examine common vertebrate species distributions across the Hubbard Brook valley in the White Mountain National Forest, including resident and migratory songbirds and small mammals, in relation to historic spring temperature and precipitation. We found no directional change in distributions through time for any of the species. However, we show that the majority of low-elevation bird species in our study area respond to warm spring temperatures by shifting upslope. All bird species that shifted were long-distance migrants. Each low-elevation migrant species responded differently to warm spring temperatures, through upslope distribution expansion, downslope distribution contraction, or total distribution shift upslope. In contrast, we found a majority of high-elevation bird species and both high- and low-elevation mammal species did not shift in response to spring temperature or precipitation and may be subject to more complex climate trends. The heterogeneous response to climate change highlights the need for more comprehensive studies on the subject and careful consideration for appropriate species and habitat management plans in northeastern montane regions.

Keywords: climate change; temperature; precipitation; Hubbard Brook; elevational shifts; mountains

---

1. Introduction

Mounting evidence suggests plants and animals are responding to ongoing climate change in numerous ways, including through significant distributional shifts [1–3]. A simple paradigm asserts that species will respond to rising temperatures by shifting their distributions poleward (e.g., References [4–6]) or up in elevation [2,3,7]. However, as time goes on, studies are finding some species are either not shifting or are shifting downslope through time, even as temperatures increase [8–11]. Yet, it is not clear what is driving montane species to shift heterogeneously and how montane species are ultimately moving in the face of climate change.

Montane species can be particularly susceptible to the effects of climate change, because their distributions are elevationally constrained, leaving them vulnerable to range restriction or local extirpation through an inability to react sufficiently to a changing climate [12–14]. Montane songbird species and some small mammal species often occupy narrow niches [15,16] and therefore may be sensitive to climate induced habitat alteration [17,18]. The elevational gradient of montane regions may also exacerbate the severity of climate change, as climate regimes are not changing uniformly at all elevations [19–21]. Most notably, precipitation rates are typically greater at high elevations, and precipitation is expected to increase more rapidly at high elevations with ongoing climate change [19–21]. As a result, precipitation may affect high-elevation species disproportionately [10,21], and in some cases, species may shift down to find areas with more favorable precipitation [12,21]. Temperature may also increase more rapidly at some higher elevation sites [20,22], putting further stress on montane communities. Migratory bird species may be particularly impacted, as individuals must recolonize areas yearly and may therefore shift greater distances than non-migratory species [23]. Songbirds and small mammals in northeastern montane forests may be especially vulnerable, as these regions are highly threatened by climate change [19,20]. However, with few studies of climate induced distributional shifts in northeastern forests, it is relatively unknown how climate change will impact songbird and small mammal species in these regions.

Long-term studies of songbird and small mammal communities are crucial to understanding how these communities are responding to a changing climate. The songbird and small mammal communities have been systematically surveyed annually since 1999 at the Hubbard Brook Experimental Forest (HBEF) within the White Mountain National Forest. Here, we used these long-term survey data to test whether changing climate has affected songbird and small mammal distributions in northeastern hardwood and boreal forests. We hypothesized that both temperature and precipitation contribute to fine scale distributional shifts within the songbird and small mammal communities and that songbird migratory status would affect the magnitude of the distribution shifts. If songbird and small mammal distributions are governed by climate, we predicted low-elevation species would shift upslope with warm temperatures, while high-elevation species would predominantly shift downslope in response to increased high-elevation precipitations. Additionally, we predicted that migratory songbird species that recolonize breeding grounds within HBEF yearly would exhibit larger distribution shifts than resident species.

2. Materials and Methods

2.1. Study Site

Hubbard Brook Experimental Forest (43°56′N, 71°45′W, NH, USA; HBEF) is a 3600 ha watershed located within the White Mountain National Forest, ranging from 200–1000 m above sea-level (Figure 1). HBEF is forested and comprised of northern hardwoods found predominantly at low to middle elevations, transitioning to boreal spruce-fir forests at elevations above 800 m [24]. The climate within HBEF is temperate, with long, cold winters and mild, wet summers [25]. Over 100 bird species regularly breed within HBEF, the majority of which are Neotropical migrants that spend temperate winters in the tropics. Eastern chipmunks (*Tamias striatus* Linnaeus; EACH) and red squirrels (*Tamiasciurus hudsonicus* Erxleben; RESQ) are also common in the valley and are frequent nest predators of these bird species [26,27].

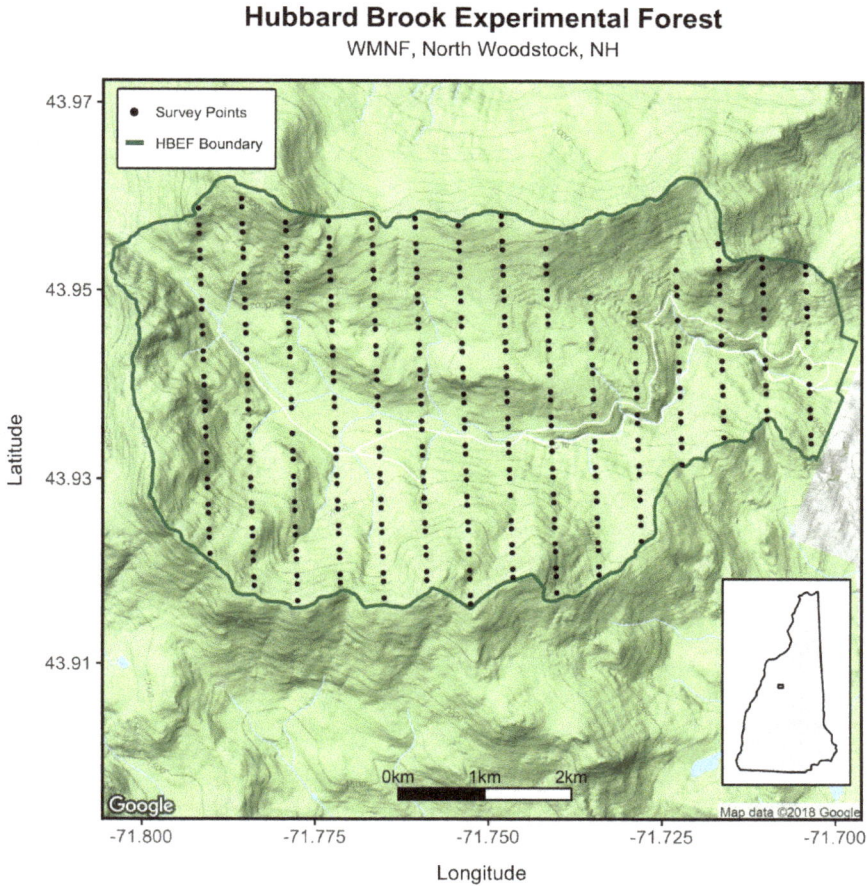

Figure 1. Map of Hubbard Brook Experimental Forest (HBEF) with survey locations in relation to New Hampshire.

2.2. Survey Methods

Point count surveys were used to collect avian and mammal occupancy data and were conducted annually from 1999 to 2016, excluding 2003 and 2004. Counts were conducted along 15 north–south transects, separated by 500 m that span the elevation gradient within HBEF. Survey locations along each transect were spaced either 100 or 200 m apart. Each survey location ($n = 373$) was surveyed at least three times during the breeding season (May through July), by a different trained surveyor each time. Point counts were conducted between 0530 and 1000 EDT (Eastern Daylight Time). Counts were not conducted in conditions that could hinder the surveyors' ability to detect individuals (rain, high winds, canopy drip, fog, etc.). During the ten-minute survey, all bird species, EACH, and RESQ seen or heard within 50 m of the point were recorded. Birds and mammals assessed to be outside 50 m were not used in this study to avoid accidental double counting of individuals at adjacent points.

2.3. Surveyed Species

We selected five songbird species to represent low-elevation bird species and five songbird species to represent high-elevation bird species, based on our prior knowledge of their breeding distributions within HBEF and to remain consistent with species selected in other regional studies [12,13]. Species designations were then confirmed using species occupancy curves over the elevation gradient in our study site, as outlined in the statistical methods. The low-elevation species we selected were the black-capped chickadee (*Poecile atricapillus* Linnaeus; BCCH), black-throated blue warbler (*Setophaga caerulescens* Gmelin; BTBW), hermit thrush (*Catharus guttatus* Pallas; HETH), ovenbird (*Seiurus aurocapilla* Linnaeus; OVEN), and red-eyed vireo (*Vireo olivaceus* Linnaeus; REVI). The high-elevation species we selected were the blackpoll warbler (*Setophaga striata* Forster; BLPW), magnolia warbler (*Setophaga magnolia* Wilson; MAWA), dark-eyed junco (*Junco hyemalis* Linnaeus; DEJU), Swainson's thrush (*Catharus ustulatus* Nuttall; SWTH), and winter wren (*Troglodytes hiemalis* Viellot; WIWR).

Red squirrel (RESQ) data were available for all years, excluding the two years during which surveys were not conducted, 2003 and 2004. The eastern chipmunk (EACH) data was missing one additional year of data (2002). No other mammals were surveyed systematically during our study period. We categorized mammal species into low- and high-elevations using the same criteria we used for birds. EACH were designated as a low-elevation species and RESQ as a high-elevation species because of their association with conifers [16].

2.4. Environmental Variables

For this study, we focused on climate variables with long-term datasets available through our study period (1999–2016). We were primarily interested in how temperature and precipitation during the months when surveys were conducted affected species distributions within HBEF. Most migratory species arrive back at HBEF to breed in May. Therefore, we used mean May temperature and precipitation as potential drivers of distribution shifts since they coincide with the arrival of migratory bird species [28] and when RESQ and EACH pups first typically become active [29,30] (hereafter called "spring temperature" and "spring precipitation", unless specified). For migratory bird species that arrive at the breeding grounds in late May, and as such may not be influenced by early May weather, we chose to test whether mean June temperature and precipitation influenced their distribution shifts. Daily temperature and precipitation data from 1999 to 2014 were downloaded from the Long Term Ecological Research Network website (https://portal.lternet.edu). Additional temperature data for 2015 and 2016 were provided by the US Forest Service. All temperature and precipitation data were collected from the weather station at the USDA Forest Service Headquarters building at HBEF (252 m above sea level). Daily mean temperature and precipitation were averaged each year over May and June to generate mean annual May and June temperature and precipitation values.

2.5. Statistical Methods

Imperfect detection during animal surveys can lead to biased occupancy and abundance estimates and is common in point count data [31]. To account for imperfect species detection in our analyses, we used single and multi-season occupancy models (*unmarked package* v0.12-0, [32]) to predict true species occupancy in relation to elevation. Using these adjusted occupancy estimates, we confirmed our a priori songbird and small mammal elevation designations by assessing how their multiyear site occupancy probability varied over the elevation gradient within HBEF (Figure 2). Species occupancy curves that increased to 100% predicted the occupancy above 800 m (approximately where the ecotone between deciduous and boreal forests reside in HBEF) were confirmed as high-elevation species, and species with a predicted occupancy that peaked and then declined before 800 m in elevation were confirmed as low-elevation species.

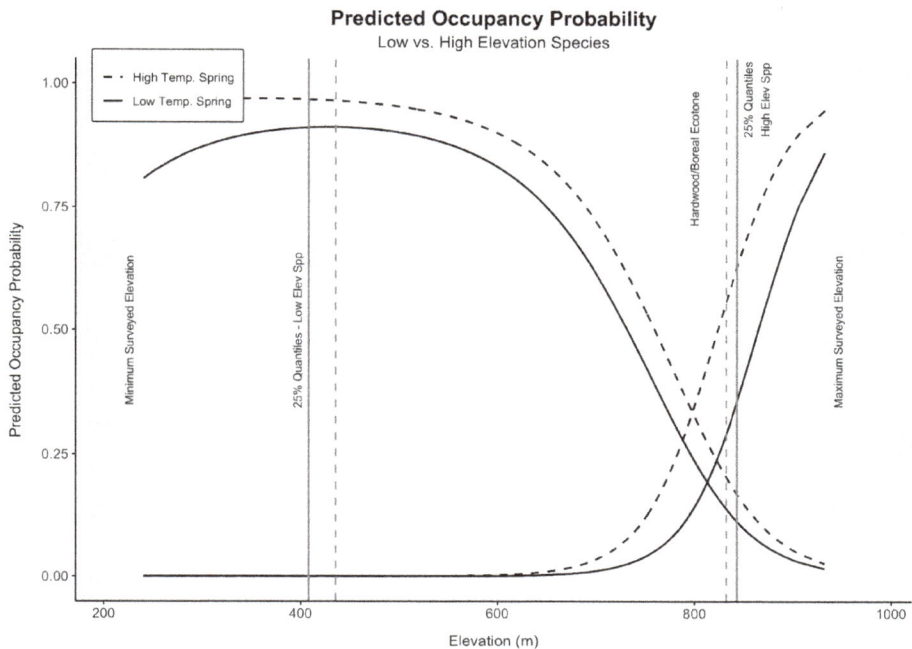

Figure 2. The simulated occupancy probabilities of high- and low-elevation species occupying a given elevation within HBEF, New Hampshire in relation to varying mean spring temperatures based on survey data corrected for imperfect detection. The solid lines represent species occupancy in a year with low spring temperatures, and the dashed lines represent occupancy in a year with high spring temperatures. The quantiles show how species distribution changes within their yearly range with mean spring temperatures.

The elevational gradient within HBEF (approximately 200 to 1000 m) likely does not encompass the full elevational distribution of some of the species included in our analyses, particularly those of high-elevation species. Following DeLuca and King [12], we accounted for partial elevational distributions by segmenting the predicted occupancy for each species into quantiles (2.5%, 5%, 25%, 50%, 75%, 95%, and 97.5%) to allow us to assess how species distributions were shifting and whether species distributions shifted uniformly, contracted, or expanded. Separate linear models for each predicted occupancy quantile were used to assess whether elevational distributions shifted through time within our study period. Linear models for each quantile were then used to assess whether the observed elevational distribution shifts were in response to average annual spring temperatures or average annual spring precipitations. We used Akaike's Information Criterion for limited sample sizes (AICc, MuMIn package v1.42.1, [33]) to identify the most parsimonious models among our candidate set of models. For low-elevation bird species, we examined the relationship between environmental covariates and the elevation of the 97.5% quantile across all low-elevation species, as the upper distributional range appears more sensitive to changes in climate [2,21]. For high-elevation bird species, we focused on the 2.5% quantile across all high-elevation species because HBEF is at the lower elevation band of their distribution. For our mammal species, we tested model appropriateness at the 50% quantile across both mammal species because our mammals include both high-elevation and low-elevation species.

All data are presented as mean $\pm$ 95% confidence interval, and results were considered significant at $p < 0.05$. All data were analyzed in R (v3.5.1, [34]).

3. Results

3.1. Environmental Variables

Historic mean annual temperature (df = 59, R^2 = 0.36, p < 0.001; Figure 3) and precipitation rates (df = 35, R^2 = 0.09, p = 0.044; Figure 3) have increased within HBEF, mirroring the trends observed in the northeastern North America over the past 100 years [19,20]. However, there was no change in the mean May temperature (df = 16, R^2 = 0.06, p = 0.165) or precipitation (df = 14, R^2 = −0.07, p = 0.909) over the 18-year period (1999 to 2016) that coincides with our survey data. The same 18-year period had substantial annual variability in both temperature and precipitation (Figure 3).

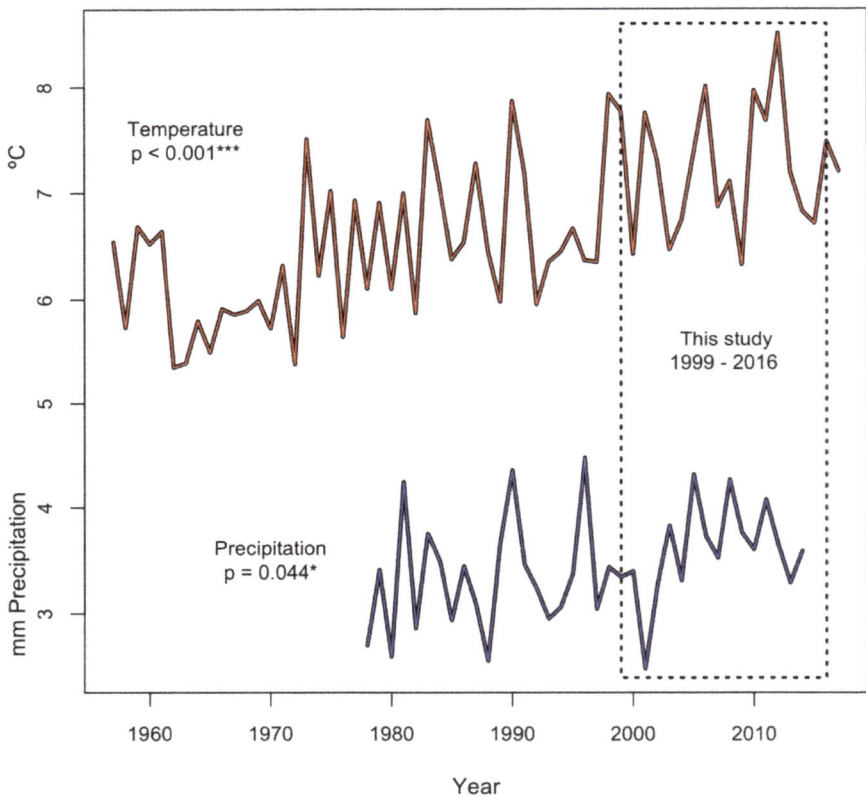

Figure 3. The mean annual temperature (1957–2017) and precipitation (1978–2014) trends in HBEF, New Hampshire. All data were collected from the USDA Forest Service Headquarters building at HBEF at 252 m above sea level.

3.2. Low-Elevation Birds

We found no evidence that low-elevation songbird species distributions shifted significantly with time over the study period (Table S1). However, three of five low-elevation bird species shifted upslope significantly with warm spring temperatures (BTBW, OVEN, and REVI; Table 1 and Table S2), while no-low elevation species shifted as a result of mean spring precipitation (Table S3). BTBWs contracted the bottom half of their distribution upslope by an average of 9.11 m/°C, while their upper distribution above 50% remained stable. In contrast to BTBWs, OVENs expanded their upper distribution (95 to 97.5% quantiles) upslope by an average of 10.58 m/°C, while their lower distribution remained stable. REVIs shifted nearly their entire elevational distribution (5% to 97.5% predicted occupancy quantiles) upslope significantly with warm spring temperature, moving the top 50% of their distribution upslope an average of 16.94 m/°C, with the top 5% shifting an average of 20.48 m/°C. All (100%) of the species that shifted significantly were species that migrate to wintering grounds outside of the United States [35–37] (hereafter "long-distance migrants"), while the remaining two species were residents or migrants within the United States [38,39] (hereafter "short-distance migrants").

The mean spring temperature alone was the most parsimonious of our candidate models for low-elevation bird species (Table S4), although two other models (temperature + precipitation and time + temperature) had ΔAICc values of <2.0.

Table 1. HBEF songbird and small mammal species distribution shifts in relation to the mean spring temperature. The mean May temperature was used for all species, excluding BLPW, where the mean June temperature was used. Asterisks (*) denote statistically significant p-values (* $p \leq 0.05$; ** $p \leq 0.01$; *** $p < 0.001$). Periods (.) denote near significant p-values (0.05 < $p \leq 0.10$). Abbreviations: Black-capped chickadee (BCCH), black-throated blue warbler (BTBW), hermit thrush (HETH), ovenbird (OVEN), red-eyed vireo (REVI), blackpoll warbler (BLPW), magnolia warbler (MAWA), dark-eyed junco (DEJU), Swainson's thrush (SWTH), winter wren (WIWR), red squirrel (RESQ), and eastern chipmunk (EACH).

| | | Mean Spring Temperature (1999–2016) | | | | |
| | | Low-Elevation Species | | | | |
Spp.	% Occu	Intercept	R^2	p	Significance
	2.5%	8.71	0.30	0.016	*
	5%	10.29	0.38	0.007	**
	25%	10.40	0.61	0.000	***
BTBW	50%	7.06	0.38	0.007	**
	75%	3.77	−0.01	0.370	
	95%	3.43	−0.04	0.545	
	97.5%	4.39	−0.03	0.446	
	2.5%	−0.90	−0.01	0.390	
	5%	−1.00	−0.04	0.534	
	25%	0.40	−0.07	0.882	
OVEN	50%	2.78	−0.01	0.361	
	75%	5.78	0.10	0.129	
	95%	10.12	0.22	0.040	*
	97.5%	11.05	0.21	0.041	*
	2.5%	0.77	0.14	0.082	.
	5%	1.62	0.21	0.040	*
	25%	6.43	0.53	0.001	***
REVI	50%	10.95	0.67	0.000	***
	75%	15.86	0.70	0.000	***
	95%	21.55	0.55	0.001	***
	97.5%	19.41	0.42	0.004	**
		High-Elevation Species				
	2.5%	14.81	−0.06	0.656	
	5%	9.47	−0.01	0.392	
	25%	4.93	0.21	0.041	*
BLPW	50%	3.47	0.34	0.011	*
	75%	1.83	0.33	0.012	*
	95%	0.42	0.47	0.002	**
	97.5%	0.16	0.23	0.034	*

3.3. High-Elevation Birds

We found that most high-elevation songbird species distributions did not shift significantly with time. Two species shifted portions of their distributions upslope with time (SWTH and DEJU) but only for one predicted occupancy quantile for each (25% and 75% respectively; Table S1). One of three high-elevation, long-distance migrant species [40] shifted distributions with the mean spring temperature (BLPW; Table 1), while the remaining long-distance migrants [41,42] and resident species [43,44] did not (Table S2). Contrary to our predictions, no high-elevation species shifted downslope, with the exception of the BLPW. The upper edge of the BLPW's distribution, at the maximum elevation of our study site, contracted slightly downslope with warm May temperature (97.5% quantile; −0.14 m/°C). However, BLPW distribution from the 25% to 97.5% quantiles expanded upslope with the mean June temperature (Table 1), with an average upslope shift of 2.16 m/°C.

No high-elevation bird distributions shifted as a result of the mean spring precipitation (Table S3). Our most parsimonious model of our candidate high-elevation species models was spring precipitation (Table S4), although models for both temperature and time had ΔAICc values of <2.0.

3.4. Mammals

We found no evidence that time, mean spring temperature, or mean spring precipitation had any significant effect on RESQ or EACH distributions through our study period (Tables S1–S3). The mean spring temperature was the most parsimonious model of our candidate models for mammals (Table S4) but only slightly more so than precipitation (ΔAICc = 0.05) and time (ΔAICc = 0.11).

4. Discussion

Overall, we did not find that songbird or small mammal elevational distributions were shifting through time within our study period. This is perhaps because temperature and precipitation did not increase significantly over our study period. Instead, our study species responded proximately to annual variations in climatic variables, with the majority of low-elevation songbird species responding to warm spring temperatures by shifting upslope (Figure S1a). However, most high-elevation songbird species and both small mammal species did not respond to temperature or precipitation (Figure S1a,b), perhaps because shifts seen in other studies occurred at higher elevations than those within our study site and climate constraints may be more extreme at higher elevations [8,11,12].

The distributional shifts we observed in low-elevation songbirds may be related to migratory status. Avian migratory status has been assessed by a handful of other studies (e.g. References [21,23]) but is not commonly assessed by studies looking at distributional shifts. Walther et al. [23] argue that migratory species are more likely to shift than resident species because they recolonize areas yearly. Within HBEF, first time breeders often must colonize new territories and are therefore subject to distribution shifts through changes in abundance and reactions to climatic variables and territory quality. This is in contrast with the resident species that take the slower, more stable route of local population extinction and colonization. They therefore exhibit smaller and slower distribution shifts [23], and is perhaps why we did not see shifts in resident species. Similar to Walther's [23] findings, we found that all three of our low-elevation songbird species that shifted upslope with warm spring temperatures were long-distance migrants, while the two species that did not shift significantly were a resident species (BCCH) and a short-distance migrant (HETH). Notably, our low-elevation migrant with the longest migration route and latest spring arrival time of our low-elevation songbirds, the REVI, was our most responsive species to warm spring temperatures. REVIs shifted nearly their entire distribution upslope with warm spring temperatures and shifted upslope farther than any of our other low-elevation species, at almost +20 m per degree of temperature increase. The other two low-elevation species that responded to warm spring temperature (BTBW and OVEN) responded to a lesser degree, shifting shorter distances and only portions of their distribution (Table 1). Long distance migrants have no way of assessing specific climate conditions at their summer breeding grounds, and they may instead use cues other than temperature, including day length [45,46] and favorable flying conditions at wintering grounds [47] to determine when to begin spring migration. Plastic responses to habitat quality upon arrival at summer breeding grounds may help buffer against phenological mismatches inherent with having a relatively set migratory schedule. Species with the longest migration routes, such as the REVI, which winters in South America [37], may benefit from having plastic settlement patterns, and these long migration routes may be one of the underlying reasons we see REVIs shift more than other species.

Contrary to our predictions, low-elevation mammals did not shift significantly with time, temperature, or precipitation. Although EACH is a frequent nest predator of songbird eggs and nestlings [26,27], their diet consists primarily of tree seeds, including those from the exceptionally long-lived [48] American beech trees (*Fagus grandifolia* Ehrhart) and sugar maple trees (*Acer saccharum* Marshall) [29,49]. EACH are therefore highly reliant on seed producing tree species as a food source, especially for caching food for consumption over the winter months [29,49]. In non-mast years, when tree seeds are scarce, EACH still rely heavily on vegetative food sources [29]. This reliance on seed producing tree species and other vegetation may be driving distribution shifts, or the lack thereof, as EACH may be tracking the slower changes in distribution of long-lived tree species over changes in temperature and precipitation [50].

In contrast with low-elevation songbird species, we found no distributional shifts in high-elevation songbirds (Figure S1a), with the exception of the BLPW, which shifted downslope with warm May temperatures. However, with an average shift of -0.14 m/°C, BLPW distributional shifts as a result of the mean May temperature were much smaller than those seen in low-elevation species. Others have also documented high-elevation birds shifting downslope [8,12,21], although the proposed causes vary. Some potential explanations include shifting vegetation [8,12] and increased high-elevation precipitation [12,21]. Neither of these explanations appear to be occurring within HBEF, as HBEF forest structure appears stable [51] and BLPWs did not respond to the mean spring precipitation. However, BLPW distributional shifts may be responding to the mean June temperature. Contrary to shifts related to the May temperatures, BLPWs showed a significant upslope expansion through all but the lowest reaches of their distribution with warm June temperatures (Figure S1a). BLPWs are long-distance migrants [40] and are typically the latest birds to arrive in HBEF out of the species we assessed in this study. They typically arrive to HBEF near the end of May and begin nesting shortly after [40]. As a result, they likely influenced most by the temperature at the end of May, which, within HBEF, are more typical of temperatures we would expect to see in June. So, why is the BLPW shifting while other high-elevation species are not? Like the REVI, BLPWs are one of the longest distance migrants of our high-elevation species [40] and may be more plastic in their response to environment as a way to buffer against phenological mismatches when arriving at breeding grounds.

Overall, the lack of observed distribution shifts by high-elevation songbird and mammal species at HBEF, with the exception of BLPWs, may be due in part to the elevation range in our study site. Most of our high-elevation sites are lower than those in numerous other studies that observed birds and small mammals move downslope (elevational maximums: this study, 903 m; Archaux [8], 3099 m; DeLuca and King [12], 1470 m; etc.), and climate constraints on our high-elevation species may not be as extreme as those at higher elevation sites [19,20]. Boreal habitats within HBEF are found primarily above 800 m in elevation and are therefore restricted to the highest ridges in the valley. As a result, we also may be missing reactions to precipitation by high-elevation species due to the small elevational range of boreal spruce-fir within HBEF. Alternately, high-elevation species in HBEF may be experiencing the effects of increased precipitation but may not be shifting significantly out of necessity for specific breeding habitats. Archaux [8] found that avian abundance and distribution within two study sites in the French Alps was closely tied with habitat distribution shifts, despite significant warming. Similar studies of high-elevation mammals like RESQ have found they may also be tracking habitat distribution shifts over changes in climatic conditions, much like their low-elevation counterparts [11,50]. RESQ are heavily reliant on conifer seeds as a food source and typically hold territories in or near stands of conifers [16,52]. Elevationally restricted habitat may cause increased competition by habitat-constrained species for breeding or food caching territory [12,15,29], and as a result, high-elevation species may be using all of the available habitat that suits their needs, regardless of quality. If this is the case, any evidence of habitat tracking would likely only be seen over longer time periods, as vegetation distribution shifts are typically slower than temperature change, due to the long lifespans of trees [1,20].

4.1. Future Implications

Distributional shifts will likely impact montane passerines and small mammals in several ways. With temperature and precipitation rates expected to increase faster at higher elevations, species shifting upslope may encounter novel combinations of climate and vegetation. Also, because some species are shifting while others are not, the avian community composition will likely change as a result, and novel species interactions may arise [53]. As birds shift in elevation, they are also likely to encounter non-avian species and habitats they have not interacted with before [53,54]. While these novel interactions may be beneficial for some species, it is unknown how this will ultimately affect sensitive populations and ecosystem dynamics. For instance, nest predation is a major factor in avian reproductive success or failure [27,49,55], and as bird species distributions shift, birds moving into areas where RESQ and EACH overlap may encounter higher rates of nest predation, potentially impacting their ability to produce young. Additionally, species that are not shifting may encounter southern and lower-elevation species that are expanding into areas left by species moving upslope or poleward [6,13], like red-bellied woodpeckers (*Melanerpes carolinus*), tufted titmice (*Baeolophus bicolor*), and eastern gray squirrels (*Sciurus carolinensis*). Species with contracting distributions, like the BTBW, may also face increased intraspecies competition during the breeding season, as the preferred breeding habitat decreases.

Small yearly distributional shifts may not be ecologically significant alone, and over short timescales, the impacts of elevational shifts may not be obvious. However, as temperatures increase over the long term, these small shifts may lead to larger scale shifts that will affect ecosystems and individual species alike. The mean annual temperature in New England is expected to increase 1.7 to 4.4 °C or more by the year 2100 [19,20], and under the worst-case warming scenario, we could see REVI and other highly plastic species shift upslope an average of almost 100 m by the year 2100. For species occupying sites near the maximum elevation within HBEF, an upslope shift of 100 m will likely extirpate them from the valley entirely. Precipitation and temperature increases are expected to be more severe at higher elevations [20,21]. As a result, species at higher elevations in the region may shift even farther than the species detailed here [12,21,54], thus exacerbating the effects of elevational shifts and novel community interactions.

4.2. Further Research

Long-term datasets are invaluable for measuring species distribution shifts. Therefore, the continued collection of species occupancy and abundance coupled with climate variables is essential to understanding the impacts of climate change on montane species as well as impacts on lower elevation species that may eventually colonize montane habitats. It is likely other local and regional variables in addition to temperature and precipitation influence elevational distribution shifts in our study species. Other variables should be assessed to increase our understanding of how species distributions are changing, including vegetation shifts at transition zones, changes in winter snowpack depth, intra and interspecies interactions, and prey distribution shifts.

HBEF supports a wide variety of avian species, including high-elevation species, despite HBEF being at the lower end of many of their elevational distributions. Yet, as climate continues to change and species continue to shift upslope, we may begin to lose high-elevation species within HBEF as they move to higher elevation sites elsewhere. Mountaintop extirpation as a result of climate change has been the focus of only a few studies in the northeast and should be explored further to increase our understanding of how species distributions shift with climate change.

5. Conclusions

We found almost no directional change in distributions through time for any of our songbird or small mammal species. However, we found three of our five (60%) low-elevation bird species (BTBW, OVEN, and REVI) responded to warm spring temperatures by shifting upslope. All low-elevation songbird species that shifted were long-distance migrants, while those that did not shift were year-round residents of HBEF or short-distance migrants (BCCH and HETH, respectively). Each low-elevation migrant species responded differently to warm spring temperatures, through upslope distribution expansion, downslope distribution contraction, or total distribution shift upslope. BLPW was the only high-elevation songbird species that shifted with warm spring temperature (downslope with the mean May temperature and upslope with the mean June temperature). The remaining high-elevation bird species (MAWA, DEJU, SWTH, and WIWR) and both high- and low-elevation mammal species (RESQ and EACH, respectively) did not shift in response to spring temperature or precipitation. This may be due to the limited elevation range in our study site, which may not experience the severe climate constraints found at higher elevations.

Supplementary Materials: The following are available online at http://www.mdpi.com/1999-4907/10/2/84/s1, Figure S1: Elevational Shifts in Target Species, Table S1: Hubbard Brook Experimental Forest (HBEF) songbird and small mammal species distribution shifts in relation to time in years, Table S2: HBEF songbird and small mammal species distribution shifts in relation to the mean spring temperature, Table S3: HBEF songbird and small mammal species distribution shifts in relation to the mean spring precipitation.

Author Contributions: Conceptualization, A.V.T., E.F., and M.H.; methodology, A.V.T., E.F., and M.H.; software, A.V.T. and M.H.; validation, M.H.; formal analysis, A.V.T. and M.H.; investigation, N.R. and T.S.S.; resources, N.R. and T.S.S.; data curation, M.H.; writing—original draft preparation, A.V.T. and M.H.; writing—review and editing, M.H., E.F., T.S.S. and N.R.; visualization, A.V.T.; supervision, M.H.; project administration, N.R., T.S.S. and M.H.; funding acquisition, N.R. and T.S.S.

Funding: This manuscript is a contribution of the Hubbard Brook Ecosystem Study. Hubbard Brook is part of the Long-Term Ecological Research (LTER) network, which is supported by the U.S. National Science Foundation. The Hubbard Brook Experimental Forest is operated and maintained by the USDA Forest Service, Northern Research Station, Newtown Square, PA. This research was funded by the National Science Foundation, grant numbers 9810221 and 0423259.

Acknowledgments: A special thanks to Amey Bailey at the USDA Forest Service headquarters at HBEF for compiling missing years of temperature data on our behalf. We also thank our two anonymous reviewers whose helpful suggestions strengthened our manuscript considerably, Bill DeLuca for the helpful statistics suggestions, and all of the dedicated field technicians who collected data for our avian and mammal dataset.

Conflicts of Interest: The authors declare no conflict of interest.

References

1.   Beckage, B.; Osborne, B.; Gavin, D.G.; Pucko, C.; Siccama, T.; Perkins, T. A rapid upward shift of a forest ecotone during 40 years of warming in the Green Mountains of Vermont. *Proc. Natl. Acad. Sci. USA* **2008**, *105*, 4197–4202. [CrossRef] [PubMed]

2.   Freeman, B.G.; Class Freeman, A.M. Rapid upslope shifts in New Guinean birds illustrate strong distributional responses of tropical montane species to global warming. *Proc. Natl. Acad. Sci. USA* **2014**, *111*, 4490–4494. [CrossRef] [PubMed]

3.   Zuckerberg, B.; Woods, A.M.; Porter, W.F. Poleward shifts in breeding bird distributions in New York State. *Glob. Chang. Biol.* **2009**, *15*, 1866–1883. [CrossRef]

4.   Hitch, A.T.; Leberg, P.L. Breeding Distributions of North American Bird Species Moving North as a Result of Climate Change. *Conserv. Biol.* **2007**, *21*, 534–539. [CrossRef] [PubMed]

5.   La Sorte, F.A.; Thompson, F.R. Poleward shifts in winter ranges of North American birds. *Ecology* **2007**, *88*, 1803–1812. [CrossRef] [PubMed]

6. Myers, P.; Lundrigan, B.L.; Hoffman, S.M.G.; Haraminac, A.P.; Seto, S.H. Climate-induced changes in the small mammal communities of the Northern Great Lakes Region. *Glob. Chang. Biol.* **2009**, *15*, 1434–1454. [CrossRef]

7. Chen, I.-C.; Shiu, H.-J.; Benedick, S.; Holloway, J.D.; Chey, V.K.; Barlow, H.S.; Hill, J.K.; Thomas, C.D. Elevation increases in moth assemblages over 42 years on a tropical mountain. *Proc. Natl. Acad. Sci. USA* **2009**, *106*, 1479–1483. [CrossRef]

8. Archaux, F. Breeding upwards when climate is becoming warmer: No bird response in the French Alps. *IBIS* **2004**, *146*, 138–144. [CrossRef]

9. Lenoir, J.; Gegout, J.-C.; Guisan, A.; Vittoz, P.; Wohlgemuth, T.; Zimmermann, N.E.; Dullinger, S.; Pauli, H.; Willner, W.; Svenning, J.-C. Going against the flow: Potential mechanisms for unexpected downslope range shifts in a warming climate. *Ecography* **2010**, *33*, 295–303. [CrossRef]

10. McCain, C.M.; Colwell, R.K. Assessing the threat to montane biodiversity from discordant shifts in temperature and precipitation in a changing climate: Climate change risk for montane vertebrates. *Ecol. Lett.* **2011**, *14*, 1236–1245. [CrossRef]

11. Wen, Z.; Wu, Y.; Ge, D.; Cheng, J.; Chang, Y.; Yang, Z.; Xia, L.; Yang, Q. Heterogeneous distributional responses to climate warming: Evidence from rodents along a subtropical elevational gradient. *BMC Ecol.* **2017**, *17*, 17. [CrossRef] [PubMed]

12. DeLuca, W.V.; King, D.I. Montane birds shift downslope despite recent warming in the northern Appalachian Mountains. *J. Ornithol.* **2017**, *158*, 493–505. [CrossRef]

13. Rodenhouse, N.L.; Matthews, S.N.; McFarland, K.P.; Lambert, J.D.; Iverson, L.R.; Prasad, A.; Sillett, T.S.; Holmes, R.T. Potential effects of climate change on birds of the Northeast. *Mitig. Adapt. Strategies Glob. Chang.* **2008**, *13*, 517–540. [CrossRef]

14. Sekercioglu, C.H.; Schneider, S.H.; Fay, J.P.; Loarie, S.R. Climate Change, Elevational Range Shifts, and Bird Extinctions: Elevation, Climate Change, and Bird Extinctions. *Conserv. Biol.* **2008**, *22*, 140–150. [CrossRef] [PubMed]

15. Sabo, S.R.; Holmes, R.T. Foraging Niches and the Structure of Forest Bird Communities in Contrasting Montane Habitats. *Condor* **1983**, *85*, 121–138. [CrossRef]

16. Saunders, D.A. *Adirondack Mammals*; College of Environmental Science and Forestry, State University of New York: New York, NY, USA, 1988; 216p.

17. Crick, H.Q.P. The impact of climate change on birds: Impact of climate change on birds. *IBIS* **2004**, *146*, 48–56. [CrossRef]

18. Jetz, W.; Wilcove, D.S.; Dobson, A.P. Projected Impacts of Climate and Land-Use Change on the Global Diversity of Birds. *PLoS Biol.* **2007**, *5*, e157. [CrossRef]

19. IPCC Climate Change 2014: Synthesis Report. In *Contribution of Working Groups I, II and III to the Fifth Assessment Report of the Intergovernmental Panel on Climate Change*; Core Writing Team; Pachauri, R.K.; Meyer, L.A. (Eds.) IPCC: Geneva, Switzerland, 2014; p. 151.

20. Janowiak, M.K.; D'Amato, A.W.; Swanston, C.W.; Iverson, L.; Thompson, F.R.; Dijak, W.D.; Matthews, S.; Peters, M.P.; Prasad, A.; Fraser, J.S.; et al. *New England and Northern New York Forest Ecosystem Vulnerability Assessment and Synthesis: A Report from the New England Climate Change Response Framework Project*; U.S. Department of Agriculture, Forest Service, Northern Research Station: Newtown Square, PA, USA, 2018.

21. Tingley, M.W.; Koo, M.S.; Moritz, C.; Rush, A.C.; Beissinger, S.R. The push and pull of climate change causes heterogeneous shifts in avian elevational ranges. *Glob. Chang. Biol.* **2012**, *18*, 3279–3290. [CrossRef]

22. Seidel, T.M.; Weihrauch, D.M.; Kimball, K.D.; Pszenny, A.A.P.; Soboleski, R.; Crete, E.; Murray, G. Evidence of Climate Change Declines with Elevation Based on Temperature and Snow Records from 1930s to 2006 on Mount Washington, New Hampshire, U.S.A. *Arct. Antarct. Alpine Res.* **2009**, *41*, 362–372. [CrossRef]

23. Walther, G.-R.; Post, E.; Convey, P.; Menzel, A.; Parmesan, C.; Beebee, T.J.C.; Fromentin, J.-M.; Hoegh-Guldberg, O.; Bairlein, F. Ecological responses to recent climate change. *Nature* **2002**, *416*, 389–395. [CrossRef]

24. Hubbard Brook Research Foundation. *Chapter 01: The Hubbard Brook Ecosystem Study: Site, History, and Research Approaches*; HBEF: North Woodstock, NH, USA, 2018. Available online: https://hubbardbrook.org/online-book (accessed on 31 December 2018).

25. Holmes, R.T.; Sherry, T.W.; Sturges, F.W. Bird Community Dynamics in a Temperate Deciduous Forest: Long-Term Trends at Hubbard Brook. *Ecol. Monogr.* **1986**, *56*, 201–220. [CrossRef]
26. Reitsma, L.R.; Holmes, R.T.; Sherry, T.W. Effects of Removal of Red Squirrels, *Tamiasciurus hudsonicus*, and Eastern Chipmunks, *Tamias striatus*, on Nest Predation in a Northern Hardwood Forest: An Artificial Nest Experiment. *Oikos* **1990**, *57*, 375. [CrossRef]
27. Sherry, T.W.; Wilson, S.; Hunter, S.; Holmes, R.T. Impacts of nest predators and weather on reproductive success and population limitation in a long-distance migratory songbird. *J. Avian Biol.* **2015**, *46*, 559–569. [CrossRef]
28. Rodewald, P. *The Birds of North America*; Cornell Laboratory of Ornithology: Ithaca, NY, USA, 2015.
29. Elliott, L. Social behavior and foraging ecology of the eastern chipmunk (*Tamias striatus*) in the Adirondack Mountains. *Smithson. Contrib. Zool.* **1978**, 1–107. [CrossRef]
30. Rusch, D.A.; Reeder, W.G. Population Ecology of Alberta Red Squirrels. *Ecology* **1978**, *59*, 400–420. [CrossRef]
31. Chandler, R.B.; Royle, J.A.; King, D.I. Inference about density and temporary emigration in unmarked populations. *Ecology* **2011**, *92*, 1429–1435. [CrossRef]
32. Fiske, I.; Chandler, R. unmarked: An R Package for Fitting Hierarchical Models of Wildlife Occurrence and Abundance. *J. Stat. Softw.* **2011**, *43*, 1–23. [CrossRef]
33. Bartoń, K. *MuMIn: Multi-Model Inference*, R package Version 1.42.1; 2018. Available online: https://CRAN.R-project.org/package=MuMIn (accessed on 31 December 2018).
34. R Core Team R. *A Language and Environment for Statistical Computing*; R Foundation for Statistical Computing: Vienna, Austria, 2018.
35. Holmes, R.T.; Kaiser, S.A.; Rodenhouse, N.L.; Sillett, T.S.; Webster, M.S.; Pyle, P.; Patten, M.A. Black-Throated Blue Warbler (*Setophaga caerulescens*), version 3.0. In *The Birds of North America*; Rodewald, P.G., Ed.; Cornell Lab of Ornithology: Ithaca, NY, USA, 2017. [CrossRef]
36. Porneluzi, P.; Van Horn, M.A.; Donovan, T.M. Ovenbird (*Seiurus aurocapilla*), version 2.0. In *The Birds of North America*; Poole, A.F., Ed.; Cornell Lab of Ornithology: Ithaca, NY, USA, 2011. [CrossRef]
37. Cimprich, D.A.; Moore, F.R.; Guilfoyle, M.P. Red-eyed/Chivi Vireo (*Vireo olivaceus/chivi*), version 2.0. In *The Birds of North America*; Poole, A.F., Gill, F.B., Eds.; Cornell Lab of Ornithology: Ithaca, NY, USA, 2000. [CrossRef]
38. Dellinger, R.; Wood, P.B.; Jones, P.W.; Donovan, T.M. Hermit Thrush (*Catharus guttatus*), version 2.0. In *The Birds of North America*; Poole, A.F., Ed.; Cornell Lab of Ornithology: Ithaca, NY, USA, 2012. [CrossRef]
39. Foote, J.R.; Mennill, D.J.; Ratcliffe, L.M.; Smith, S.M. Black-capped Chickadee (*Poecile atricapillus*), version 2.0. In *The Birds of North America*; Poole, A.F., Ed.; Cornell Lab of Ornithology: Ithaca, NY, USA, 2010. [CrossRef]
40. DeLuca, W.; Holberton, R.; Hunt, P.D.; Eliason, B.C. Blackpoll Warbler (*Dendroica striata*). In *The Birds of North America*; Poole, A.F., Ed.; Cornell Lab of Ornithology: Ithaca, NY, USA, 2013. [CrossRef]
41. Mack, D.E.; Yong, W. Swainson's Thrush (*Catharus ustulatus*), version 2.0. In *The Birds of North America*; Poole, A.F., Gill, F.B., Eds.; Cornell Lab of Ornithology: Ithaca, NY, USA, 2000. [CrossRef]
42. Dunn, E.H.; Hall, G.A. Magnolia Warbler (*Setophaga magnolia*), version 2.0. In *The Birds of North America*; Poole, A.F., Ed.; Cornell Lab of Ornithology: Ithaca, NY, USA, 2010. [CrossRef]
43. Nolan, V., Jr.; Ketterson, E.D.; Cristol, D.A.; Rogers, C.M.; Clotfelter, E.D.; Titus, R.C.; Schoech, S.J.; Snajdr, E. Dark-eyed Junco (*Junco hyemalis*), version 2.0. In *The Birds of North America*; Poole, A.F., Gill, F.B., Eds.; Cornell Lab of Ornithology: Ithaca, NY, USA, 2002. [CrossRef]
44. Hejl, S.J.; Holmes, J.A.; Kroodsma, D.E. Winter Wren (*Troglodytes hiemalis*), version 2.0. In *The Birds of North America*; Poole, A.F., Gill, F.B., Eds.; Cornell Lab of Ornithology: Ithaca, NY, USA, 2002. [CrossRef]
45. Wolfson, A. Day Length, Migration, and Breeding Cycles in Birds. *Sci. Mon.* **1952**, *74*, 191–200.
46. Bauer, S.; Nolet, B.A.; Giske, J.; Chapman, J.W.; Åkesson, S.; Hedenström, A.; Fryxell, J.M. Cues and decision rules in animal migration. In *Animal Migration*; Milner-Gulland, E.J., Fryxell, J.M., Sinclair, A.R.E., Eds.; Oxford University Press: Oxford, UK, 2011; pp. 68–87, ISBN 978-0-19-956899-4.
47. Richardson, W.J. Timing and Amount of Bird Migration in Relation to Weather: A Review. *Oikos* **1978**, *30*, 224. [CrossRef]
48. Loehle, C. Tree life history strategies: The role of defenses. *Can. J. For. Res.* **1988**, *18*, 209–222. [CrossRef]
49. Clotfelter, E.D.; Pedersen, A.B.; Cranford, J.A.; Ram, N.; Snajdr, E.A.; Nolan, V.; Ketterson, E.D. Acorn mast drives long-term dynamics of rodent and songbird populations. *Oecologia* **2007**, *154*, 493–503. [CrossRef] [PubMed]

50. Ulateig, G. *Vegetation Changes and Small Rodent Responses along Alpine Gradients in Oceanic and Continental Climate*; Norwegian University of Life Sciences: Ås, Norway, 2010.

51. van Doorn, N.S.; Battles, J.J.; Fahey, T.J.; Siccama, T.G.; Schwarz, P.A. Links between biomass and tree demography in a northern hardwood forest: A decade of stability and change in Hubbard Brook Valley, New Hampshire. *Can. J. For. Res.* **2011**, *41*, 1369–1379. [CrossRef]

52. Smith, M.C. Red Squirrel Responses to Spruce Cone Failure in Interior Alaska. *J. Wildl. Manag.* **1968**, *32*, 305–317. [CrossRef]

53. Blois, J.L.; Zarnetske, P.L.; Fitzpatrick, M.C.; Finnegan, S. Climate Change and the Past, Present, and Future of Biotic Interactions. *Science* **2013**, *341*, 499–504. [CrossRef] [PubMed]

54. Moritz, C.; Patton, J.L.; Conroy, C.J.; Parra, J.L.; White, G.C.; Beissinger, S.R. Impact of a Century of Climate Change on Small-Mammal Communities in Yosemite National Park, USA. *Science* **2008**, *322*, 261–264. [CrossRef] [PubMed]

55. Bradley, J.E.; Marzluff, J.M. Rodents as Nest Predators: Influences on Predatory Behavior and Consequences to Nesting Birds. *Auk* **2003**, *120*, 1180–1187. [CrossRef]

© 2019 by the authors. Licensee MDPI, Basel, Switzerland. This article is an open access article distributed under the terms and conditions of the Creative Commons Attribution (CC BY) license (http://creativecommons.org/licenses/by/4.0/).

forests

MDPI

Article

Damage Diversity as a Metric of Structural Complexity after Forest Wind Disturbance

Chris J. Peterson

Department of Plant Biology, University of Georgia, Athens, GA 30602, USA; chris@plantbio.uga.edu;
Tel.: +011-706-542-3754

Received: 31 December 2018; Accepted: 21 January 2019; Published: 23 January 2019

Abstract: This study presents a new metric for quantifying structural complexity using the diversity of tree damage types in forests that have experienced wind disturbance. Structural complexity studies of forests have to date not incorporated any protocol to address the variety of structural damage types experienced by trees in wind disturbances. This study describes and demonstrates such a protocol. Damage diversity, defined as the richness and evenness of types of tree damage, is calculated analogously to species diversity using two common indices, and termed a 'Shannon Damage Heterogeneity Index' (Sh-DHI) and an inverse Simpson Damage Heterogeneity Index (iSi-DHI). The two versions of the DHI are presented for >400 plots across 18 distinct wind disturbed forests of eastern North America. Relationships between DHI and pre-disturbance forest species diversity and size variability, as well as wind disturbance severity, calculated as the fraction of basal area downed in a wind disturbance event, are examined. DHIs are only weakly related to pre-disturbance tree species diversity, but are significantly positively related to pre-disturbance tree size inequality (size diversity). Damage diversity exhibits a robust curvilinear relationship to severity; both versions of the DHI show peaks at intermediate levels of wind disturbance severity, suggesting that in turn structural complexity may also peak at intermediate levels of severity.

Keywords: species diversity; structural complexity; legacies; wind damage; uprooting; trunk breakage

---

1. Introduction

Diversity in forests can be of several types, such as compositional, functional, and structural [1–3], although historically the great majority of discussion of diversity has focused on compositional diversity (usually at the species level). In the past two decades, however, interest in functional and structural diversity has greatly increased [3–5]. These three types of diversity may interact in complex and interesting ways, such that they might be considered either cause or effect, depending on circumstance.

Most obviously, compositional diversity may result in both functional and structural diversity. For example, a larger suite of species will—simply because of the sampling effect—be likely to have more variety of ecophysiological responses to resources (functional diversity) as well as a wider range of growth forms and architectures (structural diversity) [6–8].

Conversely, if a suite of species that includes potential members of a community are functionally diverse (e.g., in drought or shade tolerance), there is greater chance for coexistence, thus facilitating compositional diversity: the classical niche partitioning explanation for species diversity [9,10].

A third type of interaction, rooted in the seminal work by MacArthur [11], hypothesizes that structural diversity provides more habitat niches and thereby facilitates higher species diversity. MacArthur showed that different warbler species utilized different feeding locations (niches) in spruce and fir trees, and proposed that this spatial partitioning allowed the warblers to coexist. More recently, this idea has been reformulated and presented with the label structural complexity (e.g., [12–14]). The closely-related concept of biological legacies has much overlap with structural complexity, although legacies are most often discussed in the context of post-disturbance conditions [15–18]. Regardless of the terminology, the idea is that forests with greater structural variability or heterogeneity should provide for more niche diversity (either plant or animal), and therefore greater species diversity; this facilitation has been documented numerous times for various faunal groups [19–23]. Throughout this manuscript, the term "structural complexity" will be used to describe this idea.

Note that a distinct and important developing idea applies at the ecosystem level, wherein structural complexity is defined as canopy rugosity; higher complexity is correlated with, and possibly facilitating, higher net primary productivity [24,25]. Moreover, other studies show a positive correlation between structural complexity and aboveground carbon stocks [26,27]; the latter idea is outside the scope of this paper, although reconciliation of these two approaches (community-level and ecosystem-level) would seem to benefit multiple areas of ecological understanding.

Given that the various types of diversity (compositional, functional, structural) may influence each other as either cause or effect, especially in light of the third example (structural diversity facilitating species diversity), it is pertinent to ask if structural or compositional diversity before a disturbance might influence structural diversity after disturbance. While a variety of studies examine structural diversity or complexity after fire, insect outbreaks, or harvesting [13], only a few have examined structural complexity after wind disturbance [28–30]. This paper aims to characterize one component of structural complexity after wind disturbance, and test for associations with two types of pre-disturbance forest diversity. In particular, analyses will test for potential influence of pre-disturbance species diversity or size variation on post-disturbance damage diversity (described below).

Many different metrics have been employed by researchers seeking to quantify structural complexity, often in an attempt to establish some benchmark based on conditions in old-growth or unmanaged forests, that can subsequently serve as a target of management actions [31–33]. While some complexity metrics address species diversity [13], spatial distribution patterns of trees [4,34–38], or size variation [28,34,39–41], an entirely distinct class of complexity metrics seek to capture the variety of structural categories present in a stand. Examples of the latter are measures of coarse woody debris or snag abundance [29,30]. This paper introduces an apparently novel metric of structural complexity: damage diversity. This metric is most suited to quantifying the variety of distinct tree structural categories on the basis of the damage experienced in wind disturbances. The rationale of the proposed damage diversity metric is that uprooted trees provide very distinct habitat niches from trees that experienced trunk breakage, and these categories further differ in obvious ways from intact or lightly damaged trees. For example, abundant research has examined the influence of uprooted tree root mounds and pits on fine scale patterns of plant species composition [42–51]. Trunk-broken trees simultaneously deposit tree crowns on the ground (which can greatly influence spatial variation in fire intensity [52]) as well as creating standing snags of various heights [53]. Uprooted and trunk-broken trees also differ in the amount of tree bole deposited on the soil surface—boles can serve as fine-scale fire breaks [52] and can impede movement of ungulate herbivores and thus reduce browsing [54–57], and as they decay can provide suitable microsites for seedling establishment [58,59]. Lighter types of damage (e.g., bent, leaning or light crown damage) will, in hardwood forests, prompt release of advanced regeneration (suppressed saplings) and prolific sprouting of the damaged trees, particularly by smaller trees. These responses often result in limited new seedling colonization and thus continuance of site dominance by the pre-disturbance individuals [30,48,60–67]. Clearly the distinct types of tree damage have potentially great influence on post-disturbance dynamics of both composition and diversity of vegetation: at the extremes, light damage will perpetuate existing composition and diversity, while uprooting will facilitate new seedling establishment on treefall microsites and potentially shift composition to an early-seral suite of species. It is expected, therefore, that a greater diversity of damage types after wind disturbance will foster a higher species diversity during forest regeneration.

Expectations for how pre-disturbance compositional and structural diversity might influence post-disturbance damage diversity are developed as follows. Tree species differ in their vulnerability to wind damage [68], thus it is expected (Hypothesis 1) that as pre-disturbance species richness or diversity increases, tree damage diversity should increase. Similarly, it is well-established that larger trees are more likely to suffer trunk breakage or uprooting (e.g., [68]); in contrast, small trees and saplings may be sufficiently flexible that they bend, a phenomenon rarely witnessed in large trees [69], and one that may allow small trees to avoid complete trunk breakage or uprooting. Thus smaller trees may be prone to different types of damage than large trees; as a consequence the second expectation (Hypothesis 2) is that while types of tree damage may shift with increasing tree size, damage diversity will be roughly similar across stands of different tree sizes. Following this logic, stands with limited tree size inequality (more uniform sizes) should show low damage diversity (although the actual types of damage may differ depending on the mean tree size). Conversely, if tree size inequality is high, a greater variety of types of tree damage are expected, and therefore the third expectation (Hypothesis 3) is that damage diversity will increase with increasing size inequality.

A fourth hypothesis emerges from the recent recognition that intermediate-severity wind disturbances may be an important source of structural complexity in forests [30], although the sites available for sampling did not allow those authors to explicitly address how complexity might vary with wind disturbance severity. There are several reasons to expect that post-disturbance structural complexity might exhibit a unimodal relationship to wind disturbance severity, with a peak at intermediate levels; some of this reasoning is a logical result of the definition of severity as the proportion of trees broken or uprooted. First, at low levels of severity, the more drastic types of tree damage (e.g., trunk breakage and uprooting) are likely to be uncommon while lighter types of tree damage will be most common, thus, damage diversity should be low in low severity situations. At the very highest levels of severity, where most or all trees are fallen, the lighter types of damage are obviously rare or absent and therefore damage diversity is again low. In contrast, at intermediate levels of severity, trees are likely to suffer a variety of types of damage, resulting in maximal damage diversity when severity is intermediate. This straightforward logic produces an expectation (Hypothesis 4) that damage diversity will be maximal at intermediate levels of wind damage severity. An additional subsidiary expectation is that the predominant type of tree damage will change with changing severity, such that uprooting and trunk breakage predominate at high severity, while lighter types of damage predominate at lower severity.

Specific objectives of this study are to (1) demonstrate a protocol for developing a damage diversity metric in recently wind-disturbed stands; (2) apply this protocol to describe the damage diversity in 18 separate wind disturbed forests of eastern and central North America; and (3) test the four hypotheses presented above, to explore the influence of pre-disturbance compositional and structural diversity, as well as disturbance severity, on the resulting damage diversity.

2. Materials and Methods

2.1. Study Sites

From the study sites from which data are reported here, several previous publications report other aspects of damage and regeneration [43,44,63,68–75]. Descriptions below are presented chronologically based on date of disturbance event; locations shown in Figure 1. Dominant woody species are presented in decreasing order of cumulative basal area.

Figure 1. General location of 18 study sites within Eastern U.S. Multiple locations close together shown as a single box with multiple numbers. 1 = Mingo; 2 = Tionesta 1994; 3 = Gould Farm; 4 = Texas Hill; 5 = Rapid River; 6 = Taylor; 7 = Hattons; 8 = Fishhook; 9 = Meditation Lake; 10 = Shirttail; 11 = Twin; 12 = ThreeMile Island; 13 = Natchez Trace; 14 = Gum Road; 15 = Boggs Creek; 16 = Martin Branch; 17 = Timpson; 18 = Smokies.

Mingo site: Mingo National Wildlife Refuge is located in southeastern Missouri (36.974° N, 90.176° W; Table 1). On 4 June, 1993, an F1-rated tornado passed over the Refuge; lower hillslope and bottomland forests were damaged along a 20-km path. Tree damage was sampled in August 1994, in 9 survey plots distributed along transects; the 6 on hillslopes were 20 m × 20 m, while the 3 in bottomlands were 30 m × 30 m to adjust for lower tree density. Forests are of low density and consist of a high diversity of *Quercus* species (oaks) and other mesic hardwoods.

Tionesta 1994 site: This site is a primary forest preserve within Allegheny National Forest of northwestern Pennsylvania (41.638° N, 78.940° W; Table 1). Dominant woody vegetation was *Fagus grandifolia* Ehrh. (greatly diminished since the 1990s because of beech bark disease) and *Tsuga canadensis* L., with smaller components of *Betula alleghaniensis* Britton, *Betula lenta* L., *Prunus serotina* Ehrh., and *Acer rubrum* L. The site was disturbed by a small F2-rated tornado on 10 July, 1994, which damaged a patch of roughly 2 ha. (tornadoes in the United States are rated according to severity of damage, on a six-level scale from F0 to F5 [prior to 2007] or EF0 to EF5 [2007 and later]; Storm Prediction Center Fujita Tornado Damage Scale). The entire damaged patch was mapped with surveying equipment in summer 1996, and subsequently partitioned into 20 m × 20 m contiguous plots [Peterson 2000]. This site has been free of major disturbance (prior to 1994) for at least 3 centuries [76].

Gould Farm and Texas Hill sites: These two sites, separated by ~15 km, were disturbed by the same F4-rated tornado on May 29, 1995. Gould Farm is in western Massachusetts (42.175° N, 73.249° W), and had *Pinus strobus* L., *T. canadensis*, and *Acer saccharum* Marshall as dominant species. Texas Hill is in far eastern New York (42.207° N, 73.606° W; Table 1), and *Quercus montana* Willd., *Quercus rubra* L., and *T. canadensis* were dominants. Each site was sampled in 1997 with ten 20 m × 20 m plots distributed as 5 plots along each of 2 transects. Both are secondary stands with unknown disturbance history.

Rapid River site: This site in the central Upper Peninsula of Michigan (45.944° N, 86.658° W; Table 1), is covered in secondary forest dominated by *A. rubrum*, *Betula papyrifera* Marshall and *Populus grandidentata* Michx. Forests were damaged by a 5 October, 1997 thunderstorm downburst, which impacted an area roughly 5 km wide and 35 km long. Eleven 20 m × 20 m plots distributed along 2 parallel transects were sampled in July of 1999.

Hattons and Taylor sites: These sites, separated by ~10 km, are located in secondary forests of northeastern Pennsylvania (Table 1). Hattons (41.398° N, 75.126° W) was dominated by *A. rubrum* and *P. strobus*, while Taylor (41.266° N, 75.128° W) was a mix of several species of *Quercus* and *Carya*. Both were damaged by an F2-rated tornado on 30 May, 1998, which caused an 85-ha damage patch at Hattons, and a 4-ha damage patch at Taylor. At the Hattons site, 8 contiguous 20 m × 20 m plots were inventoried in July 1999 for tree damage. At the Taylor site, roughly 4 ha of damaged forest was mapped with surveying equipment, in June–July 2000, and subsequently subdivided into 77 contiguous 20 m × 20 m plots.

Boundary Waters Canoe Area sites: Fishhook, Shirttail, Meditation Lake, Threemile and Twin Island are the five study sites located with the Boundary Waters Canoe Area, a wilderness that consists of nearly 500,000 ha of primary sub-boreal forests in northeastern Minnesota, entirely within Superior National Forest. On 4 July, 1999, a large derecho event damaged >150,000 ha of forest in the Boundary Waters Canoe Area Wilderness [77]. Five sites in the vicinity of Seagull Lake were sampled for tree damage in 2000 and 2001 [71]; (Table 1). At Fishhook, 16 plots were sampled along 5 transects; at Meditation Lake, 9 contiguous plots (3 rows of 3) were sampled; at Shirttail, an area slightly less than 2 ha was mapped and partitioned into 18 contiguous plots; at Twin Island 20 plots were arranged along two transects; and at Threemile, 22 plots were distributed on five transects. Woody vegetation at Fishhook (48.138° N, 90.890° W) was predominantly *Picea* species, *Pinus banksiana* Lamb., and *B. papyrifera*, while that at Shirttail (48.118° N, 90.939° W) was a mixture of *Abies balsamea* (L.) Mill, *Thuja occidentalis* L., and *Pinus resinosa* Sol ex Aiton. Twin Island (48.120° N, 90.934° W) was an *A. balsamea-B. papyrifera-T. occidentalis* stand. Dominant species at Meditation Lake (48.134° N, 90.883° W) were *A. balsamea*, *Populus tremuloides* Michx., and *B. papyrifera*; while at Threemile (48.137° N, 90.899° W) the dominants were *P. resinosa*, *T. occidentalis*, and *Pinus banksiana* Lamb. The study sites are all of post-fire origin.

NTSF site: Natchez Trace State Park and Forest (NTSF) is located in west-central Tennessee (35.717° N, 88.300° W; Table 1), and occupied by secondary forests that originated in the 1930s. Dominant tree species are several *Quercus* species, *Carya* species, and *Pinus taeda* L. On 5 May 1999, a straight-line windstorm (downburst) with sustained winds >90 km/h and gusts >145 km/h struck NTSF and damaged ~3000 ha. Thirty-two plots (30 m × 30 m) were distributed haphazardly in 2 areas (16 in each area), and were inventoried for damage to all trees >10 cm trunk diameter in the summers of 2000 and 2001.

Gum Road was a secondary *Q. rubra-Q. alba-Fraxinus americana* L. stand in central Tennessee (35.667° N, 75.148° W; Table 1) disturbed in April 2002 by an F2 tornado, and was sampled in 2004. Twenty-eight plots were distributed as 4 contiguous 20 m × 20 m plots in each of 7 larger 40 m × 40 m plots. Disturbance history at this site is unknown.

Boggs Creek, Timpson, and Martin Branch are all located within the Chattahoochee National Forest in northern Georgia (Table 1). They were disturbed on April 27, 2011 by an EF-3 tornado that ultimately had a total track length of 64 km. At Boggs Creek (34.699° N, 83.883° W) *Q. montana*, *Quercus coccinea* Munchh., *Pinus virginiana* Mill., *Q. alba*, *P. strobus*, and *Oxydendrum arboreum* (L.) DC were the dominant tree species. The Timpson (34.876° N, 83.481° W) site's woody vegetation was predominantly *P. strobus*, *Q. alba*, and *Q. coccinea*. At Martin Branch (34.775° N, 83.796° W), dominant tree species were *P. strobus*, *A. rubrum* and *O. arboreum*. These 3 sites were sampled in the summer and fall of 2011, and summer of 2012. At Boggs Creek, 38 plots were distributed along 8 transects; at Martin Branch 18 plots were arranged along 6 transects; and at Timpson, 14 plots were distributed among 5 transects. These stands originated after clearcutting in the late 19th and early 20th centuries.

The Smokies study site was located in the northwestern corner of Great Smoky Mountains National Park in southeastern Tennessee (35.621° N, 83.882° W; Table 1). The disturbance was also on April

27, 2011, and was an EF-4 tornado with a 33 km track length. *A. rubrum*, *T. canadensis*, and *P. strobus* were the dominant tree species. It was sampled in 2012 and 2013, using 22 plots distributed along park trails in 3 areas. Histories of the locations of actual plots are unknown, although most stands in this area are of post-fire origin and the area is generally considered primary forest.

Table 1. Study site names, pre-disturbance vegetation characteristics, number of trees sampled, and disturbance severity. Values are means ± standard deviation across plots.

Site	Pre-Dist Species Richness	Pre-Dist. H' (Density)	Pre-Dist 1/λ (Density)	Pre-Dist. Evenness (Density)	Severity (prop. Basal Area Lost)	Num. Trees	Pre-Dist. Diam. (cm)
Mingo *[a,5]	7.67 ± 2.34	1.56 ± 0.28	3.72 ± 1.12	0.78 ± 0.09	0.58 ± 0.23	192	17.08 ± 3.48
Mingo *[b,5]	8.67 ± 1.15	1.80 ± 0.14	4.84 ± 0.88	0.83 ± 0.03	0.75 ± 0.23	92	28.58 ± 3.59
Tionesta 1994 [a,5]	3.48 ± 1.46	1.04 ± 0.31	2.55 ± 0.75	0.82 ± 0.12	0.30 ± 0.28	643	27.00 ± 8.09
Gould Farm [a,5]	7.50 ± 1.84	1.54 ± 0.30	3.80 ± 1.47	0.77 ± 0.11	0.85 ± 0.19	360	22.02 ± 3.21
Texas Hill [a,5]	5.60 ± 1.84	1.23 ± 0.37	2.78 ± 1.07	0.72 ± 0.12	0.97 ± 0.04	255	21.84 ± 2.23
Rapid River [a,5]	6.09 ± 1.97	1.27 ± 0.30	2.91 ± 0.74	0.72 ± 0.10	0.28 ± 0.21	573	15.99 ± 0.86
Taylor [a,5]	4.96 ± 2.23	1.36 ± 0.37	3.46 ± 1.20	0.86 ± 0.09	0.80 ± 0.25	1224	23.73 ± 5.41
Hattons [a,5]	3.88 ± 0.64	0.90 ± 0.21	2.04 ± 0.58	0.67 ± 0.12	0.73 ± 0.27	158	22.57 ± 4.61
Fishhook [a,5]	3.31 ± 0.79	0.79 ± 0.22	1.93 ± 0.39	0.67 ± 0.16	0.47 ± 0.16	611	15.35 ± 1.92
Med. Lake [a,5]	4.44 ± 0.73	0.85 ± 0.21	1.85 ± 0.40	0.58 ± 0.13	0.63 ± 0.14	554	11.01 ± 2.83
Shirttail [a,5]	5.17 ± 1.04	1.32 ± 0.26	3.28 ± 0.82	0.81 ± 0.11	0.86 ± 0.16	548	16.26 ± 2.61
Twin [a,5]	5.35 ± 1.27	1.17 ± 0.22	2.64 ± 0.73	0.72 ± 0.12	0.83 ± 0.14	878	15.19 ± 3.94
ThreeMile [a,10]	3.86 ± 1.39	1.00 ± 0.35	2.48 ± 0.84	0.77 ± 0.14	0.29 ± 0.24	404	22.40 ± 3.89
Natchez Trace [b,5]	12.07 ± 2.38	2.02 ± 0.27	5.87 ± 1.75	0.82 ± 0.06	0.38 ± 0.20	1637	19.07 ± 2.56
Gum Road [a,5]	7.89 ± 2.21	1.86 ± 0.32	5.92 ± 1.87	0.91 ± 0.04	0.33 ± 0.26	674	23.24 ± 3.23
Boggs Creek [a,10]	8.24 ± 2.07	1.79 ± 0.32	5.12 ± 1.77	0.86 ± 0.08	0.66 ± 0.29	894	25.71 ± 5.29
Smokies [a,10]	7.14 ± 1.96	1.66 ± 0.30	4.66 ± 1.74	0.86 ± 0.09	0.58 ± 0.24	487	25.60 ± 3.03
Martin Branch [a,10]	7.28 ± 1.32	1.73 ± 0.17	4.87 ± 0.96	0.88 ± 0.06	0.74 ± 0.24	383	26.72 ± 5.49
Timpson [a,10]	7.71 ± 1.07	1.77 ± 0.18	4.91 ± 1.20	0.87 ± 0.06	0.72 ± 0.16	274	27.47 ± 3.17

* Mingo site sampled with 6 20 × 20 m plots, and 3 30 × 30 m plots. [a,b] = Plot size; a = 20 m × 20 m; b = 30 m × 30 m; [5,10] = minimum tree diameter (in cm) sampled; 5 = 5 cm; 10 = 10 cm.

2.2. Methods

Field methods For all of the above sites, field sampling consisted of an inventory of all trees above a minimum size threshold (either 5 cm or 10 cm trunk diameter at 1.4 m) within sample plots, in which species, type of damage, and dbh (cm) were recorded, as well as other variables not relevant to this report. Type of damage was categorized as: intact (undamaged), light crown damage (LCD; <50% of crown missing), heavy crown damage (HCD; >50% of crown missing), bent (trunk curved), leaning (trunk straight but >20° from vertical), trunk broken, or uprooted, for a total of 7 possible damage categories.

Sampling plots were typically 20 m × 20 m in size; and distributed non-adjacently (typically spaced along transects) or as contiguous plots such as at Meditation Lake, Shirttail, Tionesta 1994, and Taylor. Two sites were exceptions. Natchez Trace was sampled with 32 non-adjacent 30 m × 30 m plots [73]. At Mingo, 3 out of the 9 total plots were also 30 m × 30 m in size [74].

Statistical analyses Plot-level damage severity was calculated as percent of pre-disturbance tree basal area that was felled (trunk broken or uprooted) by the wind disturbance. This is a better representation of the change in site conditions than using percent of individuals felled by the wind disturbance, because basal area better captures the size variation among individuals, and is therefore more closely correlated with the amount of canopy openness in a closed-canopy stand. A more nuanced measure of severity might include the partial damage categories by using some fraction of a tree's pre-disturbance basal area in the total subtracted from the plot's pre-disturbance total; however, it is not immediately obvious how such partial basal areas would be assigned to the lighter damage categories.

The term 'damage heterogeneity index' or DHI, is introduced as a measure of the diversity of damage categories; it is entirely analogous to typical measures of diversity of species. Two versions of the DHI were calculated, one called Sh-DHI and using the Shannon index ($H' = \Sigma_i p_i \ln p_i$) and another called iSI-DHI and using the inverse Simpson index ($D = 1/\lambda$, where $\lambda = \Sigma_i(p_i^2)$). In both of the versions of DHI, damage categories were used instead of species as the classification variable, and p_i defined as the proportion of total individuals or total basal area represented by a given species. Evenness was also calculated along with the two primary indices, as $J = H'/H_{max}$ [78]. Note that while the great majority of uses of the Shannon or Simpson index use the number of individuals as the measure of abundance, this is by no means necessary, and may indeed by misleading when individuals differ by orders of magnitude in size. Using the summed basal area for a given damage category in a given plot mitigates this extreme variation and gives a better indicator of the actual dominance of a damage category in the sample. While the seven damage categories place an upper limit on the values Sh-DHI (max = 1.946) and iSi-DHI (max = 7.00) thus calculated, it still provides a quantitative measure of the diversity of damage types across the tree samples.

To explore any potential relationship between the DHIs and pre-disturbance forest characteristics and damage severity, the plot-level values of DHI (Sh-DHI and iSi-DHI) were regressed against plot-level forest characteristics (species richness and diversity, mean and inequality of tree diameter) and wind damage severity. Both linear and quadratic regressions were tested; if the secondary term in the quadratic regression significantly improved the regression, the relationship between predictor and response variable is curvilinear. Tree diameter inequality was quantified as the Gini coefficient [79]. All regressions were performed in Sigma-Plot 11.0 (Systat Software, San Jose, CA, USA).

Nonmetric multidimensional scaling ordination was used to visualize how type of tree damage varied among sample plots. In this novel ordination approach, damage types are used analogously to how species are used in traditional ordinations; the distribution of plots in "damage type ordination space" reveals the most prevalent type of tree damage for those plots. PC-Ord 7 (Wild Blueberry Media LLC, Corvalis, OR, USA) was used to perform the ordination. The following options were used in the ordination: Autopilot = off; distance index = Sorensen; Axes = 2; Runs with real data = 50; Stability criterion = 0.00001; Iterations to evaluate stability = 1; Maximum number of iterations = 250; Initial step length= 0.20; starting coordinates = Random. The multiple runs with real data were utilized to avoid local minima. Preliminary output was rotated with a varimax rigid rotation to improve separation; a varimax rotation moves output points simultaneously around the centroid to obtain the greatest spread of scores along the first axis. Final stress of the two-dimensional solution was 17.46. The input matrix was 340 rows × 7 columns of data; the columns define attributes—in this case, types of treefall (e.g., intact is one column; branches broken is next column, etc.) and the rows define entities—in this case 20 × 20 m sample plots. Entries in the input data matrix were number of trees experiencing that category of damage in that plot. After the ordination, plots were divided into those with 5 cm dbh as the minimum tree size (n = 237 plots) and those with 10 cm dbh as the minimum tree size (n = 111 plots) to show on separate graphs; the 35 plots that were 30 m × 30 m in size were excluded from the ordinations due to the small sample size.

3. Results

A total of 413 plots were sampled, encompassing 10,762 trees. Twenty-eight plots had five or fewer trees pre-disturbance and were subsequently excluded from analyses, resulting in 385 plots and 10,675 trees. Among the seven damage categories, uprooting was the most common overall, with 27.8% of trees (2973 trees), while trunk breakage occurred in 17.9% of trees (1907 trees). Lighter damage categories accounted for 33.0% of trees (3519 trees), while 25.9% of trees (2769 trees) remained intact (undamaged).

The relative abundances of damage categories varied immensely, even among plots within the same study site; many plots had no representation of one or more of the seven damage categories, while others had all seven categories in broadly similar abundances. Figure 2 presents the nine plots of Meditation Lake as just one example. In this site, plot C2 is overwhelmingly composed (62.1% of basal area) of uprooted trees; plot C1 has damage dominated by trunk breakage (54.1% of basal area); while in plot B2 intact or branches broken were the most abundant damage categories (36.9% and 31.9% of basal area, respectively).

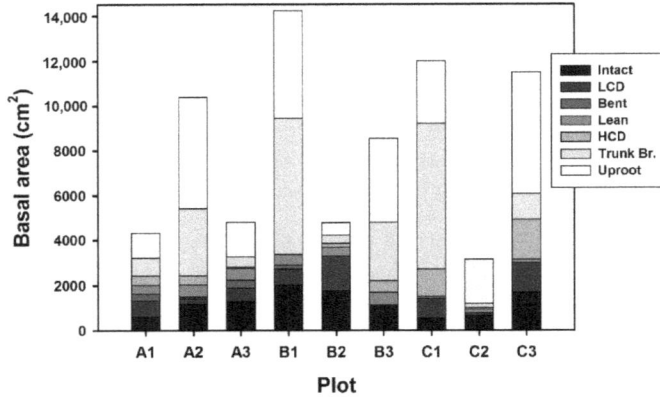

Figure 2. Relative basal area of seven tree damage categories at the Meditation Lake study site. In the figure legend, 'LCD' = light crown damage and 'HCD' = heavy crown damage; other damage categories are words, not acronyms.

Mean Sh-DHI (based on basal area) was close to half of the maximum for both the plots sampled with 5 cm dbh minimum, and those sampled with a 10 cm dbh minimum (Figure 3). For iSi-DHI, the corresponding levels were roughly 35% and 37% of the maximum. Larger plots (30 m × 30 m plot size) exhibited higher diversity than smaller 20 m × 20 m plots in both the Sh-DHI (roughly 84% of maximum) and iSi-DHI (50.3% of maximum) indices. A ranked one-way ANOVA (Kruskal-Wallis test) found that the three plot types differed significantly ($H = 47.55$, $p < 0.001$ for Sh-DHI, and $H = 42.69$, $p < 0.001$ for iSi-DHI). In both cases, Dunn's pairwise comparisons revealed that the 5 cm dbh minimum plots and the 10 cm dbh minimum plots did not differ from one another, while the 30 m × 30 m plots significantly differed from the 5 cm dbh minimum plots. The 30 m × 30 m plots were sampled with a 5 cm diameter minimum, thus are not tested against the 20 m × 20 m plots that were sampled with a 10 cm diameter minimum.

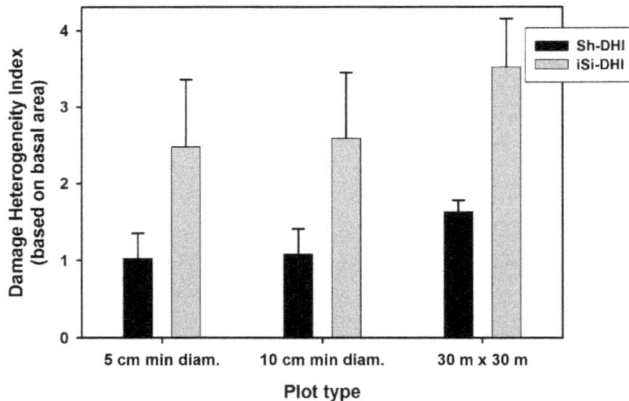

Figure 3. Mean and standard deviation of two damage heterogeneity indices, for three types of sampling plots. Each plot's damage heterogeneity was calculated using the basal area in each damage category as the abundance values input into the diversity formula. The '5 cm minimum diameter and '10 cm minimum diameter' plot types were all 20 m × 20 m plots, while the '30 m × 30 m' plot type had a 5 cm diameter minimum.

Hypothesis 1: For plots sampled with a 5-cm diameter minimum, neither species richness nor species diversity of pre-disturbance trees was significantly related to post-disturbance Sh-DHI or iSi-DHI (data not shown).

For plots sampled with a 10-cm diameter minimum, the DHIs showed no significant relationship to pre-disturbance tree species richness (Figure 4a; Sh-DHI results shown; results for iSi-DHI are similar). There was a weak positive correlation between pre-disturbance tree species diversity and Sh-DHI (Figure 4b), although the R^2 was very small.

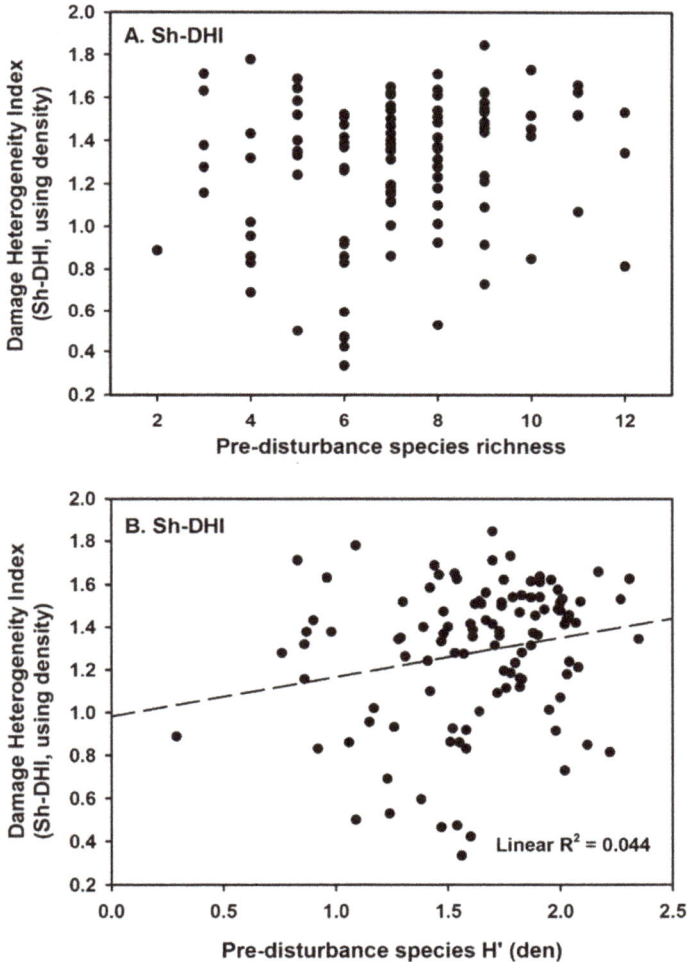

Figure 4. Damage heterogeneity index (DHI) vs. pre-disturbance plot species richness or diversity, for plots with 10-cm diameter minimum. (**A**) DHI calculated with Shannon formula, vs. species richness; (**B**) DHI calculated with Shannon formula, vs. pre-disturbance species diversity calculated using density and the Shannon formula. Only significant regressions shown.

Hypothesis 2: For plots sampled with 5-cm diameter minimum, there was a weak linear relationship between pre-disturbance mean tree diameter and post-disturbance Sh-DHI and iSi-DHI (data not shown). In both cases, the linear regression was significant (for Sh-DHI: $p < 0.001$, $R^2 = 0.048$; for iSi-DHI: $p = 0.036$, $R^2 = 0.019$) and had a negative slope. However, note that again the linear regressions both had very low R² values. For these plots, the secondary term of the quadratic regression was not significant ($p > 0.05$ in both cases).

For plots sampled with 10-cm diameter minimum, Sh-DHI and iSi-DHI both showed significant relationship to the pre-disturbance mean tree diameter (Figure 5). Both linear regressions were significant (for Sh-DHI: $p = 0.014$, $R^2 = 0.054$; for iSi-DHI: $p = 0.025$, $R^2 = 0.045$), and had negative slopes. For this data set, however, the secondary term in the quadratic regressions was also significant and negative for both types of DHI. Even with the improvement provided by the curvilinear term, though, the overall R^2 values were again quite small (Figure 5).

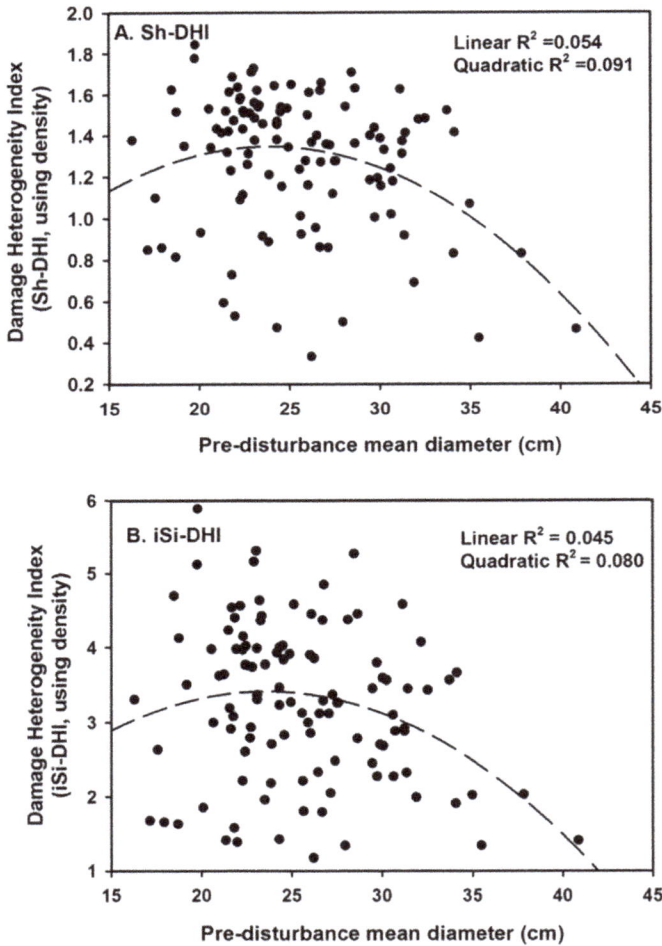

Figure 5. Damage heterogeneity index vs. pre-disturbance plot mean trunk diameter, for plots with 10-cm diameter minimum; $n = 111$. (**A**) DHI calculated with Shannon formula; (**B**) DHI calculated with inverse Simpson formula. Only significant regressions shown; quadratic regression shown if it had greater R^2 than linear regression.

Hypothesis 3: For plots sampled with 5-cm minimum diameter, both the Sh-DHI and the iSi-DHI increased significantly with greater pre-disturbance tree size inequality (Figure 6). In both cases, the linear regression was significant (for Sh-DHI: $p < 0.001$, $R^2 = 0.147$; For iSi-DHI, $p < 0.001$, $R^2 = 0.114$)

but the secondary term in the quadratic regression was not. Note that in these regressions the variance explained is somewhat better than in previous regressions.

For plots sampled with 10-cm minimum diameter (data not shown), none of the regressions between pre-disturbance tree size inequality and post-disturbance DHIs were significant.

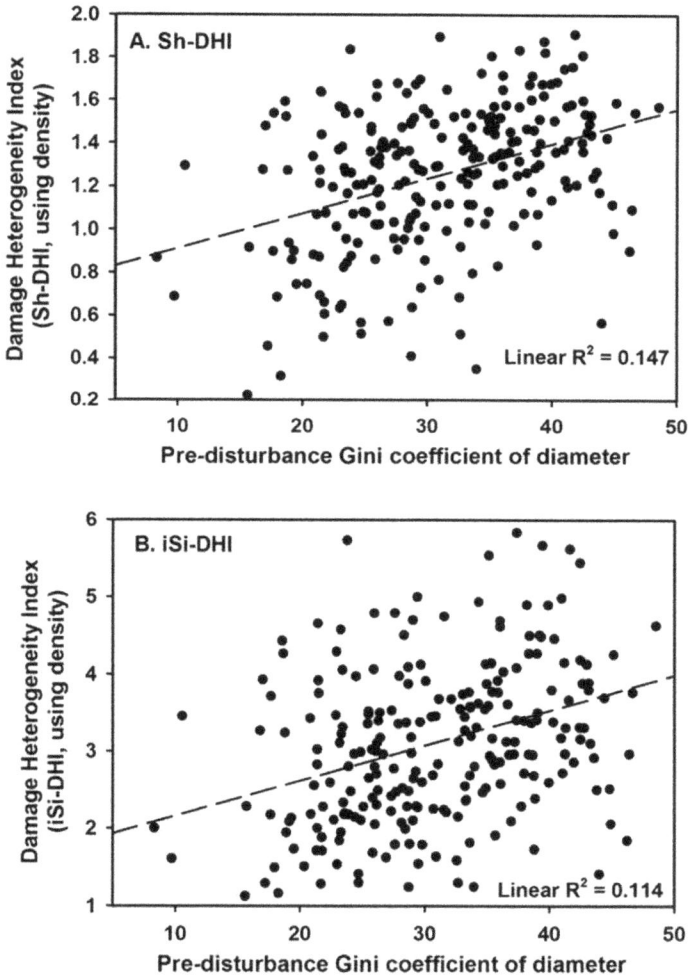

Figure 6. Damage heterogeneity index vs. pre-disturbance tree size inequality, for plots with 5-cm diameter minimum; *n* = 111. (**A**) DHI calculated with the Shannon formula; (**B**) DHI calculated with the inverse Simpson formula. Only significant ($p < 0.05$) regression results shown.

Hypothesis 4 The Nonmetric Multidimensional Scaling ordination produced a distribution of plots in the reduced 2-dimensional space (Figure 4) in which uprooted trees are lower and to the right of the centroid of the distribution, while intact trees are nearer the top of the graph. In the both groups of plots, severity of damage is strongly positively correlated with Axis 1 (5-cm minimum plots: $r = 0.792$, $p < 0.0001$; 10-cm minimum plots: $r = 0.732$, $p < 0.0001$), and strongly negatively correlated with Axis 2 (5-cm minimum plots: $r = -0.537$, $p < 0.0001$; 10-cm minimum plots: $r = -0.766$, $p < 0.0001$). Thus in the ordination graphs, highest severity damage is located in the lower right of the graph. Since the 'uproot' category is low and furthest to the right, this indicates that uprooting was the dominant type of damage in plots that experienced severe damage. It is noteworthy that the trends in Figure 7a,b are largely consistent, indicating that the results shown are not an artifact of one particular minimal tree size threshold.

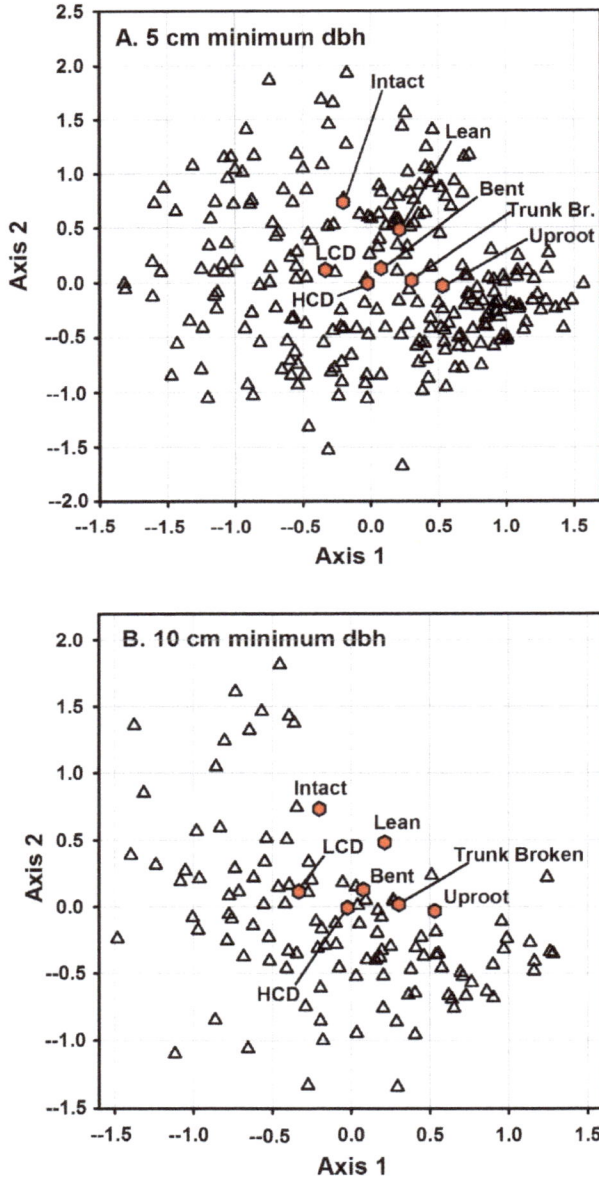

Figure 7. Nonmetric Multidimensional Scaling ordination of plots based on damage type categories. (**A**) 5-cm minimum diameter, *n* = 237 plots; (**B**) 10-cm minimum diameter, *n* = 111 plots. LCD = Light crown damage; HCD = Heavy crown damage.

Graphing the damage diversity (whether calculated with Sh-DHI or iSi-DHI) against severity of plot damage revealed a clear and consistent pattern: damage diversity peaks at intermediate levels of severity (Figures 8 and 9). In all four of the datasets presented in Figures 8 and 9, the quadratic regressions had much higher R^2 values than the linear relationship, which was sometimes not significant at all. This confirms existence of the peak at intermediate severity.

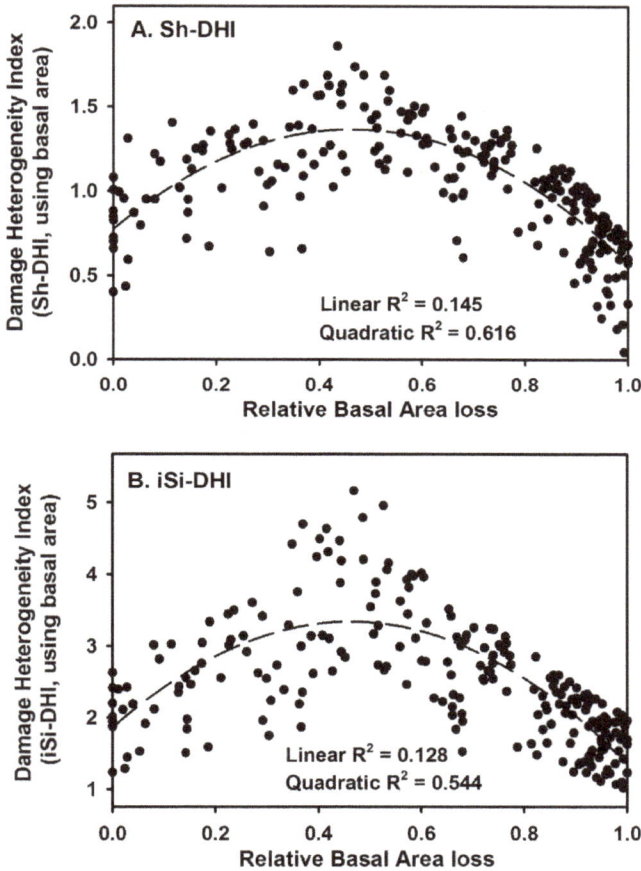

Figure 8. Damage heterogeneity index in relation to severity of damage, using plots sampled with 5-cm diameter minimum. (**A**) Sh-DHI, calculated with the Shannon index; (**B**) iSi-DHI, calculated with the inverse Simpson index. In all calculations of damage heterogeneity above, the measure of abundance used was the cumulative basal area in each damage category for that plot. Variance explained (R^2) by linear and quadratic regressions was compared to determine if the relationship between DHI and severity was significantly curved.

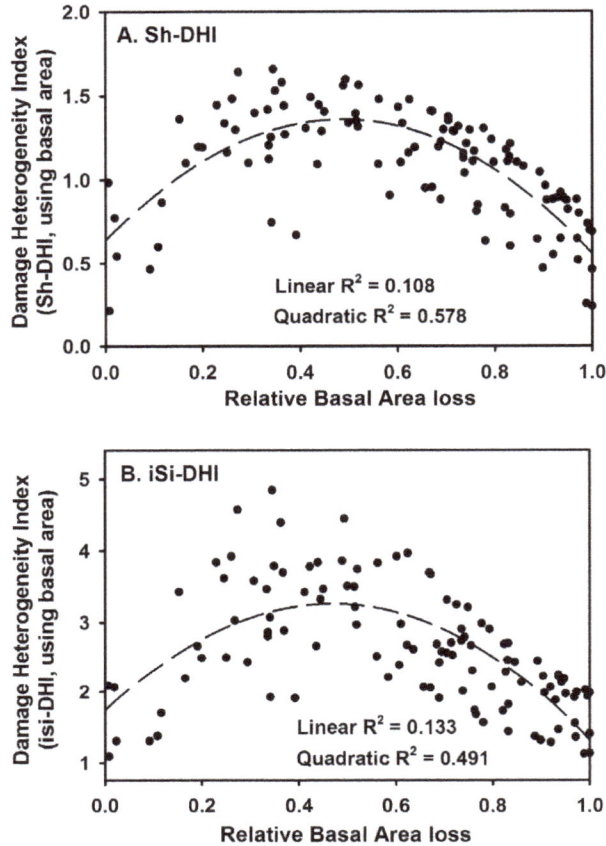

Figure 9. Damage heterogeneity index in relation to severity of damage, using plots sampled with 10-cm diameter minimum. (**A**) Sh-DHI, calculated with the Shannon index; (**B**) iSi-DHI, calculated with the inverse Simpson index. In all calculations of damage heterogeneity above, the measure of abundance used was the number of trees in each damage category for that plot. Variance explained (R^2) by linear and quadratic regressions were compared to determine if the relationship between DHI and severity was significatly curved.

All of the relationships shown in Figures 8 and 9 are based on the two DHIs calculated using basal area as the measure of abundance of the damage categories; analogous DHI calculations using counts of individuals rather than basal area as the measure of abundance (not shown) produced similar curves with peaks at intermediate severity, although the relationships were noisier. In all cases, the quadratic regressions again indicated a significant curvature with a DHI peak at intermediate levels of wind disturbance severity.

Finally, the intermediate peak in the DHIs is not an artifact of pooling multiple sites. Figure 10 shows results from two of the study sites with larger sample sizes—Boggs Creek and Tionesta 1994. In both cases, Sh-DHI peaks at intermediate levels of damage severity, and the quadratic regression has much higher R^2 than the linear, indicating that the peak is statistically real.

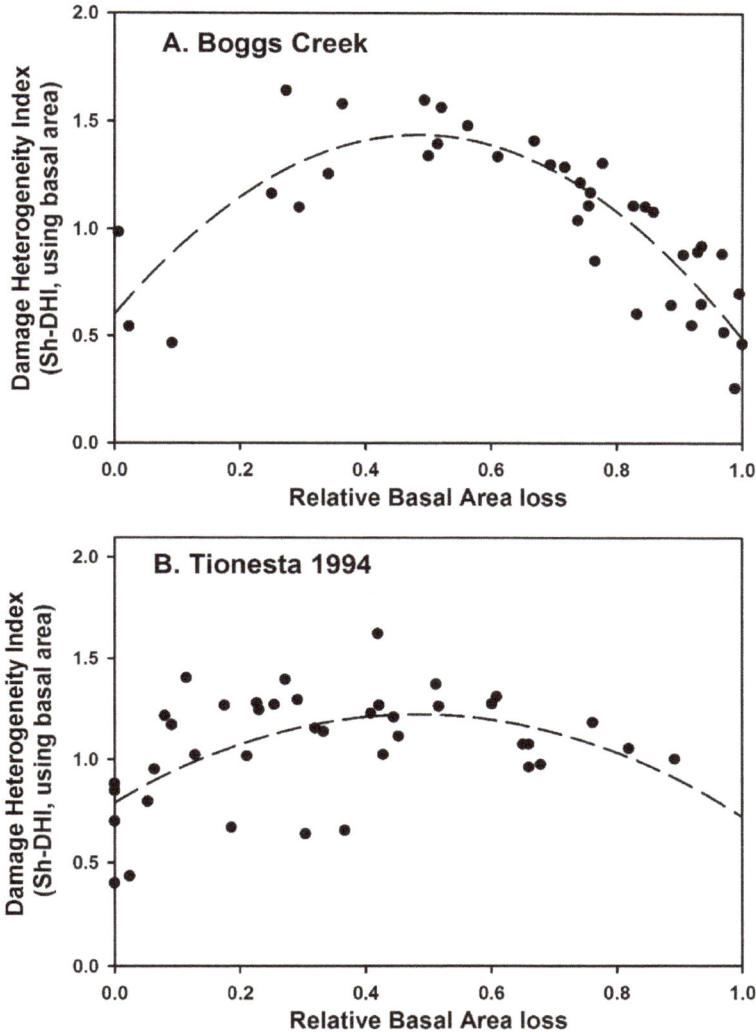

Figure 10. Damage heterogeneiety index (Sh-DHI) in relation to severity of damage, from two of the study sites. (**A**) Boggs Creek (linear $R^2 = 0.124$, quadratic $R^2 = 0.726$); and (**B**) Tionesta 1994 (linear $R^2 = 0.106$, quadratic $R^2 = 0.298$). In both panels, quadratic regression line is shown. In all calculations of damage heterogeneity above, the measure of abundance used was the cumulative basal area in each damage category for that plot.

4. Discussion

This work demonstrates for the first time that tree damage categories can be used to construct a structural complexity metric based on damage diversity. Just as some early work on structural complexity used size categories in a diversity index calculation [12,13,80], the same logic is here applied to distinct categories of tree damage. The similarity of mean DHI values for plots with 5-cm vs. 10-cm diameter minima (Figure 3), as well as the similar distribution of the two types of plots in the ordination (Figure 7), suggest that this approach is not dramatically affected by the minimum tree size sampled, at least for these two common size thresholds. However, it is clear that plot size may influence the quantitative values obtained in this type of damage diversity study (Figure 3), as evidenced by the significant difference in mean diversity index value for 20 m × 20 m plots vs. 30 m × 30 m plots. Also, although the actual values obtained with the Sh-DHI and the iSi-DHI differed (as expected, given the different formulas used), the trends were remarkably similar, again suggesting that the approach proposed here is robust to the actual diversity index employed, as well as to the minimum tree size sampled.

The pre-disturbance forest characteristics examined for Hypotheses 1–3 differed substantially in their influence on post-disturbance damage diversity. Surprisingly, pre-disturbance species richness and diversity appeared to have little effect on damage diversity (Figure 4), therefore Hypothesis 1 must be rejected. Because a number of studies have documented among-species differences in tree damage during high winds [81,82], having a greater variety of trees would be expected to result in greater damage diversity. However, as pointed out in Peterson [68], the species effect is often small, suggesting that in this analysis, the numerous other influencing variables (e.g., tree sizes, site conditions, storm conditions) may have obscured any potential species effect. Thus in this analysis, pre-disturbance compositional diversity appears to have little causative effect on post-disturbance structural complexity.

Mean tree size significantly influenced the two DHIs, but the variance explained was quite small (Figure 5). The extent to which this result is contingent on the distribution of tree sizes in addition to the plot mean, is unknown; expected patterns of damage might be expected to be very different between two plots with, for example a mean trunk diameter of 30 cm. In one plot, many small trees and two or three large ones (e.g., >60 cm) would likely experience damage distinct from another plot with all trees in the range of 20–40 cm diameter. Nevertheless, this weak effect can be interpreted as a confirmation of Hypothesis 2, which expected no trend in damage diversity with changes in mean tree sizes.

Potentially more interesting is the stronger relationship between pre-disturbance size inequality (measured via the Gini coefficient) and damage diversity (Figure 6). While noisy, this relationship was stronger than the previous ones, and confirms Hypothesis 3. The confirmation of Hypothesis 3, in combination with rejection of Hypothesis 1, is consistent with several previous studies [68,81] that conclude that tree size is a greater influence than species on wind damage patterns. Pre-disturbance size variability, then, is indeed positively associated with higher values of one component of structural complexity—diversity begets diversity, in this case. Such a continuation of structural diversity through severe wind damage echoes other studies that suggest positive feedbacks in vegetation composition, in which species composition becomes entrained and spatial patchiness in composition is maintained through time via positive neighborhood effects [83].

Visualizing the "damage category composition" through ordination reveals that there is a clear gradient from predominantly intact plots to predominantly uprooted plots as severity of damage increases (Figure 7). This confirms the subsidiary expectation to Hypothesis 4, namely that as severity changes, the predominant type of tree damage changes as well. At first glance, this is perhaps expected, but further consideration suggests that such a finding is not a foregone conclusion since trees in very severely damage plots could fail primarily by trunk breakage. Although post-windthrow circumstances are known in which most tree failure is via trunk breakage (Gum Road as described in [68]), these data sets suggest that for wind damage in eastern North America, high severity wind damage will mostly produce uprooted trees. The accompanying root pits and mounds have important implications for regeneration of vegetation, driving fine scale spatial patterns [42–51], as well as providing an important establishment site for pioneers that typically do not establish on intact forest floor [72] and thus enriching vegetation diversity. Treefall pits and mounds are also a source of soil turnover and mixing [84,85].

Perhaps the most notable result of this study is the clear relationship between damage diversity (as quantified in the two types of DHI) and damage severity. In hindsight, such a relationship may be unsurprising in light of the logic presented for Hypothesis 4, but this appears to nevertheless be the first documentation of such a pattern. At the lowest severities, most trees are intact or have broken branches, and on the left sides of Figures 8–10, increasing severity means adding the more serious damage categories, thus driving the diversity index upward. On the right sides of Figures 8–10, an analogous process occurs, except that as severity increases from intermediate to very high, the light damage categories disappear and most or all of the trees are snapped or uprooted, thus driving down diversity. Therefore, considering damage diversity as a component of multidimensional concepts of structural complexity, it may be expected that structural complexity overall may also show a peak at intermediate levels of damage severity, at least after wind disturbances. At least three previous studies have examined structural complexity after wind disturbances [28–30]. Szmyt and Dobrowolska [28] specifically reported higher structural complexity in intermediate-severity sites, although their metrics did not include damage diversity. Meigs and Keeton [30] did not quantitatively relate complexity to severity; they did, however, note that the intermediate-severity wind disturbance can augment structural complexity. Since these interesting trends reported in previous work [28,30] do not employ damage diversity in their calculations of structural complexity, there could be a variety of components of complexity that are maximal at intermediate levels of wind damage severity.

The intermediate peak in damage diversity appears to be a very robust relationship, as affirmed by several lines of evidence. First, using two different indices of diversity (Figures 8 and 9), and with datasets employing two different tree size minima, there was a consistent peak in damage diversity at intermediate levels of disturbance severity. Second, very similar—albeit noisier—relationships were revealed when using counts of individuals in the damage diversity calculations, showing that this trend is also robust to the actual measure of abundance (counts vs. basal area). And third, these trends are revealed within individual sites as well as across the pooled datasets (Figure 10), quelling concerns that the intermediate peak relationship could be some sort of artifact of the pooled analysis. Thus, very similar trends are seen across two indices of diversity, two measures of abundance, two tree size minima, two plot sizes, and for individual sites as well as pooled sites.

The results reported here have striking parallels in the suggestion by Gough et al. [25] that intermediate severity disturbances may, via enhanced complexity, provide a mechanism for aging mature forests to maintain net ecosystem production. Their reasoning is that disturbances that are frequent but that are not fully stand-replacing will maximize canopy rugosity (a single-metric measure of complexity) which in turn will promote higher-than-expected net primary productivity that can offset increasing heterotrophic respiration, and thus maintain net ecosystem productivity as forests age (see similar reasoning in [14]). If broadly confirmed, their hypothesis, along with results reported here, together suggest that low- to intermediate-severity disturbances may play vital roles in both ecosystem and community ecology via their influence on complexity.

Limitations As with any correlative study, the relationships revealed here are intertwined with multiple other causal factors and cannot be convincingly separated without complex, controlled experiments. For example, while this work revealed little influence of species richness or diversity on damage diversity, it is quite likely the species composition influenced the types of tree damage and therefore the level of damage diversity. One striking species effect on wind damage is the phenomenon of interlocking or extensively grafting root systems [86], which can result in exceedingly strong rooting and therefore a virtual absence of uprooting in the face of high winds for such trees. Alternatively, Canham et al. [82] suggest that certain species may preferentially 'sacrifice' major limbs to maintain an intact trunk and root system and thereby assure the ability to regrow from an elevated position; such species might rarely suffer uprooting or trunk breakage even in extreme winds. It is likely that species differences influenced the results seen here, but that influence may not compromise the validity of these results. The trends reported appear to be consistent across many forest types, and under a wide variety of sampling constraints and methodologies; even if species effects were hidden influences on the patterns in one or a few locations, it seems unlikely that trends observed across such a large collection of data sets would be vulnerable to such influences. Moreover, even single sites (Figure 10) show the same trends as the larger ensemble of observations; since compositional variation within a given site is modest, the peaked damage diversity vs severity curve would be unlikely to be driven by species effects.

Management While the primary objective of this research was to understand how wind disturbance might contribute to structural complexity, there may be management insights to be gained from considering these findings. Creation, by human intervention, of the types of damage discussed herein, would be likely to be laborious and expensive, in part because several different tasks would need to be employed, and the damage types discussed are the result of application of large forces high in a tree. In some circumstances, managers may choose to apply novel techniques, such as use of explosives to remove the top of a tree and create a snag with live lower branches [18]; such actions, though, are likely to be financially affordable only when applied to a modest number of selected individual trees. However, managers can exploit the well-known phenomenon [87] that trees in recently thinned or harvested stands are much more likely to fall. Application of variable density thinning or newer harvesting methods to retain scattered trees will parallel an intermediate severity wind event in terms of canopy opening, and will by default result in the occurrence of many of these types of damage discussed here. While such actions are an imperfect mimic of natural wind disturbance, they can serve to facilitate creation of some structural complexity attributes while still allowing resource extraction [87].

5. Conclusions

Damage diversity is presented as a potentially useful metric of structural complexity in forests following wind disturbance. Characterization of damage diversity in 18 distinct study sites in eastern North America shows that the metric is readily quantified and robust to differing indices of diversity, differing minimal tree sizes, and differing measures of abundance. Visualization via ordination showed that uprooting tends to become the predominant damage category as damage severity increases. A noteworthy result was that damage diversity peaks at intermediate levels of damage severity, and when viewed in combination with previous research, there may be an emerging trend that structural complexity overall is maximal when wind disturbance is of intermediate severity.

This research is a first step on a novel conceptual pathway, and suggests a variety of ways the damage diversity concept may be extended. Considering emphases of past research on spacing and size variation as additional major dimensions of complexity, it may be informative to merge the damage diversity approach presented here with tree size categories to have a larger suite of size-damage categories, each potentially with distinct implications for fostering species diversity in the recovering forest. Damage categories were examined here in an entirely non-spatial manner, but could easily be analyzed using measures of spatial autocorrelation, clustering, or neighborhood intermingling. It remains to be seen if other components of complexity, such as metrics based on spacing or size variation, show similar relationships to severity. With an eye to generalizing to other disturbance types, this approach might easily be applied to ice storm damage in forests. Considering fire, although the particular damage categories presented here obviously would not apply to a post-fire context, a similar perspective might examine categories of fire damage such as intact, partial foliage scorch, all foliage scorched, and larger branches consumed.

Funding: Funding for this work was provided by grants DEB 114-3511 from the Population & Community Ecology program, and AGS 114-1926 from the Physical & Dynamic Meteorology program, of the U.S. National Science Foundation. Sampling at Natchez Trace was funded by the USDA Forest Service Southern Research Station, cooperative agreement # SRS 01-CA-11330136-405. Additional funding was provided by Sigma Xi, the Cary Institute, Rutgers University, University of Georgia, and the Andrew Mellon Foundation.

Acknowledgments: I thank the following agencies or offices for permission to sample at the field sites: Chattahoochee National Forest, Blue Ridge Ranger District; Allegheny National Forest, Bradford Ranger District; Mingo National Wildlife Refuge; Superior National Forest; Hiawatha National Forest; Natchez Trace State Forest of the Tennessee Department of Natural Resources; Great Smoky Mountains National Park, and the Blooming Grove Hunting Club. Many field assistants over the period from 1994–2014 labored in often-difficult conditions to make this synthesis possible; I am grateful to all of them.

Conflicts of Interest: The author declares no conflict of interest. The funders had no role in the design of the study; in the collection, analyses, or interpretation of data; in the writing of the manuscript, or in the decision to publish the results.

References

1. Swanson, M.E.; Franklin, J.F.; Beschta, R.L.; Crisafulli, C.M.; DellaSala, D.A.; Hutto, R.L.; Lindenmayer, D.B.; Swanson, F.J. The forgotten stage of forest succession: Early-successional eocsystems on forest sites. *Front. Ecol. Environ.* **2011**, *9*, 117–125. [CrossRef]
2. D'Amato, A.W.; Bradford, J.B.; Fraver, S.; Palik, B.J. Forest management for mitigation and adaptation to climate change: Insights from long-term silviculture experiments. *For. Ecol. Manag.* **2011**, *262*, 803–816. [CrossRef]
3. Fahey, R.T.; Alveshere, B.C.; Burton, J.I.; D'Amato, A.W.; Dickinson, Y.L.; Keeton, W.S.; Kern, C.C.; Larson, A.J.; Palik, B.J.; Puettmann, K.J.; et al. Shifting conceptions of complexity in forest management and silviculture. *For. Ecol. Manag.* **2018**, *421*, 59–71. [CrossRef]
4. Zenner, E.K. Does old-growth condition imply high live-tree structural complexity? *For. Ecol. Manag.* **2004**, *195*, 243–258. [CrossRef]
5. Lutz, J.A.; Larson, A.J.; Freund, J.A.; Swanson, M.E.; Bible, K.J. The importance of large-diameter trees to forest structural heterogeneity. *PLoS ONE* **2013**, *8*, e82784. [CrossRef] [PubMed]
6. Kinzig, A.P.; Pacala, S.W.; Tilman, D. *The Functional Consequences of Biodiversity*; Princeton University Press: Princeton, NJ, USA, 2001.
7. Tilman, D.; Isbell, F.; Cowles, J.M. Biodiversity and ecosystem functioning. *Annu. Rev. Ecol. Syst.* **2014**, *45*, 471–493. [CrossRef]
8. Isbell, F.; Craven, D.; Connolly, J.; Loreau, M.; Schmid, B.; Beierkuhnlein, C.; Bezemer, T.M.; Bonin, C.; Bruelheide, H.; de Luca, E.; et al. Biodiversity increases the resistance fo ecosystem productivity to climate extremes. *Nature* **2015**, *526*, 574–577. [CrossRef] [PubMed]
9. Hutchinson, G.E. Homage to Santa Rosalia, or why are there so many kinds of animals? *Am. Nat.* **1959**, *93*, 145–159. [CrossRef]
10. Tilman, D.; Pacala, S. The maintenance of species richness in plant communities. In *Species Diversity in Ecological Communities*; Ricklefs, R.E., Schluter, D., Eds.; University of Chicago Press: Chicago, IL, USA, 1993; pp. 13–25.
11. MacArthur, R.H. Population ecology of some warblers of northeastern coniferous forests. *Ecology* **1958**, *39*, 599–619. [CrossRef]
12. Neumann, M.; Starlinger, F. The significance of different indices for stand structure and diversity in forests. *For. Ecol. Manag.* **2001**, *145*, 91–106. [CrossRef]
13. McElhinny, C.; Gibbons, P.; Brack, C.; Bauhus, J. Forest and woodland stand structural complexity: Its definition and measurement. *For. Ecol. Manag.* **2005**, *218*, 1–24. [CrossRef]
14. Danescu, A.; Albrecht, A.T.; Bauhus, J. Structural diversity promotes productivity of mixed, uneven-aged forests in southwestern Germany. *Oecologia* **2016**, *182*, 319–333. [CrossRef] [PubMed]
15. Foster, D.R.; Knight, D.H.; Franklin, J.F. Landscape patterns and legacies resulting from large, infrequent forest disturbances. *Ecosystems* **1998**, *1*, 497–510. [CrossRef]
16. Franklin, J.F.; Lindenmayer, D.B.; MacMahon, J.A.; McKee, A.; Magnuson, J.; Perry, D.A.; Waide, R.; Foster, D. Threads of continuity: Ecosystem disturbance, recovery, and the theory of biological legacies. *Conserv. Pract.* **2000**, *1*, 8–16. [CrossRef]
17. Keeton, W.S.; Franklin, J.F. Do remnant old-growth trees accelerate rates of succession in mature Douglas-fir forests? *Ecol. Monogr.* **2005**, *75*, 103–118. [CrossRef]
18. Franklin, J.F.; Mitchell, R.J.; Palik, B.J. *Natural Disturbance and Stand Development Principles for Ecological Forestry*; General Technical Report NRS-19; USDA Forest Service: Newtown Square, PA, USA, 2007.
19. Bouget, C.; Duelli, P. The effects of windthrow on forest insect communities: A literature review. *Boil. Conserv.* **2004**, *118*, 281–299. [CrossRef]
20. Janssen, P.; Fortin, D.; Hebert, C. Beetle diversity in a matrix of old-growth boreal forest: Influence of habitat heterogeneity at multiple scales. *Ecography* **2009**, *32*, 423–432. [CrossRef]

21. Goetz, S.J.; Steinberg, D.; Betts, M.G.; Holmes, R.T.; Doran, P.J.; Dubayah, R.; Hofton, M. Lidar remote sensing variables predict breeding habitat of a Neotropical migrant bird. *Ecology* **2010**, *91*, 1569–1576. [CrossRef]

22. Arnan, X.; Bosch, J.; Comas, L.; Gracia, M.; Retana, J. Habitat determinants of abundance, structure and composition of flying Hymenoptera communities in mountain old-growth forests. *Insect Conserv. Divers.* **2011**, *4*, 200–211. [CrossRef]

23. Huth, N.; Possingham, H.P. Basic ecological theory can inform habitat restoration for woodland birds. *J. Appl. Ecol.* **2011**, *48*, 293–300. [CrossRef]

24. Hardiman, B.S.; Gough, C.M.; Halperin, A.; Hofmeister, K.L.; Nave, L.E.; Bohrer, G.; Curtis, P.S. Maintaining high rates of carbon storage in old forests: A mechanism linking canopy structure to forest function. *For. Ecol. Manag.* **2013**, *298*, 111–119. [CrossRef]

25. Gough, C.M.; Curtis, P.S.; Hardiman, B.S.; Scheuermann, C.M.; Bond-Lamberty, B. Disturbance, complexity and succession of net ecosystem production in North America's temperate deciduous forests. *Ecosphere* **2016**, *7*, e01375. [CrossRef]

26. Wang, W.; Lei, X.; Ma, Z.; Kneeshaw, D.D.; Peng, C. Positive relationship between aboveground carbon stocks and structural diversity in spruce-dominated forest stands in New Brunswick, Canada. *For. Sci.* **2011**, *57*, 506–515.

27. Ford, S.E.; Keeton, W.S. Enhanced carbon storage through management for old-growth characteristics in northern hardwood-conifer forests. *Ecosphere* **2017**, *8*, e01721. [CrossRef]

28. Szmyt, J.; Dobrowolska, D. Spatial diversity of forest regeneration after catastrophic wind in northeastern Poland. *iForest* **2016**, *9*, 414–421. [CrossRef]

29. Fraver, S.; Dodds, K.J.; Kenefic, L.S.; Morrill, R.; Seymour, R.S.; Sypitkowski, E. Forest structure following tornado damage and salvage logging in northern Maine, USA. *Can. J. For. Res.* **2017**, *47*, 560–564. [CrossRef]

30. Meigs, G.W.; Keeton, W.S. Intermediate-severity wind disturbance in mature temperate forests: Legacy structure, carbon storage, and stand dynamics. *Ecol. Appl.* **2018**, *28*, 798–815. [CrossRef]

31. Kuuluvainen, T. Forest management and biodiversity conservation based on natural ecosystem dynamics in northern Europe: The complexity challenge. *Ambio* **2009**, *38*, 309–315. [CrossRef]

32. Palik, B.J.; Mitchell, R.J.; Hiers, J.K. Modelling silviculture after natural disturbance to sustain biodiversity in the longleaf pine (*Pinus palustris*) ecosystem: balancing complexity and implementation. *For. Ecol. Manag.* **2002**, *155*, 347–356. [CrossRef]

33. Jogiste, K.; Korjus, H.; Stanturf, J.A.; Frelich, L.E.; Baders, E.; Donis, J.; Jansons, A.; Kangur, A.; Koster, K.; Learman, D. Hemiboreal forest: Natural disturbances and the importance of ecosystem legacies to management. *Ecosphere* **2017**, *8*, e02503. [CrossRef]

34. Donato, D.C.; Campbell, J.L.; Franklin, J.F. Multiple successional pathways and precocity in forest development: Can some forests be born complex? *J. Veg. Sci.* **2012**, *23*, 576–584. [CrossRef]

35. Bace, R.; Svoboda, M.; Janda, P.; Morrissey, R.C.; Wild, J.; Clear, J.L.; Cada, V.; Donato, D.C. Legacy of pre-disturbance spatial pattern determines early structural diversity following severe disturbance in montane spruce forests. *PLoS ONE* **2015**, *10*, e0139214. [CrossRef] [PubMed]

36. Franklin, J.F.; Van Pelt, R. Spatial aspects of structural complexity. *J. For.* **2004**, *102*, 22–28.

37. Bachofen, H.; Zingg, A. Effectiveness of structure improvement thinning on stand structure in subalpine Norway spruce (Picea abies (L.) Karst.) stands. *For. Ecol. Manag.* **2001**, *145*, 137–149. [CrossRef]

38. Schneider, E.E.; Larson, A.J. Spatial aspects of structural complexity in Sitka spruce–western hemlock forests, including evaluation of a new canopy gap delineation method. *Can. J. For. Res.* **2017**, *47*, 1033–1044. [CrossRef]

39. Meigs, G.W.; Morrissey, R.C.; Bace, R.; Chaskovskyy, O.; Cada, V.; Despres, T.; Donato, D.C.; Janda, P.; Labusova, J.; Seedre, M.; et al. More ways than one: Mixed-severity disturbance regimes foster structural complexity via multiple developmental pathways. *For. Ecol. Manag.* **2017**, *406*, 410–426. [CrossRef]

40. Peck, J.E.; Zenner, E.K.; Brang, P.; Zingg, A. Tree size distribution and abundance explain structural complexity differentially within stands of even-aged and uneven-aged structure types. *Eur. J. For. Res.* **2014**, *133*, 335–346. [CrossRef]

41. Zenner, E.K.; Peck, J.E. Floating neighborhoods reveal contribution of individual trees to high sub stand scale heterogeneity. *For. Ecol. Manag.* **2018**, *412*, 29–40. [CrossRef]

42. Beatty, S.W. Influence of microtopography and canopy species on spatial patterns of forest understory plants. *Ecology* **1984**, *65*, 1406–1419. [CrossRef]

43. Peterson, C.J.; Carson, W.P.; McCarthy, B.C.; Pickett, S.T.A. Microsite variation and soil dynamics within newly created treefall pits and mounds. *Oikos* **1990**, *58*, 39–46. [CrossRef]
44. Peterson, C.J.; Pickett, S.T.A. Microsite and elevational influences on forest regeneration three years after catastrophic windthrow. *J. Veg. Sci.* **1990**, *1*, 657–662. [CrossRef]
45. Carlton, G.C.; Bazzaz, F.A. Regeneration of three sympatric birch species on experimental hurricane blowdown microsites. *Ecol. Monogr.* **1998**, *68*, 99–120. [CrossRef]
46. Waldron, K.; Ruel, J.-C.; Gauthier, S.; de Grandpre, L.; Peterson, C.J. Effects of post-windthrow salvage logging on microsites, plant composition and regeneration. *Appl. Veg. Sci.* **2014**, *17*, 323–334. [CrossRef]
47. Spicer, J.E.; Suess, K.F.; Wenzel, J.W.; Carson, W.P. Does salvage logging erase a key physical legacy of a tornado blowdown? A case study of tree tip-up mounds. *Can. J. For. Res.* **2018**, *48*, 976–982. [CrossRef]
48. Plotkin, A.B.; Schoonmaker, P.; Leon, B.; Foster, D. Microtopography and ecology of pit-mound structures in second-growth versus old-growth forests. *For. Ecol. Manag.* **2017**, *404*, 14–23. [CrossRef]
49. Sass, E.M.; D'Amato, A.W.; Foster, D.R.; Plotkin, A.B.; Fraver, S.; Schoonmaker, P.K.; Orwig, D.A. Long-term influence of disturbance-generated microsites on forest structural and compositional development. *Can. J. For. Res.* **2018**, *48*, 958–965. [CrossRef]
50. Vodde, F.; Jogiste, K.; Engelhart, J.; Frelich, L.E.; Moser, W.K.; Sims, A.; Metslaid, M. Impact of wind-induced microsites and disturbance severity on tree regeneration patterns: Results from the first post-storm decade. *For. Ecol. Manag.* **2015**, *348*, 174–185. [CrossRef]
51. Simon, A.; Gratzer, G.; Sieghardt, M. The influence of windthrow microsites on tree regeneration and establishment in an old growth mountain forest. *For. Ecol. Manag.* **2011**, *262*, 1289–1297. [CrossRef]
52. Cannon, J.B.; O'Brien, J.J.; Loudermilk, E.L.; Dickinson, M.B.; Peterson, C.J. The influence of experimental wind disturbance on forest fuels and fire characteristics. *For. Ecol. Manag.* **2014**, *330*, 294–303. [CrossRef]
53. Cline, S.P.; Berg, A.B.; Wight, H.M. Snag characteristics and dynamics in Douglas-fir forests, western Oregon. *J. Wildl. Manag.* **1980**, *44*, 773–786. [CrossRef]
54. Tilghman, N.G. Impacts of white-tailed deer on forest regeneration in northwestern Pennsylvania USA. *J. Wildl. Manag.* **1989**, *53*, 524–532. [CrossRef]
55. Ripple, W.J.; Larsen, E.J. The role of postfire coarse woody debris in aspen regeneration. *West. J. Appl. For.* **2001**, *16*, 61–64.
56. De Chantal, M.; Granstrom, A. Aggregations of dead wood after wildfire act as browsing refugia for seedlings of *Populus tremula* and *Salix Caprea*. *For. Ecol. Manag.* **2007**, *250*, 3–8. [CrossRef]
57. Relva, M.A.; Westerholm, C.L.; Kitzberger, T. Effects of introduced ungulates on forest understory communities in northern Patagonia are modified by timing and severity of stand mortality. *Plant Ecol.* **2009**, *201*, 11–22. [CrossRef]
58. Cornett, M.W.; Reich, P.B.; Puettmann, K.J.; Frelich, L.E. Seedbed and moisture availability determine safe sites for early *Thuja occidentalis* (Cupressaceae) regeneration. *Am. J. Bot.* **2000**, *87*, 1807–1814. [CrossRef] [PubMed]
59. Harmon, M.E.; Franklin, J.F.; Swanson, F.J.; Sollins, P.; Gregory, S.V.; Lattin, J.D.; Anderson, N.H.; Cline, S.P.; Aumen, N.G.; Sedell, J.R.; et al. Ecology of coarse woody debris in temperate ecosystems. *Adv. Ecol. Res.* **1986**, *15*, 133–302.
60. Webb, S.L. Disturbance by wind in temperate-zone forests. In *Ecosystems of Disturbed Ground*; Walker, L.R., Ed.; Elsevier: Amsterdam, The Netherlands, 1999; pp. 187–222.
61. Cooper-Ellis, S.; Foster, D.R.; Carlton, G.; Lezberg, A. Forest response to catastrophic wind: Results from an experimental hurricane. *Ecology* **1999**, *80*, 2683–2696. [CrossRef]
62. Girard, F.; De Grandpre, L.; Ruel, J.-C. Partial windthrow as a driving process of forest dynamics in old-growth boreal forests. *Can. J. For. Res.* **2014**, *44*, 1165–1176. [CrossRef]
63. Peterson, C.J. Damage and recovery from two different tornadoes in the same old Growth forest: A comparison of infrequent events. *For. Ecol. Manag.* **2000**, *135*, 237–252. [CrossRef]
64. Vandermeer, J.; Mallona, M.A.; Boucher, D.; Yih, K.; Perfecto, I. Three years of ingrowth following catastrophic hurricane damage on the Caribbean coast of Nicaragua: Evidence in support of the direct regeneration hypothesis. *J. Trop. Ecol.* **1995**, *11*, 465–471. [CrossRef]
65. Harcombe, P.A.; Leipzig, L.E.M.; Elsik, I.S. Effects of Hurricane Rita on three long-term forest study plots in east Texas, USA. *Wetlands* **2009**, *29*, 88–100. [CrossRef]

66. Nagel, T.A.; Svoboda, M.; Diaci, J. Regeneration patterns after intermediate wind disturbance in an old-growth *Fagus-Abies* forest in southeastern Slovenia. *For. Ecol. Manag.* **2006**, *226*, 268–278. [CrossRef]

67. Dodet, M.; Collet, C.; Frochot, H.; Wehrlen, L. Tree regeneration and plant species diversity responses to vegetation control following a major windthrow in mixed broadleaved stands. *Eur. J. For. Res.* **2011**, *130*, 41–53. [CrossRef]

68. Peterson, C.J. Consistent influence of tree diameter and species on damage in nine Eastern North America tornado blowdowns. *For. Ecol. Manag.* **2007**, *250*, 96–108. [CrossRef]

69. Peterson, C.J.; Pickett, S.T.A. Stem damage and resprouting following catastrophic windthrow in an old-growth hemlock-hardwoods forest. *For. Ecol. Manag.* **1991**, *42*, 205–217. [CrossRef]

70. Peterson, C.J.; Pickett, S.T.A. Forest reorganization: A case study in an old-growth forest catastrophic blowdown. *Ecology* **1995**, *76*, 763–774. [CrossRef]

71. Peterson, C.J. Within-stand variation in windthrow in southern-boreal forests of Minnesota: Is it predictable? *Can. J. For. Res.* **2004**, *34*, 365–375. [CrossRef]

72. Peterson, C.J.; Carson, W.P. Generalizing forest regeneration models: The dependence of propagule availability on disturbance history and stand size. *Can. J. For. Res.* **1996**, *26*, 45–52. [CrossRef]

73. Peterson, C.J.; Leach, A.D. Limited salvage logging effects on forest regeneration after moderate-severity windthrow. *Ecol. Appl.* **2008**, *18*, 407–420. [CrossRef]

74. Peterson, C.J.; Rebertus, A.J. Tornado damage and initial recovery in three adjacent, lowland temperate forests in Missouri. *J. Veg. Sci.* **1997**, *8*, 559–564. [CrossRef]

75. Sobhani, V.M.; Barrett, M.J.; Peterson, C.J. Robust prediction of treefall pit and mound sizes from tree size across 10 forest blowdowns in eastern North America. *Ecosystems* **2014**, *17*, 837–850. [CrossRef]

76. Bjorkbom, J.C.; Larson, R.G. *The Tionesta Scenic and Research Natural Areas*; General Technical Report NE-31; USDA Forest Service: Washington, DC, USA, 1977.

77. U.S. Department of Agriculture Forest Service. *After the Blowdown: A Resource Assessment of the Boundary Waters Canoe Area Wilderness, 1999-2003*; General Technical Report NRS-7; U.S. Department of Agriculture Forest Service: Washington, DC, USA, 2007.

78. Pielou, E.C. Measurement of diversity in different types of biological collections. *J. Theor. Boil.* **1966**, *13*, 131–135. [CrossRef]

79. Weiner, J.; Solbrig, O.T. The meaning and measurement of size hierarchies in plant populations. *Oecologia* **1984**, *61*, 334–336. [CrossRef] [PubMed]

80. Stoudhammer, C.L.; LeMay, V.M. Introduction and evaluation of possible indices of stand structural complexity. *Can. J. For. Res.* **2001**, *31*, 1105–1115. [CrossRef]

81. Everham, E.M.; Brokaw, N.V.L. Forest damage and recovery from catastrophic wind. *Bot. Rev.* **1996**, *62*, 113–185. [CrossRef]

82. Canham, C.D.; Papaik, M.J.; Latty, E.F. Interspecific variation in susceptibility to windthrow as a function of tree size and storm severity for northern temperate tree species. *Can. J. For. Res.* **2001**, *31*, 1–10. [CrossRef]

83. Frelich, L.E.; Sugita, S.; Reich, P.B.; Davis, M.B.; Friedman, S.K. Neighborhood effects in forests: Implications for within-stand patch structure. *J. Ecol.* **1998**, *86*, 149–161. [CrossRef]

84. Ulanova, N.G. The effects of windthrow on forests at different spatial scales: A review. *For. Ecol. Manag.* **2000**, *135*, 155–167. [CrossRef]

85. Phillips, J.D.; Marion, D.A.; Turkington, A.V. Pedologic and geomorphic impacts of a tornado blowdown event in a mixed pine-hardwood forest. *Catena* **2008**, *75*, 278–287. [CrossRef]

86. Basnet, K.; Scatena, F.N.; Likens, G.E.; Lugo, A.E. Ecological consequences of root grafting in Tabonuco (*Dacryodes excelsa*) trees in the Luquillo Experimental Forest, Puerto Rico. *Biotropica* **1993**, *25*, 28–35. [CrossRef]

87. Bauhus, J.; Puettmann, K.; Messier, C. Silviculture for old-growth attributes. *For. Ecol. Manag.* **2009**, *258*, 525–537. [CrossRef]

© 2019 by the author. Licensee MDPI, Basel, Switzerland. This article is an open access article distributed under the terms and conditions of the Creative Commons Attribution (CC BY) license (http://creativecommons.org/licenses/by/4.0/).

forests

MDPI

Review

Excess Nitrogen in Temperate Forest Ecosystems Decreases Herbaceous Layer Diversity and Shifts Control from Soil to Canopy Structure

Frank S. Gilliam

Department of Biology, University of West Florida, Pensacola, FL 32514, USA; fgilliam@uwf.edu;
Tel.: +1-850-474-2750

Received: 28 November 2018; Accepted: 12 January 2019; Published: 15 January 2019

Abstract: Research Highlights: Excess N from atmospheric deposition has been shown to decrease plant biodiversity of impacted forests, especially in its effects on herbaceous layer communities. This work demonstrates that one of the mechanisms of such response is in N-mediated changes in the response of herb communities to soil resources and light availability. Background and Objectives: Numerous studies in a variety of forest types have shown that excess N can cause loss of biodiversity of herb layer communities, which are typically responsive to spatial patterns of soil resource and light availability. The objectives of this study were to examine (1) gradients of temporal change in herb composition over a quarter century, and (2) spatial patterns of herb cover and diversity and how they are influenced by soil resources and canopy structure. Materials and Methods: This study used two watersheds (WS) at the Fernow Experimental Forest, West Virginia, USA: WS4 as an untreated reference and WS3 as treatment, receiving 35 kg N/ha/yr via aerial application. Herb cover and composition was measured in seven permanent plots/WS from 1991 to 2014. In 2011, soil moisture and several metrics of soil N availability were measured in each plot, along with measurement of several canopy structural variables. Backwards stepwise regression was used to determine relationships between herb cover/diversity and soil/canopy measurements. Results: Herb diversity and composition varied only slightly over time on reference WS4, in contrast to substantial change on N-treated WS3. Herb layer diversity appeared to respond to neither soil nor canopy variables on either watershed. Herb cover varied spatially with soil resources on WS4, whereas cover varied spatially with canopy structure on WS3. Conclusions: Results support work in many forest types that excess N can decrease plant diversity in impacted stands. Much of this response is likely related to N-mediated changes in the response of the herb layer to soil N and light availability.

Keywords: herbaceous layer; excess nitrogen; canopy structure; temperate forests

---

1. Introduction

The herbaceous layer is increasingly acknowledged for its significant contribution to the integrity of the structure and function of forest ecosystems [1–3]. This has likely contributed, in part, to the notable increase, in recent decades, in research activity of plant ecologists investigating the dynamics of forest herb layer communities. Based on the number of publications, the new millennium has witnessed an unprecedented increase in research efforts. The number of papers published on the herb layer in the current, incomplete decade alone exceeds the total number of papers published in the 20th century by nearly 100% [2].

Along with growing awareness of the essential role that the herb layer plays in forest ecosystems is an increase in studies investigating the ecological factors that influence its dynamics, including spatial and temporal variation in species composition and aboveground cover/biomass. A review of this expanding literature reveals sharp contrasts among the forest types studied, including widely varying land use history, that preclude broad generalizations regarding specific ecological drivers that help shape forest herb communities. Certainly, any list of abiotic factors affecting these communities would include light and moisture, especially as mitigated by the overstory canopy, as well as temperature regimes [4]. Temperature responses are of particular interest with their implications for the future of forests in the context of climate change [5,6].

Although the utility of considering stratification of vegetation in forest communities has been questioned (see Parker and Brown [7] and discussion therein), it is clear that solar radiation attenuates through a forest canopy in ways that that alter both the quantity and quality of light reaching the forest floor. Light availability is the most spatially and temporally variable component of the environment of the forest floor, varying at numerous levels over time and space, in a manner described as a *dynamic mosaic* [8], penetrating the canopy and reaching the forest floor in sunflecks, i.e., mosaics of discrete patches of light varying in size and distribution over time scales from the diurnal to the seasonal. In addition, soil resources, including nutrients and moisture, are often spatially quite heterogeneous in ways that influence the dynamics of the herb layer [9–11].

Given the increased interest in the importance of the herb layer to maintaining the structure and function of forest ecosystems, it is not surprising that considerable research has been devoted to understanding what environmental factors most sensitively affect herb layer composition and cover. Some of the more extensive earlier work was carried out by Rogers [12–16] in the 1980s in conifer and hardwood forests of the northern U.S. and Southern Canada. This work revealed that there can be great interannual variability in herb cover. He also concluded that soil fertility was more important than climate variables in influencing forest herb communities. In one of the more complete studies of vegetation recovery following disturbance, however, Reiners [17] showed that community reorganization was largely driven by light availability, as *Rubus* spp. increased and then sharply decreased during canopy redevelopment. Using a database of studies investigating effects of resources on herb layer diversity, Bartels and Chen [9] concluded neither resource availability nor heterogeneity solely influences herb layer dynamics. Reich et al. [18] found that spatial heterogeneity in the light environment most directly affect herb richness in southern boreal forests. This was confirmed for both richness and herb cover by Kumar et al. [19] studying similar stand types, with the same response found for biomass [20].

In addition to these components of the ambient environment of the forest floor are anthropogenic influences, virtually all of which represent threats to forest health, particularly regarding the herb layer of impacted forests, including the effects of excess N from atmospheric deposition [21–23]. Gilliam et al. [24] examined a quarter century of experimental additions of N on an entire watershed (WS) at the Fernow Experimental Forest (FEF), West Virginia, to study the effects of excess N on temporal and spatial dynamics of the herb layer of a Central Appalachian hardwood forest. This study involved nearly annual monitoring of herb layer composition and cover of permanent plots from 1991 to 2014 on both an N-treated watershed and the long-term reference watershed at FEF (WS4), a ~100 year old mixed hardwood forest. They found a pronounced shift in herb layer composition over the 25 years of experimental N additions to the treatment watershed, and determined that such change arose from increases in a nitrophilic species (i.e., *Rubus allegheniensis* Porter that competitively excluded numerous N-efficient herbaceous species, ultimately decreasing plant diversity.

This response supported the N homogeneity hypothesis, which predicts that excess N deposition to forest ecosystems increases the spatial homogeneity of N by decreasing natural patchiness of N availability essentially by filling in the low-N matrix within which discrete high-N patches occur [25]. As a result, temporal increases in atmospheric inputs of N should increase N availability within this matrix to approach that within the patches of high fertility. Nitrophilic plant species of the forest herbaceous layer then increase in dominance, outcompeting the more numerous N-efficient species and decreasing biodiversity of the forest, up to 90% of which is represented by the herb layer [1].

To further investigate this response, Walter et al. [26] carried out field studies on the response of *R. allegheniensis* to variation in light and N availability. They compared relative cover of *R. allegheniensis* in N-treated WS3 and another untreated watershed at FEF (WS7) and among N-fertilized and unfertilized experimental plots; both approaches utilized canopy openness as a covariate. The ex situ experiment used a two-way factorial design, measuring leaf area with two levels of N and three of light. Results of both approaches were consistent in revealing that the effects of N availability on cover were significantly mitigated by availability of light.

The objectives of this paper are two-fold. The first is to further analyze data from the Gilliam et al. [24] study to more specifically focus on gradients of change in herb layer composition and temporal variation in indices of biodiversity in the context of forest response to excess N. The second includes previously unpublished canopy structural data to examine spatial patterns of herb cover and diversity and how they are influenced by several metrics of canopy structure (many of which are indicative of light availability to the forest floor [27]) and soil resources in both watersheds, with a particular focus on how excess N might alter these relationships.

2. Methods

2.1. Study Site

This study was carried out at the Fernow Experimental Forest (FEF), Tucker County, West Virginia (39°03′15″ N, 79°49′15″ W), as part of a long-term study on the effects of chronic additions of N on the structure and function of central Appalachian hardwood forest ecosystems. Fernow Experimental Forest is a ~1900 ha area of the Allegheny Mountain section the unglaciated Allegheny Plateau. Precipitation for FEF averages ~1430 mm yr^{-1}, with precipitation generally increasing through the growing season and with higher elevations [24].

Two watersheds were used for the location of sample plots: WS3 and WS4, with WS3 serving as the treatment watershed, receiving aerial additions of $(NH_4)_2SO_4$, and WS4 serving as reference watershed. Applications of $(NH_4)_2SO_4$ to WS3 began in 1989 and are currently on-going. Aerial applications of $(NH_4)_2SO_4$ are made three times per year, and historically have been administered by either helicopter or fixed-wing aircraft. March and November applications are 33.6 kg/ha of fertilizer, or 7.1 kg/ha of N. July applications are 100.8 kg/ha fertilizer (21.2 kg/ha N). Stands on WS3 were ~45 yr-old at the time of most recent sampling in this study (2014); these are even-aged and developed following clearcutting. WS4 currently supports even-aged stands >100 yr old.

Study watersheds support mixed hardwood stands. Overstory dominant species include sugar maple (*Acer saccharum* Marsh.), sweet birch (*Betula lenta* L.), American beech (*Fagus grandifolia* Ehrh.), yellow poplar (*Liriodendron tulipifera* L.), black cherry (*Prunus serotina* Ehrh.), and northern red oak (*Quercus rubra* L.). In 1991, species composition of the herbaceous layer was quite similar between watersheds, despite differences in stand age, including species of *Viola*, *Rubus*, mixed ferns, and seedlings of striped maple *Acer pensylvanicum* L. and red maple *A. rubrum* L.

2.2. Field Methods

The herbaceous layer was sampled in five circular 1 m^2 sub-plots within each of seven circular 0.04 ha permanent sample plots using methods described in Gilliam et al. [24]. Briefly, all vascular plants ≤1 m in height in each subplot were identified to species (sometimes to genus) and visually estimated for cover (%); see Walter et al. [28] for detailed description of this method. This was carried out in the first week of July each of the sample years 1991, 1992, 1994, 2003, and annually from 2009 to 2014, for a total of 10 sample years over a 24-year period representing 26 years of N treatment on WS3. When used for regression analyses, soil data were taken from Gilliam et al. [29], including soil moisture (%), extractable NH_4^+ and NO_3^-, and net N mineralization and nitrification. Soil moisture was determined gravimetrically, NH_4^+ and NO_3^- were measured colorimetrically following 1N KCl extraction, and net N mineralization and nitrification were determined using in situ incubations. These were for mineral soil only (O horizon excluded) that was taken to a 5-cm depth, with measurements made on a monthly basis. For this analysis, data from July 2011 were used to align with the herb layer and forest canopy measurements.

Forest canopy measurements were made in July 2011 with a Riegl LD90-3100VHS-FLP laser rangefinder (operating in first-return mode at 890 nm and 2 kHz, laser safety class I) mounted to the front of a frame at 1 m above the ground and manually pointed upward, making 2000 measurements per second. Data were transferred through a serial cable to laptop [30]. Using constant walking speed, locations of each range measurement were estimated from its sequence in the data file. Generally, distances between measurements were <1 cm, with the spot size of the laser beam being 4–6 cm at the ranges measured.

The data files were edited to identify values that were out-of-range (e.g., when penetrating canopy openings to the sky) and remove spurious values. The edited files were processed through a program customized for grouping ranges horizontally, calculating vertical profiles (using methodology of MacArthur and Horn (1969) [31]), estimating surface area density using the overlap transformation, and assigning coordinates to each estimate. Bins used were 1 m in the horizontal and 1 m in the vertical. Resulting estimates refer to cube-shaped voxels of $1 \times 1 \times 1$ m in the x, y, and z dimensions, respectively.

2.3. Data Analysis

Herb layer data for the entire study period were subjected to detrended correspondence analysis (DCA), using CANOCO 4.5, for each watershed separately to assess temporal change in herb community composition. For graphical purposes, two dimensional means (from individual plot axis 1 and axis 2 values) were calculated for each year as centroids. Part of the output of DCA on CANOCO 4.5 (Microcomputer Power, Ithaca, NY, USA) are several metrics of biodiversity, including species richness and evenness and Hill and Shannon diversity indices. Changes in metrics of biodiversity of the herb layer over time, including species richness, evenness, and diversity, were assessed via Pearson product–moment correlation [32].

For the purpose of this study, the following canopy structural variables were determined: canopy area index (CAI), local outer canopy height (LOCH), rugosity, and gap fraction. Canopy area index is the sum of surface area density across all levels in a column. Local outer canopy height is the maximum surface height in a column—across all columns together these define the outer canopy surface. Rugosity is a measure of the "roughness" of the forest canopy and is the standard deviation of the mean outer canopy height. The gap fraction is the fraction of horizontal locations without any canopy surface area directly above (one minus the "cover") [30].

Potential effects of both canopy structure and soil variables on herb cover and biodiversity metrics were assessed on data from 2011 (the only year for which canopy measurements were made) using backwards stepwise regression. This procedure eliminates variables from the proposed model sequentially until all remaining variables produce F statistics that are significant at a given level of probability, in this case $p < 0.05$ [32]. This was used to identify which (if any) canopy and/or soil variables best explain spatial variation in herb diversity and cover with the following initial model:

$$Y = \text{moist} + NH_4^+ + NO_3^- + \text{Nmin} + \text{nit} + \text{CAI} + \text{LOCH} + \text{gapfrac} + \text{rug}$$

where, moist is soil moisture (%), NH_4^+ is extractable soil NH_4^+ (μg N/g soil), NO_3^- is extractable soil NO_3^- (μg N/g soil), Nmin is net N mineralization (μg N/g soil/d), nit is net nitrification (μg N/g soil/d), CAI is canopy area index, LOCH is local outer canopy height (m), gapfrac is gap fraction, rug is rugosity, and Y is the dependent variable. Separate runs were made for the following dependent variables: herb cover, richness, evenness, Hill diversity, and Shannon diversity.

For spatial analyses, herb layer cover, CAI, and LOCH were kriged separately in each watershed and each year using an ordinary kriging method with a spherical variogram model and global search radius in R package gstat [33]. Each model was fit using a common initial range and sill value and interpolated onto a grid with a cell resolution of 5 × 5 m. Grids were mapped in ArcGIS using 20 equal intervals that spanned the range of kriged values for cover, CAI, and LOCH, separately, for 2011.

Temporal change in metrics of biodiversity for the study watersheds was assessed via linear correlation of individual metrics versus year, from 1991 to 2014. In addition, because of its potential as a mechanism driving change in these metrics, as just discussed, cover (%) of *R. allegheniensis* was also assessed in this manner.

3. Results and Discussion

3.1. Temporal Variation in Composition of the Herbaceous Layer

For clarity, the results of DCA are presented in two figures for each of reference WS4 and treatment WS3, the first with annual centroids labeled by year and arrows depicting temporal trends as trajectories of change in ordination space. The second figure for each watershed displays the location of prominent herb layer species for each watershed, along with unlabeled annual centroids for purposes of comparison.

As WS4 is typical of a mature second-growth stand of the central Appalachian region [34], results suggest that, although there is notable inter-annual variability, changes in herb community composition is not unidirectional over time in such stands. For example, although substantial variation occurred in the nine years between 1994 and 2003, far more occurred in the following six years to 2009; indeed, this represented a return toward increased similarity with 1991, despite the 18 year duration of this period (Figure 1). Species variation along this time trend appeared to occur across two gradients. The first of these suggests a fertility gradient, from *Vaccinium* spp. (VACC), which are well-adapted to weathered, infertile soil [35], to Smilax rotundifolia (SMRO), and then *Rubus allegheniensis* (RUAL) a nitrophilic species [26,36]. The second is likely a moisture gradient, from seedlings of overstory species (e.g., *Quercus rubra, Acer saccharum, A. rubrum*) (QURU, ACSA, ACRU) to *Laportea canadensis* (LACA) and *Dryopteris intermedia* (DRIN), which are characteristic of moist forest soils [37,38] (Figure 2).

Differences between results of DCA on temporal trends of herb composition WS4 and those of WS3 suggest the profound effects of experimental additions of N on herb layer dynamics. Although there was minimal variation along Axis 1 during the initial period of the study (1991–1994), there was a substantial shift in composition from 1994 to 2003, one which never returned toward the initial period, as was observed for WS4 (Figure 3). Similar to WS4, species variation along this time trend on WS3 occurred across two gradients. The first suggests a disturbance gradient, from *S. rotundifolia*, often associated with canopy disturbances [39], to fern species (*Polystichum acrosticoides, Dennstaedia punctiloba*) (POAC, DEOU), which are better adapted to less-disturbed conditions (Figure 4).

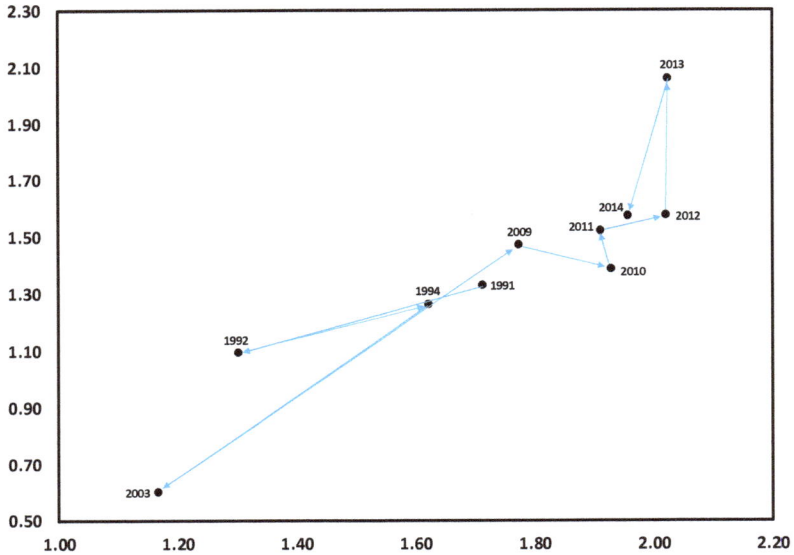

Figure 1. Detrended correspondence analysis (DCA) for herbaceous layer communities from 1991 to 2014 on reference watershed WS4. Data points shown along with years are centroids (two-dimensional means) for each year.

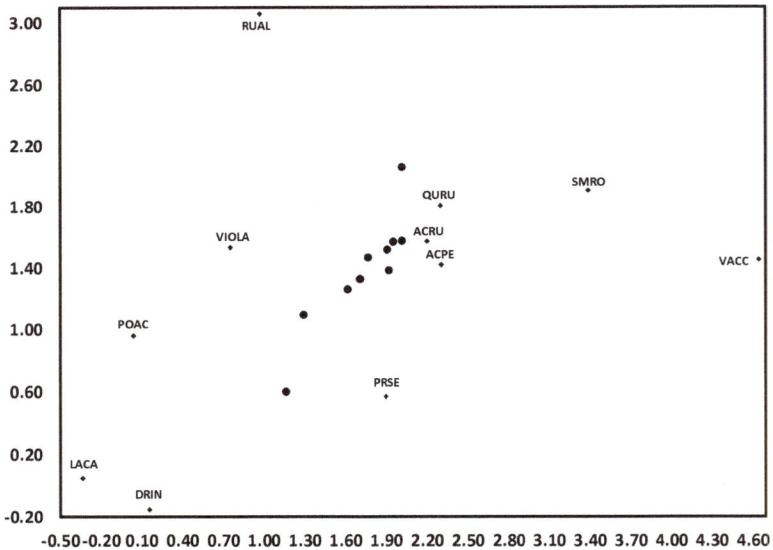

Figure 2. Detrended correspondence analysis (DCA) for herbaceous layer communities from 1991 to 2014 on reference WS4. Depicted are prominent species, along with centroids as shown in Figure 1. Species codes are as follows: ACPE (*Acer pensylvanicum*), ACRU (*A. rubrum*), DRIN (*Dryopteris intermedia*), LACA (*Laportea canadensis*), POAC (*Polystichum acrostichoides*), PRSE (*Prunus serotina*), QURU (*Quercus rubra*), RUAL (*Rubus allegheniensis*), SMRO (*Smilax rotundifolia*), VACC (*Vaccinum* spp.), and VIOLA (*Viola* spp.).

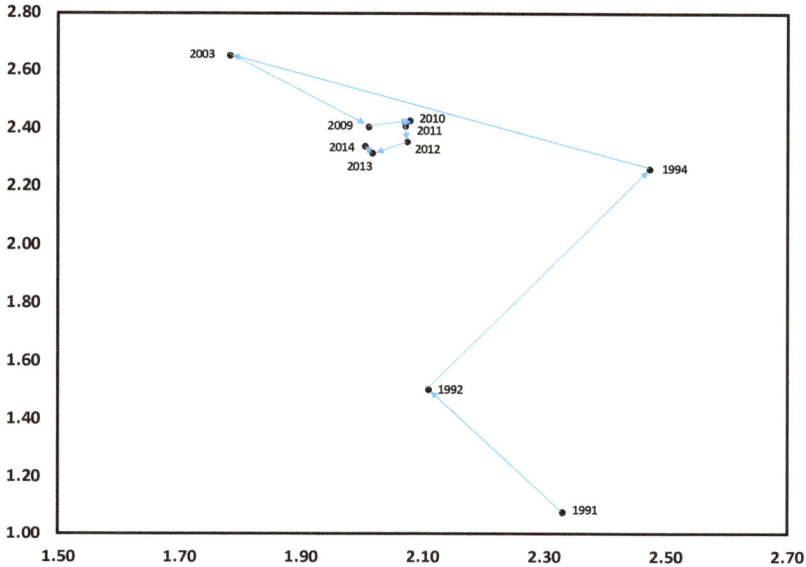

Figure 3. Detrended correspondence analysis (DCA) for herbaceous layer communities from 1991 to 2014 on N-treated WS3. Data points shown along with years are centroids (two-dimensional means) for each year per watershed.

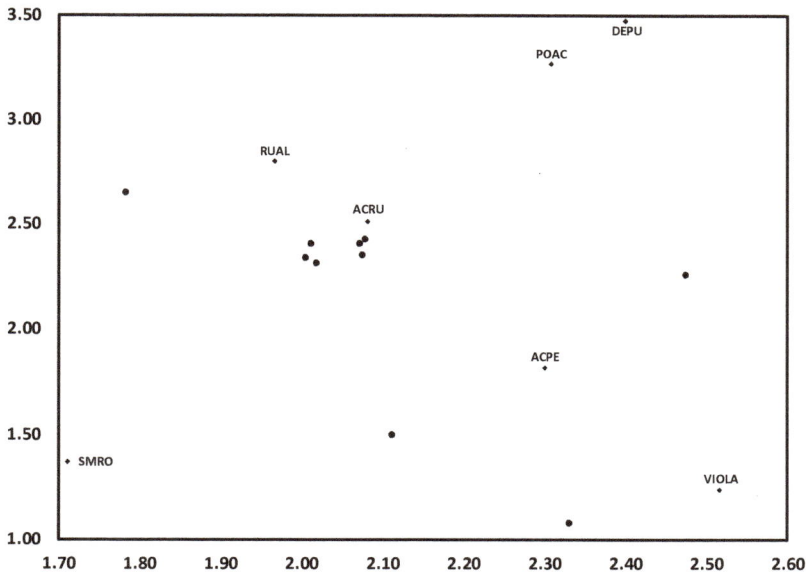

Figure 4. Detrended correspondence analysis (DCA) for herbaceous layer communities from 1991 to 2014 on N-treated WS3. Depicted are prominent species, along with centroids as shown in Figure 3. Species codes are as follows: ACPE (*Acer pensylvanicum*), ACRU (*A. rubrum*), DEPU (Dennstaedtia punctilobula), POAC (*Polystichum acrostichoides*), RUAL (*Rubus allegheniensis*), SMRO (*Smilax rotundifolia*), and VIOLA (*Viola* spp.).

The most notable gradient, and one that has important implications for effects of N on forest biodiversity, is that between *Viola spp.* and *R. allegheniensis*. In particular, both the direction and distance between those two species in ordination space reflects a change in dominance of the herb community from 1991 to 2014 (Figure 4). That is, the once-dominant and species-rich Viola group has been replaced by *R. allegheniensis*, supporting earlier predictions of the N Homogeneity Hypothesis: that increasing inputs of N would provide a competitive advantage to nitrophilic species, such as *R. allegheniensis*, at the expense of N-efficient species which comprise most of the biodiversity of the herb community [25]. This response is one of the more prominent mechanisms by which excess N can decrease forest biodiversity, considering that (1) there is far greater species richness among N-efficient species than among nitrophilic species, and (2) up to 90% of plant diversity of forest ecosystems is in the herb layer [40].

3.2. Change in Herb Diversity Metrics over Time

Species richness (S) increased significantly ($p < 0.0004$) on WS4 for the study period, suggesting that herb communities in mature second growth hardwood stands can experience the influx of new species well after 100 yr of post-disturbance succession. By contrast, species evenness (J) did not vary significantly ($p > 0.05$) with time, resulting in no change in herb layer diversity (Hill's N2). Significant increases in cover of *R. allegheniensis* did not influence any metrics of biodiversity (Table 1).

Table 1. Pearson product–moment correlation coefficients between species richness (S), evenness (J), and diversity (Hill's N2), and cover of *Rubus allegheniensis* (*Rubus*, in %) versus time (years) on reference WS4 and treatment WS3. Thus, significant positive and negative coefficients indicate increases and decreases, respectively, over time. Values shown are coefficients significant at $p < 0.05$. NS indicates no significant correlation.

Metric	WS4	WS3
S	0.91	NS
J	NS	−0.81
Hill's N2	NS	−0.64
Rubus	0.86	0.98

Contrasts between watersheds, as depicted in Table 1, suggest both the profound effect of excess N on herb diversity and the principle mechanism for such an effect. As already reported for this site [24], and consistent with findings of numerous other studies [41–43], chronic additions of N to WS3 have significantly ($p < 0.05$) decreased herb layer diversity. Although both richness and evenness contribute to species diversity, these results reveal that the N-mediated loss of diversity arose from decreased species evenness, which also declined significantly ($p < 0.005$), and not species richness, which showed no change (Table 1). Combining these contrasts with the highly significant ($p < 0.0001$) increase in *R. allegheniensis* on WS3 confirm that these N-mediated declines in herb diversity are driven primarily by increasing cover of a nitrophilic species, as initially predicted by Gilliam [25] and found in several studies in contrasting forest types [21,44–46]. It is notable that *R. allegheniensis* increased significantly on WS4 over this same period, yet appeared to have no influence on herb diversity metrics. The increase is quite likely from the chronically elevated ambient levels of N deposition for this site [24], whereas the lack of effect is the much lower degree of increase on WS4.

3.3. Influences of Canopy and Soil Variables on Herbaceous Biodiversity and Cover

When the model was run with each of richness, evenness, and diversity separately, none was significant ($p > 0.05$), suggesting that neither canopy nor soil variables measured herein influences herb layer biodiversity for either watershed, regardless of experimental treatment. By contrast, regressions for cover revealed significant, though contrasting, model outcomes for both watersheds (Table 2). Indeed, such contrasts were strongly suggestive of effects of added N on herb layer cover. Results indicate that for reference WS4, herb cover was influenced by soil resources, rather than canopy structural variables, especially pools of available NH_4^+ and NO_3^- (Table 2). Model outcome for treatment WS3, however, demonstrated that canopy structural, rather than soil resource, variables were primarily influencing herb cover. Thus, in the absence of added N, cover appears to be controlled primarily by soil N, whereas chronic additions of 35 kg N/ha/yr have brought about a shift in control wherein light availability, as influenced by canopy structure, more sensitively influences herb cover in these forest stands.

Table 2. Backward stepwise regression for study watershed at Fernow Experimental Forest, WV, beginning with the following initial model: cover = moist + NH_4^+ + NO_3^- + N_{min} + nit + CAI + LOCH + gapfrac + rug, where cover is percent herb layer cover, moist is soil moisture, NH_4^+ is extractable soil N, NO_3^- is extractable soil N, N_{min} is net N mineralization, nit is net nitrification, CAI is canopy area index, LOCH is local outer canopy height, gapfrac is gap fraction, and rug is rugosity. See Methods for explanation of canopy structural variables, stepwise regression procedure, and units for independent variables.

Watershed	Final Model	r^2
WS4	Cover = 60.8 − 5.3NH_4^+ + 12.6NO_3^-	0.85
WS3	Cover = 408.5 − 58.9CAI + 3.1LOCH − 909.6gapfrac	0.998

Among the canopy structural variables, canopy area index (CAI) and local outer canopy height (LOCH) appeared to exert the strongest influence on herb cover on WS3. Neither were significantly correlated with cover on WS4, but were negatively ($r = -0.88$, $p < 0.05$) and positively ($r = 0.88$, $p < 0.05$) correlated, respectively, on WS3. Actual spatial patterns of all three variables on both watersheds were determined via kriging (Figure 5). Because CAI is the sum of surface area density across all levels in a vertical column through the canopy, it should be negatively related to light availability to the forest floor. That is, higher CAI would indicate a greater leaf area intercepting more light. The negative relationship between CAI and cover on WS3 supports this (Figure 5). By contrast, LOCH likely positively related to light availability to the forest floor, being calculated as the maximum surface height in a column, rather than being an indicator of foliar surface area. This is consistent with the positive relationship between LOCH and herb cover on WS3 (Figure 5).

Contrasts between treated and reference watersheds in the present study suggest that a similar phenomenon to that reported by Walter et al. [26] for *R. allegheniensis* (see Introduction) may have occurred for the herb community as a whole. That is, 25 years of adding N to WS3 has likely shifted herb cover toward being more sensitive to spatial variability of light incident on the forest floor due to less dependence on available soil N.

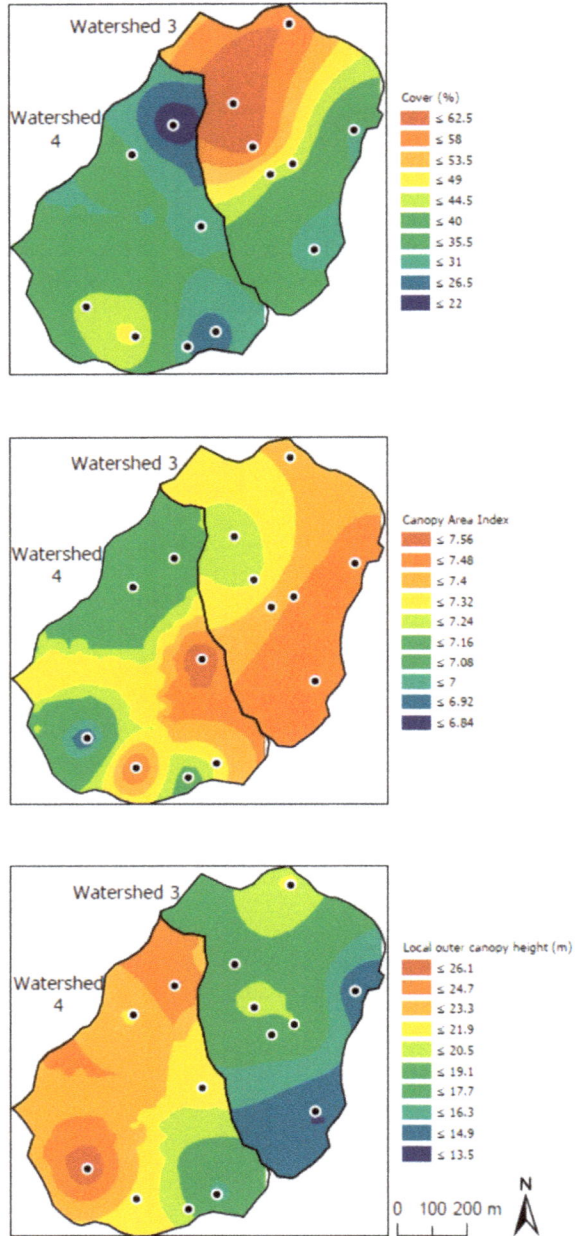

Figure 5. Spatial patterns of herbaceous layer cover, canopy area index, and local outer canopy height for study watersheds at Fernow Experimental Forest, West Virginia, July 2011. Shown also are locations of permanent sample plots on each watershed.

4. Conclusions

Differences found in this study between reference WS4 and N-treated WS3 in the nature of how the herb layer is affected by soil resources versus canopy structure highlight mechanisms of how excess N can influence forest herb communities, adding to a growing number of such studies. Results presented here further demonstrate the loss of forest biodiversity, via decreases in species diversity of herb communities, in response to simulated increases in deposition of N. In addition, this study suggests that excess N can facilitate a shift in factors controlling herb layer dynamics from variation in soil resources to variation in canopy structure.

Finally, ambient deposition of N is currently far different than that which initially created the impetus for studies of excess N on forest ecosystems. Due to the efficacy of the Clean Air Act in the United States, deposition of oxidized N has been declining in recent decades, particularly in the eastern U.S. [47,48]. In contrast, deposition of reduced N has increased over the same period [49]. Thus, future investigations on impacts of N on forest ecosystems should consider both the ways in which N-impacted forested regions recover toward pre-impact conditions and the responses of forests to changes in dominant chemical forms of N. Gilliam et al. [49] has suggested a hysteretic model for such recovery, which predicts a variable and, in some cases, considerable lag time before such changes may be detected. The question remains regarding how the herbaceous layer of N-impacted forests, and its response to soil and canopy structure, will change in a lower-N future.

Funding: This research was funded in part through United States Department of Agriculture (USDA) Forest Service, Fernow Experimental Forest, Timber and Watershed Laboratory, Parsons, W.V., under USDA Forest Service Cooperative Grants 23-165, 23-590, and 23-842 to Marshall University. Additional funding for this research was provided by USDA National Research Initiative Competitive Grants (Grant NRICGP #2006-35101-17097) to Marshall University and by the Long Term Research in Environmental Biology (LTREB) program at the National Science Foundation (Grant Nos. DEB0417678 and DEB-1019522) to West Virginia University.

Acknowledgments: I am indebted to numerous undergraduate and graduate students at Marshall University and West Virginia University for their excellent assistance in the field, and to Bill Peterjohn (West Virginia University) and Mary Beth Adams (USDA Forest Service) for field and logistical support. I am especially indebted to Jess Parker (Smithsonian Environmental Research Center) for all canopy structural measurements and to Chris Walter (University of Minnesota) for generating the kriged maps. The long-term support of the USDA Forest Service in establishing and maintaining the research watersheds is acknowledged.

Conflicts of Interest: The author declares no conflict of interest.

References

1. Gilliam, F.S. The ecological significance of the herbaceous layer in forest ecosystems. *BioScience* **2007**, *57*, 845–858. [CrossRef]
2. Gilliam, F.S. Introduction: The herbaceous layer—The forest between the trees. In *The Herbaceous Layer in Forests of Eastern North America*, 2nd ed.; Gilliam, F.S., Ed.; Oxford University Press: New York, NY, USA, 2014; pp. 1–9, ISBN 978-0-19-983765-6.
3. Thrippleton, T.; Bugmann, H.; Kramer-Priewasser, K.; Snell, R.S. Herbaceous understorey: An overlooked player in forest landscape dynamics? *Ecosystems* **2016**, *19*, 1240–1254. [CrossRef]
4. Neufeld, H.S.; Young, D.R. Ecophysiology of the herbaceous layer in temperate deciduous forests. In *The Herbaceous Layer in Forests of Eastern North America*, 2nd ed.; Gilliam, F.S., Ed.; Oxford University Press: New York, NY, USA, 2014; pp. 35–95, ISBN 978-0-19-983765-6.
5. De Frenne, P.; Rodríguez-Sánchez, F.; Coomes, D.A.; Baeten, L.; Verstraeten, G.; Vellend, M.; Bernhardt-Römermann, M.; Brown, C.D.; Brunet, J.; Cornelis, J.; et al. Microclimate moderates plant responses to macroclimate warming. *Proc. Natl. Acad. Sci. USA* **2013**, *110*, 18561–18565. [CrossRef] [PubMed]
6. Bellemare, J.; Moeller, D.A. Climate change and forest herbs of temperate deciduous forests. In *The Herbaceous Layer in Forests of Eastern North America*, 2nd ed.; Gilliam, F.S., Ed.; Oxford University Press: New York, NY, USA, 2014; pp. 460–493, ISBN 978-0-19-983765-6.
7. Parker, G.G.; Brown, M.J. Forest canopy stratification—Is it useful? *Am. Nat.* **2000**, 473–484.

8. Gilliam, F.S. The dynamic nature of the herbaceous layer: A brief synthesis. In *The Herbaceous Layer in Forests of Eastern North America*, 2nd ed.; Gilliam, F.S., Ed.; Oxford University Press: New York, NY, USA, 2014; pp. 497–508, ISBN 978-0-19-983765-6.

9. Hutchings, M.J.; John, E.A.; Wijesinghe, D.K. Toward understanding the consequences of soil heterogeneity for plant populations and communities. *Ecology* **2003**, *84*, 2322–2334. [CrossRef]

10. Bartels, S.F.; Chen, H.Y.N. Is understory plant species diversity driven by resource quantity or resource heterogeneity? *Ecology* **2010**, *91*, 1931–1938. [CrossRef] [PubMed]

11. Laliberté, E.; Grace, J.B.; Huston, M.A.; Lambers, H.; Teste, F.P.; Turner, B.L.; Wardle, D.A. How does pedogenesis drive plant diversity? *Trends Ecol. Evol.* **2013**, *28*, 331–340. [CrossRef]

12. Rogers, R.S. Hemlock stands from Wisconsin to Nova Scotia: Transitions in understory composition along a floristic gradient. *Ecology* **1980**, *61*, 178–193. [CrossRef]

13. Rogers, R.S. Mature mesophytic hardwood forest: Community transitions, by layer, from east-central Minnesota to southeastern Michigan. *Ecology* **1981**, *62*, 1634–1647. [CrossRef]

14. Rogers, R.S. Early spring herb communities in mesophytic forests of the Great Lakes Region. *Ecology* **1982**, *63*, 1050–1063. [CrossRef]

15. Rogers, R.S. Annual variability in community organization of forest herbs: Effect of an extremely warm and dry early spring. *Ecology* **1983**, *64*, 1086–1091. [CrossRef]

16. Rogers, R.S. Local coexistence of deciduous-forest groundlayer species growing in different seasons. *Ecology* **1985**, *66*, 701–707. [CrossRef]

17. Reiners, W.A. Twenty years of ecosystem reorganization following experimental deforestation and regrowth suppression. *Ecol. Monogr.* **1992**, *62*, 503–523. [CrossRef]

18. Reich, P.B.; Frelich, L.E.; Voldseth, R.A.; Bakken, P.; Adair, E.C. Understorey diversity in southern boreal forests is regulated by productivity and its indirect impacts on resource availability and heterogeneity. *J. Ecol.* **2012**, *100*, 539–545. [CrossRef]

19. Kumar, P.; Chen, H.Y.N.; Thomas, S.C.; Shahi, C. Linking resource availability and heterogeneity to understorey species diversity through succession in boreal forest of Canada. *J. Ecol.* **2018**, *106*, 1266–1276. [CrossRef]

20. Kumar, P.; Chen, H.Y.N.; Searle, E.B.; Shahi, C. Dynamics of understorey biomass, production and turnover associated with long-term overstorey succession in boreal forest of Canada. *For. Ecol. Manag.* **2018**, *427*, 152–161. [CrossRef]

21. Clark, C.M.; Morefield, P.; Gilliam, F.S.; Pardo, L.H. Estimated losses of plant biodiversity across the U.S. from historical N deposition from 1985–2010. *Ecology* **2013**, *94*, 1441–1448. [CrossRef] [PubMed]

22. Sutton, M.A.; Mason, K.E.; Sheppard, L.J.; Sverdrup, H.; Haeuber, R.; Hicks, W.K. *Nitrogen Deposition, Critical Loads and Biodiversity: Proceedings of the International Nitrogen Initiatives Workshop, Linking Experts of the Convention on Long-Range Transboundary Air Pollution and the Convention on Biological Diversity*; Springer: New York, NY, USA, 2014; ISBN 978-94-007-7938-9.

23. Simkin, S.M.; Allen, E.B.; Bowman, W.D.; Clark, C.M.; Belnap, J.; Brooks, M.L.; Cade, B.S.; Collins, S.L.; Geiser, L.H.; Gilliam, F.S.; et al. A continental analysis of ecosystem vulnerability to atmospheric nitrogen deposition. *Proc. Natl. Acad. Sci. USA* **2016**, *113*, 4086–4091. [CrossRef]

24. Gilliam, F.S.; Welch, N.T.; Phillips, A.H.; Billmyer, J.H.; Peterjohn, W.T.; Fowler, Z.K.; Walter, C.; Burnham, M.; May, J.D.; Adams, M.B. Twenty-five year response of the herbaceous layer of a temperate hardwood forest to elevated nitrogen deposition. *Ecosphere* **2016**, *7*, e01250. [CrossRef]

25. Gilliam, F.S. Response of the herbaceous layer of forest ecosystems to excess nitrogen deposition. *J. Ecol.* **2006**, *94*, 1176–1191. [CrossRef]

26. Walter, C.A.; Raiff, D.T.; Burnham, M.B.; Gilliam, F.S.; Adams, M.B.; Peterjohn, W.T. Nitrogen fertilization interacts with light to increase *Rubus* spp. cover in a temperate forest. *Plant Ecol.* **2016**, *217*, 421–430. [CrossRef]

27. Lefsky, M.A.; Cohen, W.B.; Parker, G.G.; Harding, D.J. Lidar remote sensing for ecosystem studies. *Bioscience* **2002**, *52*, 19–30. [CrossRef]

28. Walter, C.A.; Burnham, M.B.; Gilliam, F.S.; Peterjohn, W.T. A reference-based approach for measuring the cover of forest herbs. *Environ. Monit. Assess.* **2015**, *187*, 657. [CrossRef] [PubMed]

29. Gilliam, F.S.; Walter, C.A.; Adams, M.B.; Peterjohn, W.T. Nitrogen (N) dynamics in the mineral soil of a Central Appalachian hardwood forest during a quarter century of whole-watershed N additions. *Ecosystems* **2018**, *21*, 1489–1504. [CrossRef]

30. Parker, G.G.; Harding, D.J.; Berger, M. A portable LIDAR system for rapid determination of forest canopy structure. *J. Appl. Ecol.* **2004**, *41*, 755–767. [CrossRef]

31. MacArthur, R.H.; Horn, H.S. Foliage profiles by vertical measurements. *Ecology* **1969**, *50*, 802–804. [CrossRef]

32. Zar, J.H. *Biostatistical Analysis*, 5th ed.; Prentice-Hall: Upper Saddle River, NJ, USA, 2009; ISBN 0-13-100846-3.

33. Pebesma, E.J. Multivariable geostatistics in S: The gstat package. *Comput. Geosci.* **2004**, *30*, 683–691. [CrossRef]

34. Kochenderfer, J.N. Fernow and the Appalachian hardwood region. In *The Fernow Watershed Acidification Study*; Adams, M.B., DeWalle, D.R., Hom, J.L., Eds.; Springer: Dordrecht, The Netherlands, 2006; Chapter 2, pp. 17–40, ISBN 1-4020-4614-6.

35. Gilliam, F.S.; Yurish, B.M.; Adams, M.B. Temporal and spatial variation of nitrogen transformations in nitrogen-saturated soils of a Central Appalachian hardwood forest. *Can. J. For. Res.* **2001**, *31*, 1768–1785. [CrossRef]

36. Gilliam, F.S.; May, J.D.; Adams, M.B. Response of foliar nutrients of *Rubus allegheniensis* to nutrient amendments in a central Appalachian hardwood forest. *For. Ecol. Manag.* **2018**, *411*, 101–107. [CrossRef]

37. Woodland, D.W. Biosystematics of the perennial North American taxa *Urtica*. II. Taxonomy. *Syst. Bot.* **1982**, *7*, 282–290. [CrossRef]

38. Hoshizaki, B.J.; Wilson, K.A. The cultivated species of the fern genus *Dryopteris* in the United States. *Am. Fern J.* **1999**, *89*, 1–98. [CrossRef]

39. Roberts, M.R. Response of the herbaceous layer to natural disturbance in North American forests. *Can. J. Bot.* **2004**, 1273–1283. [CrossRef]

40. Gilliam, F.S. Effects of excess nitrogen deposition on the herbaceous layer of eastern North American forests. In *The Herbaceous Layer in Forests of Eastern North America*, 2nd ed.; Gilliam, F.S., Ed.; Oxford University Press: New York, NY, USA, 2014; pp. 497–508, ISBN 978-0-19-983765-6.

41. Verheyen, K.; Baeten, L.; De Frenne, P.; Bernhardt-Römermann, M.; Brunet, J.; Cornelis, J.; Decocq, G.; Dierschke, H.; Hédl, R.; Heinken, T.; et al. Driving factors behind the eutrophication signal in understorey plant communities of deciduous temperate forests. *J. Ecol.* **2012**, *99*, 352–365. [CrossRef]

42. Dirnböck, T.; Grandin, U.; Bernhardt-Römermann, M.; Beudert, B.; Canullo, R.; Forsius, M.; Grabner, M.-T.; Holmberg, M.; Kleemola, S.; Lundin, L.; et al. Forest floor vegetation response to nitrogen deposition in Europe. *Glob. Chang. Biol.* **2014**, *20*, 429–440. [CrossRef] [PubMed]

43. Roth, T.; Kohli, L.; Rihm, B.; Amrhein, V.; Achermann, B. Nitrogen deposition and multidimensional plant diversity at the landscape scale. *R. Soc. Open Sci.* **2015**, *2*, 150017. [CrossRef] [PubMed]

44. De Schrijver, A.; de Frenne, P.; Ampoorter, E.; van Nevel, L.; Demey, A.; Wuyts, K.; Verheyen, K. Cumulative nitrogen inputs drives species loss in terrestrial ecosystems. *Glob. Ecol. Biogeogr.* **2011**, *20*, 803–816. [CrossRef]

45. Ferretti, M.; Marchetto, A.; Arisci, S.; Bussotti, F.; Calderisi, M.; Carnicelli, S.; Cecchini, G.; Fabbio, G.; Bertini, G.; Matteucci, G.; et al. On the tracks of nitrogen deposition effects on temperate forests at their southern European range—An observational study from Italy. *Glob. Chang. Biol.* **2014**, *20*, 3423–3438. [CrossRef] [PubMed]

46. Walter, C.A.; Adams, M.B.; Gilliam, F.S.; Peterjohn, W.T. Non-random species loss in a forest herbaceous layer following nitrogen addition. *Ecology* **2017**, *98*, 2322–2332. [CrossRef] [PubMed]

47. Du, E.; de Vries, W.; Galloway, J.N.; Hu, X.; Fang, J. Changes in wet nitrogen deposition in the United States between 1985 and 2012. *Environ. Res. Lett.* **2014**, *9*, 095004. [CrossRef]

48. Du, E. Rise and fall of nitrogen deposition in the United States. *Proc. Natl. Acad. Sci. USA* **2016**, *113*, E3594–E3595. [CrossRef]

49. Gilliam, F.S.; Burns, D.A.; Driscoll, C.T.; Frey, S.D.; Lovett, G.M.; Watmough, S.A. Decreased atmospheric nitrogen deposition in eastern North America: Predicted responses of forest ecosystems. *Environ. Pollut.* **2019**, *244*, 560–574. [CrossRef] [PubMed]

© 2019 by the author. Licensee MDPI, Basel, Switzerland. This article is an open access article distributed under the terms and conditions of the Creative Commons Attribution (CC BY) license (http://creativecommons.org/licenses/by/4.0/).

forests

MDPI

Communication

Species-Rich National Forests Experience More Intense Human Modification, but Why?

R. Travis Belote

The Wilderness Society, Bozeman, MT 59715, USA; travis_belote@tws.org; Tel.: +1-406-581-3808

Received: 6 November 2018; Accepted: 30 November 2018; Published: 4 December 2018

Abstract: Ecologists have studied geographic gradients in biodiversity for decades and recently mapped the intensity of the "human footprint" around the planet. The combination of these efforts have identified some global hotspots of biodiversity that are heavily impacted by human-caused land cover change and infrastructure. However, other hotspots of biodiversity experience less intense modifications from humans. Relationships between species diversity and the human footprint may be driven by covarying factors, like climate, soils, or topography, that coincidentally influence patterns of biodiversity and human land use. Here, I investigated relationships between tree species richness and the degree of human modification among Forest Service ranger districts within the contiguous US. Ranger districts with more tree species tended to experience greater human modification. Using data on climate, soils, and topography, I explored mechanisms explaining the positive relationship between tree richness and human modification. I found that climate is related to both tree richness and human modification, which may be indirectly mediated through climate's role governing productivity. Ranger districts with more productive climates support more species and greater human modification. To explore potential conservation consequences of these relationships, I also investigated whether the amount of area designated within highly protected conservation lands were related to climate, productivity, and topography. Less productive ranger districts with steeper slopes tended to experience the greatest relative amounts of conservation protection. Combined, these results suggest that complex relationships explain the geographic patterns of biodiversity and the human footprint, but that climate and topography partially govern patterns of each.

Keywords: biodiversity; climate; human footprint; productivity; topography; USDA Forest Service

1. Introduction

Some forests on Earth support hundreds of different species of trees within a given-sized area (e.g., wet tropical forests), while others only support one species within the same-sized area (e.g., boreal forests) [1]. This variability in the geography of species diversity has intrigued biologists for at least two centuries, and various hypotheses have been proposed to explain gradients in diversity [2]. At more local scales, ecologists also have been studying the consequences of variability in species diversity among sites. Do sites rich with species diversity function differently than sites with fewer species? Are more diverse forests more resistant or resilient to disturbance and drought compared to less diverse forests (e.g., [3,4])? These kinds of questions form the basis of active research programs and have resulted in important insights into the effects of species diversity on ecosystem functions [5]. Acknowledging the role species diversity plays in the functioning of ecosystems influences policy and management of wildland and agricultural ecosystems. For instance, the very mission of the United States Department of Agriculture Forest Service (USFS), which manages 780,000 km^2 of land, is "to sustain the health, *diversity*, and productivity of the Nation's forests and grasslands to meet the needs of present and future generations" (emphasis added).

Human use of forests to harvest food, fuel, and fiber, as well as the transportation and infrastructure needed to support economies, has altered forested landscapes around the globe [6].

Collectively, human alteration of ecosystems has been described as the "human footprint" [7]. Use of this concept has allowed geographers and ecologists to map the degree of human footprint and by converse—wildlands [8] or wildness [9]. In most regions of the globe, the human footprint has expanded resulting in loss of the planet's wildlands at accelerating rates [10]. In some cases the human footprint is concentrated in global biodiversity hotspots [7], leaving less diverse regions more wild and undeveloped [10]. The circumboreal taiga forests and tundra, for instance, are relatively wild, but with fewer overall species compared to some places on Earth. Central America, parts of Africa, and Southeast Asia are rich with species and also subjected to an increasing human footprint. Some species-rich regions retain a relatively minimal human footprint (e.g., the Amazon River Basin), but have—in recent years—experienced an increase in human impact [7]. These patterns are concerning and indicate regions in need of conservation to monitor and mitigate human impacts to the most diverse places. The collective impacts of the human footprint on species populations and overall biodiversity result from individual and combined effects of habitat fragmentation, direct harvesting of individuals, vectors for invasive species, among others [11,12].

While relationships between gradients of biodiversity and the human footprint are clear in some regions, the reasons why species rich areas tend to experience a greater degree of human impact are intriguing and somewhat equivocal [13]. Climates favorable to many species tend to be agriculturally productive, but given the limitations of soil resource availability—this is not always the case [13]. Hot humid tropical environments rich in species also can have highly weathered soils that limit potential agricultural productivity [14]. Moreover, actual productivity of biomass and species diversity are sometimes—but not always—positively correlated within and across regions [15]. Therefore, the relationships between climate, species diversity, and the composite human footprint are complex and require further study. This is especially true given the potential conservation implications of biodiversity hotspots being more heavily modified by humans [16].

With these questions in mind, I acquired spatial data on the degree of human modification, species diversity (tree richness), climate, and soils to investigate patterns among ranger districts of USFS lands in the contiguous United States. First, I asked whether USFS ranger districts with the greatest tree species richness experience more intense impacts by humans. Or, is the human footprint randomly distributed among ranger districts with respect to richness of tree species? I then assessed relationships among climate, potential soil productivity, net primary productivity, and topography to investigate environmental conditions that may give rise to observed patterns between tree species richness and human impacts. For instance, are relationships between patterns of species richness and human impacts driven by covarying factors like climate, soils, and topography, or potentially mediated by gradients in forest productivity? Humans may tend to more intensively modify areas rich in species if climatic and topographic conditions that support species-rich areas are also more sought after for agricultural production or other human infrastructure. Finally, I investigated how ranger districts varied in their degree of conservation protection (e.g., wilderness areas) and whether the amount of protection varied along gradients in productivity and topography. I was interested in whether the percentage of ranger districts protected was related to topography or productivity. I predicted that ranger districts with steeper slopes and less productive forests may experience the greatest amount of conservation protection, as these lands are less politically contested for resource extraction (e.g., timber) than highly productive forests with gentler topography [17].

2. Materials and Methods

I focus attention on the 492 USFS ranger districts because their management is governed by a consistent set of laws and regulations, and they are distributed across varying gradients of climate, forest types, and regions (Figure 1). Understanding variability in human modification and species diversity among national forests provides a geographically dispersed and convenient case study for assessing patterns of climate, human modification, and species richness (Figure 1). Results from such a study could also help place local ranger districts into a broader national context when implementing conservation plans within and among national forests.

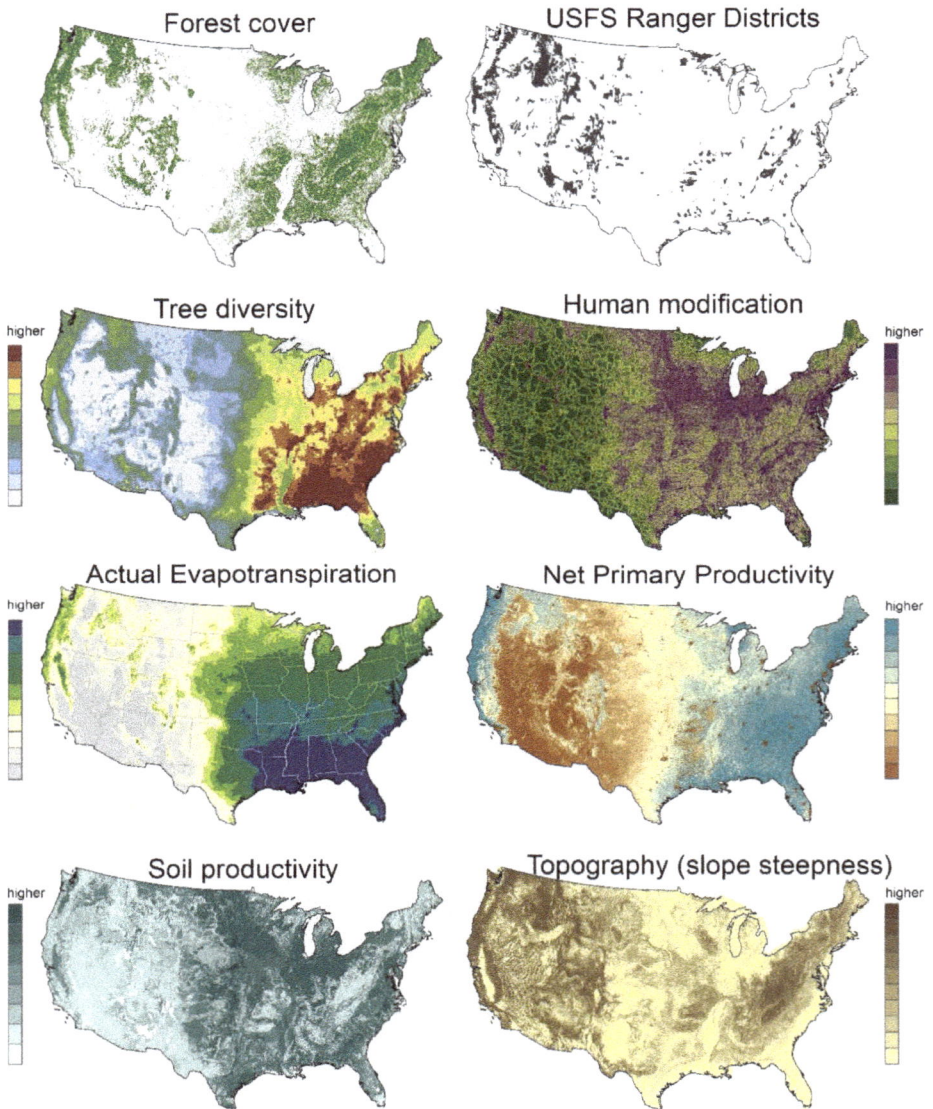

Figure 1. Spatial data used to analyze relationships between tree species richness, the degree of human modification, climate, vegetation productivity, soil productivity, and topography. Forest cover is shown for reference.

I obtained data from various sources and brought them into a geographic information system (GIS) to calculate average values of tree richness, human modification, climate, and protected areas for each ranger district (Figure 1). Data on the spatial location and distribution of administrative boundaries of ranger districts were obtained from the USFS geodata clearinghouse (https://data.

fs.usda.gov/geodata/). I obtained tree richness estimates from Jenkins et al. (2015) [18], which represents a 10-km resolution mapped gridded dataset representing the number of overlapping tree species based on distribution maps of Little (1971) [19]. Non-native tree species were not included in this dataset. I obtained human modification data from Theobald 2013 [20], which is similar to the global human footprint data [21] but is higher resolution and available for the contiguous US. This data layer is a 270-m resolution composite gridded dataset that ranges from 0 to 1 representing mapped impacts to ecosystems, including developed land cover, land use (including recent timber harvests), roads, transmission lines, and structures. Climate data were obtained from Dobrowski et al. (2013) [22]. I used estimates of average actual evapotranspiration (AET) and annual water deficit, as they integrate temperature and precipitation with topography and latitude to represent climate gradients important for plant growth and the distribution of vegetation types [23]. Net primary productivity data representing modeled annual average g C m^{-2} year^{-1} between the years 2000 and 2012 were obtained from the Numerical Terradynamic Simulation Group (NTSG) at the University of Montana [24]. Predicted soil productivity data were developed by Scott et al. (2001) [25] and used by Aycrigg et al. (2013) [26]. The dataset is an ordinal composite score based on five soil factors that influences fertility. I also used a mapped layer of estimated steepness of topography (slope in degrees) that was calculated from a 30-m resolution digital elevation model. Protected area data were obtained from the Conservation Biology Institute [27].

After bringing data into a GIS, I calculated the mean values of tree richness, human modification, AET, soil productivity, and slope steepness for each ranger district ($N = 492$). I also calculated the percentage of each ranger district that is protected in Gap Analysis Program (GAP) 1 or 2 status lands of conservation areas. GAP status values range from 1 to 4 and represent the degree of conservation protections. GAP 1 and 2 are the highest degrees of protections and include designated wilderness areas and national parks, or other designations that mandate protection of biodiversity, prevention of land cover conversion, and with strict limitations on commercial extractive activities (mining and timber harvests).

I began the analysis by producing a scatterplot of average tree richness and human modification for each ranger district and running a simple linear model to describe the relationship among ranger districts. After finding a significant positive relationship where the ranger districts with the greatest species richness tended to also be the most impacted by humans, I developed a path diagram with hypothesized relationships between climate, soils, topography, productivity, tree richness, and human impacts (Figure 2) to describe possible mechanisms explaining patterns. I considered relationships between exogenous variables (climate, soils, and topography) and endogenous variables (productivity, human modification, and species richness). I purposefully left out the connection between tree richness and human modification in the path diagram, to instead explore potential underlying explanations for the positive relationship between tree richness and human modification. I also included the percentage of land within a protected area in the path diagram to test my predictions concerning conservation protections on ranger districts with varying topographic steepness and productivity. I did not include linkages among all variables. In some cases, variables could be hypothetically linked at finer spatial scales than considered in my assessment. Human modification may, for instance, influence species richness, but I did not include this because of the coarse resolution of the species data. At stand-scales human land use via timber harvesting may influence species composition and diversity [28], but I was interested in broader biogeographic patterns and was limited to coarse data based on continental range maps of trees.

After developing the path diagram, I standardized variables by converting them to z-scores based on their distribution [29] and subjected the path diagram to a structural equation model using the R package "lavaan" [30], relying on fit metrics and parameter estimates to interpret relationships among variables. Specifically, I evaluated the X^2 model fit test statistic (where p-value > 0.05 suggest goodness of fit), comparative fit index (CFI) and Tucker-Lewis Index (TFI), the root mean square error of approximation (RMSEA), and the standardized root mean square residual (SRMS) [31]. I also evaluated the parameter estimates for each path in the model and plotted the significant results in an updated path diagram highlighting positive and negative relationships and their relative values. I first included both AET and water deficit in a latent variable "climate", which did not improve overall model fit. Instead, for my final model, I used AET alone, which is correlated with water deficit ($r = -0.62$, $p < 0.001$) and resulted in a better overall model fit.

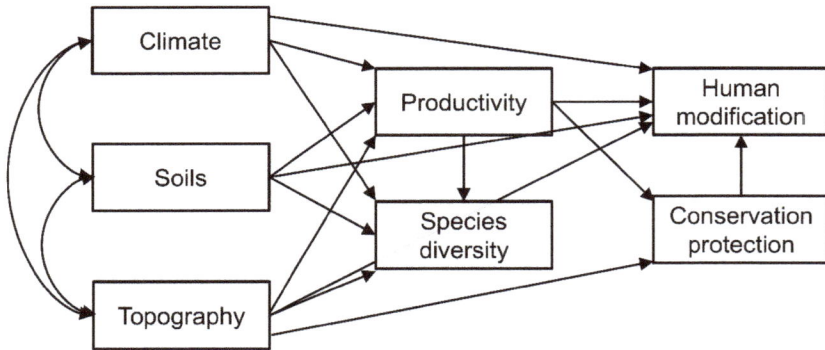

Figure 2. Hypothetical path diagram depicting relationships between climate, soils, topography and their direct or indirect influence on productivity, species diversity (tree richness), the degree of human modification, and conservation protection of United States Department of Agriculture Forest Service (USFS) ranger districts.

3. Results

Ranger districts with more predicted tree species tended to be more heavily modified by humans compared to ranger districts with fewer species (Figure 3). For instance, the predicted degree of human modification roughly doubles from ranger districts with 30 estimated species to those with 90 species (i.e., slope of relationship = 0.0034, $p < 0.0001$).

The proposed path model fit well the structure of the data ($X^2 = 1.48$; p-value = 0.687; d.f. = 3; RMSEA = 0 with 90% confidence intervals of 0 and 0.058; SRMR = 0.009; CFI = 1.0; TLI = 1.0; Figure 4; Table 1). A p-value > 0.05 from the X^2 test provides support for the proposed structure in the path model [29]. Endogenous modeled relationships varied in their degree of explanatory power with models of tree richness and human modification having the highest R^2 values of 0.815 and 0.621, respectively. R^2 values for paths to productivity and conservation protection were lower at 0.286 and 0.170, respectively.

The strength and nature (positive or negative) of relationships in the proposed path model varied from non-significant, positive, and negative (Figure 4; Table 1). For instance, AET was positively related to productivity, species richness, and the degree of human modification. Specifically, ranger districts with higher AET were more productive, species rich, and heavily modified. Not surprisingly, estimated soil productivity was positively related to observed productivity. Perhaps more surprisingly, the path coefficient between soil productivity and human modification was not related among ranger districts. Steepness of slope was positively related to productivity and conservation protection, but negatively related to human modification. Ranger districts with steeper slopes tended to be more productive. In general more mountainous districts captured more carbon. However, more topographically steep ranger districts also have more of their area in GAP 1 or 2 lands. Productivity was positively associated with species diversity and human modification, but negatively related to conservation protection. More productive ranger districts tended to be more diverse and more heavily modified and have less relative area in GAP 1 or 2 lands.

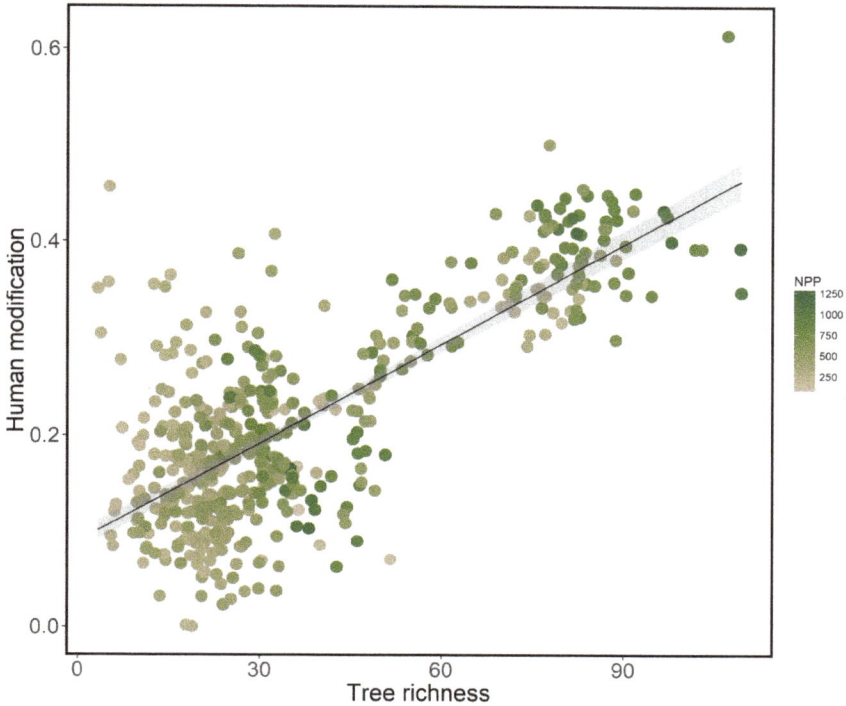

Figure 3. Relationship between average tree richness (predicted number of species based on overlapping species within mapped 10-km grid cells) and the average degree of human modification of USFS ranger districts located in the contiguous US (*N* = 492). I added vegetation net primary productivity (NPP; g C m^{-2} year^{-1}) as a color ramp to illustrate additional patterns.

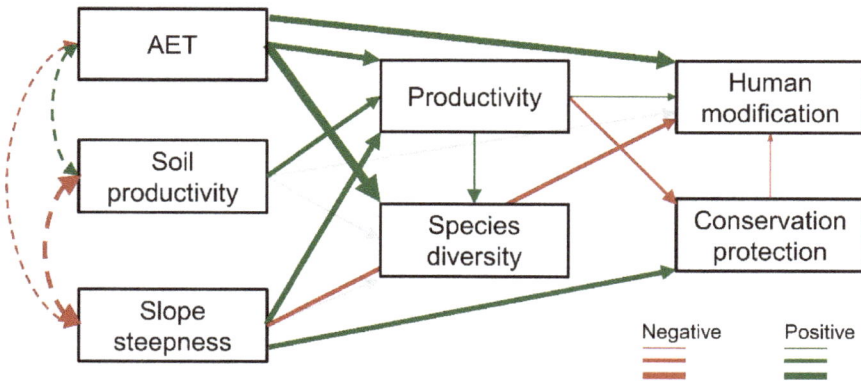

Figure 4. Results of path analysis from the hypothetical path model presented in Figure 2. Significant relationships in path coefficients are highlighted in either red (negative estimates) or green (positive estimates) with the weight of the line varying with the size of the estimates (Table 1). Grey lines represent non-significant paths. Dashed lines show correlations between exogenous variables. AET = actual evapotranspiration, an estimate of climates favorable to plant growth; NPP = net primary productivity.

Table 1. Estimates of coefficients for relationships between factors in the proposed path diagram shown in Figure 1. Results are plotted in Figure 4.

	Estimate	Standard Error	z-Value	p-Value
Human modification				
Conservation protection	−0.069	0.028	−2.425	0.015
Slope steepness	−0.248	0.045	−5.54	<0.001
Productivity	0.096	0.033	2.892	0.004
AET	0.594	0.034	17.274	<0.001
Soil productivity	−0.004	0.042	−0.095	0.925
Conservation protection				
Productivity	−0.194	0.041	−4.719	<0.001
Slope steepness	0.375	0.041	9.118	<0.001
Productivity				
AET	0.472	0.042	11.175	<0.001
Slope steepness	0.489	0.055	8.82	<0.001
Soil productivity	0.339	0.056	6.081	<0.001
Species richness				
Productivity	0.133	0.023	5.79	<0.001
AET	0.843	0.024	35.066	<0.001
Soil productivity	0.007	0.029	0.227	0.82
Slope steepness	0.019	0.03	0.617	0.537

4. Discussion

Why some places are more biologically diverse than others and why some places are more heavily impacted by humans than others are two important questions explored by ecologists, geographers, and conservation scientists. USFS ranger districts richest in tree species tended to be more modified by humans. Species-rich ranger districts are closer to concentrations of high human population density, developed lands, and infrastructure like roads and utility transmission lines compared to less species rich forests. This relationship—where the most diverse lands are most modified by humans—seems to be most strongly driven by climate. Climatic conditions related to high species richness are also associated with lands more heavily modified by humans. However, soils and topography, in some cases via their influence on productivity, also seem to partially explain the relationship.

The most species rich and modified ranger districts occur in the Southeastern US. Most of the eastern national forests were obtained by the US federal government from private land in the early 1900s after having been subjected to intensive and widespread tree harvesting by private individuals and timber companies [32]. In contrast, most western forest service lands were established from existing federal lands [33]. Therefore, the history of land use and management tenure differs between species-rich eastern and less-rich western forests. The indirect role that climate played in the history of land use is more difficult to investigate, but climate likely plays a role in geographic patterns of land use and settlement [34]. Areas with climates more favorable for crop production are typically more intensively managed with greater land cover conversions to agriculture and resulting human settlement and development of infrastructure [35]. Climate thus seems to at least partially influence patterns of human modification.

Climate also governs geographic gradients in species diversity [36]. Climate as a common variable influencing both patterns of land use and species diversity results in an indirect relationship between tree richness and the degree of human modification. The relationship between biodiversity and human footprint has been observed globally, with striking exceptions that reveal other limiting factors on human land use, namely topographic and soil-based limitations. At a global scale, climatic conditions favorable to tree diversity also have resulted in weathered soils that can be unproductive for agricultural use [13,14]. Therefore, in some areas on Earth, climate favorable to plant production and high species diversity are relatively unmodified by humans. In the case of ranger districts of national forests studied here, soil productivity—at least based on the dataset I used—was not directly related to human modification and species richness. Rather, soil productivity was indirectly related to tree richness and human modification via observed vegetation productivity. The most productive soils were more productive in terms of biomass accumulation, which has a positive relationship with tree richness and human modification. Not surprisingly, slope steepness was also related to human modification. Ranger districts with steeper slopes were less impacted by human modification. Similar patterns are observed globally, where the human footprint tends to be lower in areas characterized by soils with limited agricultural suitability [7].

Steepness of slope also predicted how much of each ranger district was protected in conservation reserves classified under GAP 1 or 2 status. The steeper the slopes across ranger districts, the higher proportion of land that was protected. This is also not surprising and has been observed in other studies [26], as steep slopes may be politically less contentious for legislative or administrative conservation protections that limit commercial resource extractions, like timber harvests. Forests with steeper slopes are more difficult and costly to harvest, and therefore represent lands more easily set-aside in reserves [17]. Interestingly, more productive ranger districts tend to have lower relative amounts of lands protected in GAP 1 or 2 conservation reserves. This pattern could be driven by the fact that less productive forests—like steep slopes—are less economically viable for commercial timber harvests. Productive forests are more likely to be sought after for active timber harvesting programs [17], making limitations to commercial extraction of those forests more politically difficult. An important area of ongoing conservation research seeks to identify lands rich with species that are currently under-represented in protected areas or areas that could serve to maintain large resilient landscapes [37,38]. There may be such opportunities in the species-rich and less well protected forests of the southeastern US [18,39].

While I have tried to uncover some biogeographic patterns of tree richness, human modification, climate, and topography, the conservation consequences of these patterns are less clear. Across national forests of the contiguous US, humans seem to have preferentially modified lands rich in species over less species rich lands. How the consequences of these modifications and their associated impacts to ecological composition, structure, and function also vary across gradients of productivity, diversity, climate, and topography may form the core of future research questions. If humans modify species-rich areas more than species poor areas, are more species exposed to detrimental impacts of human modifications, resulting in greater conservation concern? Alternatively, because species rich areas tend to be more productive, could these human-modified ecosystems have more capacity to absorb impacts and recover? Evidence supports both alternatives, depending on the metric used to assess the questions. Species-rich areas of the Southeastern US have functionally lost foundational species, like the American chestnut (*Castanea dentata* (Marshall) Borkh.), the result of an introduced fungus from Asia [40]. Tree species composition throughout the region continues to be altered by mortality from pests and pathogens, and the introduction of non-native trees [41,42]. Human pressures have also resulted in the regional extirpation of many vertebrate species [43]. The eastern forests have also shown high capacity to recover following widespread, intensive timber clearing and subsequent high-grade logging [44,45]. However, land use demands in recent years has reduced forest cover within ecoregions of the eastern US via increased development or forest harvesting [46].

While less species-rich ranger districts of the western US experience less human modification, these patterns could be the result of the history of land use that accompanied Euro-American settlement patterns, in addition to climatic influences of land use. Because western forests have not already been as modified as the eastern forests does not mean that they will not be modified if development of human infrastructure is left unmitigated. The expansion of the human footprint continues to increase across the west [47], including into climates and topographies that may have once limited development. Data presented here provide a national perspective on public forests, their degree of modification (and therefore their degree of wildness [38]), tree richness, and proportion protected in conservation reserves. Data summarized here could be used to identify valuable (wild and/or diverse) forests that may be priorities for additional conservation based on their limited levels of conservation areas. In other words, these data could identify valuable but insufficiently protected ranger districts (*sensu* [46]). In some cases, the most species rich ranger districts may be priorities for additional conservation protection [18,48]. In other cases, the least human modified (wildest) forests with minimal protection may be priorities for designating future conservation reserves [49,50]. The data summarized here can be used to identify the wildest remaining forests with relatively low levels of protections.

There are several important limitations and caveats with this analysis. First, data were summarized within USFS ranger districts, which are large (median area of ranger districts was 154,884 hectares). My aim was to explore broad biogeographic gradients in human modification, biodiversity, soils, topography, and climate among ranger districts. However, finer-scaled gradients of these factors could reveal important patterns and insights. Second, geographic patterns reflecting relationships between people and nature (i.e., social-ecological systems) are almost certainly more complex than what I investigated. Research using other data, including individual variables used to create composite scores of human modifications could yield interesting insights into which components of human modification (e.g., roads, human population density, etc.) are related to which climatic or topo-edaphic patterns. Third, I worked with the boundaries of ranger districts available through the USFS geodata clearinghouse. However, these maps represent administrative boundaries that do not always reflect ownership boundaries. Across all ranger districts, over 93% of administrative boundaries are also owned and managed by the USFS, but in the eastern US inholdings and land not yet purchased occur throughout the administrative boundaries. In some cases, private roads or homes are included in the administrative boundary for the ranger district. To explore the consequences of the different areas representing USFS lands, I produced the scatterplot between tree richness and human modification for ownership boundaries of ranger districts (from the protected areas database of the Conservation Biology Institute [27]), and found patterns matched those presented in Figure 2. Finally, I used relatively coarse estimates of tree richness based on maps of overlapping distributions of species. However, plot-level data (e.g., from the Forest Inventory and Analysis (FIA) program) could be used to estimate local to regional gradients in diversity, and whether patterns change depending on the scale that tree diversity is sampled [2,51].

5. Conclusions

I explored broad patterns of tree richness, human modification, climate, topography, soils, and conservation protection to try and uncover potential properties explaining why the most species diverse USFS ranger districts are most impacted by humans. Climate and other geophysical factors seem to simultaneously govern both tree richness and intensity of the degree of human modification. Understanding these kinds of geographic patterns in the biogeography of ecosystems and land use may provide important perspectives into the nature of social-ecological systems and insights into conservation of species-rich and relatively wild, unmodified places. The threats to biodiversity caused by intensive human modification in species-rich areas should form the basis of future global- and local-scale research. Areas rich with species that are relatively free of human modification, which I have found to be uncommon in the contiguous US, may be regarded as high priorities for additional conservation protections [18].

Funding: This research received no external funding beyond The Wilderness Society.

Acknowledgments: Thanks to Clinton Jenkins, Dave Theobald, Solomon Dobrowski, Jocelyn Aycrigg, Steve Running and NTSG, and others for making spatial data publicly available. Thanks to Aaron Ellison and Frank Gilliam for the invitation to contribute to this special issue. Finally, thanks to Greg Aplet for inspiration to explore these questions.

Conflicts of Interest: The author declares no conflict of interest.

References

1. Fine, P.V.A.; Ree, R.H. Evidence for a time-integrated species-area effect on the latitudinal gradient in tree diversity. *Am. Nat.* **2006**, *168*, 796–804. [CrossRef]
2. Keil, P.; Chase, J. Integrating global patterns and drivers of tree diversity across a continuum of spatial grains. *bioRxiv* **2018**. [CrossRef]
3. Anderegg, W.R.L.; Konings, A.G.; Trugman, A.T.; Yu, K.; Bowling, D.R.; Gabbitas, R.; Karp, D.S.; Pacala, S.; Sperry, J.S.; Sulman, B.N.; et al. Hydraulic diversity of forests regulates ecosystem resilience during drought. *Nature* **2018**, *561*, 538–541. [CrossRef] [PubMed]
4. Naeem, S.; Wright, J.P. Disentangling biodiversity effects on ecosystem functioning: Deriving solutions to a seemingly insurmountable problem. *Ecol. Lett.* **2003**, *6*, 567–579. [CrossRef]
5. Loreau, M.; Naeem, S.; Inchausti, P.; Bengtsson, J.; Grime, J.P.; Hector, A.; Hooper, D.U.; Huston, M.A.; Raffaelli, D.; Schmid, B.; et al. Biodiversity and ecosystem functioning: Current knowledge and future challenges. *Science* **2001**, *294*, 804–808. [CrossRef] [PubMed]
6. Haddad, N.M.; Brudvig, L.A.; Clobert, J.; Davies, K.F.; Gonzalez, A.; Holt, R.D.; Lovejoy, T.E.; Sexton, J.O.; Austin, M.P.; Collins, C.D.; et al. Habitat fragmentation and its lasting impact on Earth's ecosystems. *Sci. Adv.* **2015**, *1*, e1500052. [CrossRef]
7. Venter, O.; Sanderson, E.W.; Magrach, A.; Allan, J.R.; Beher, J.; Jones, K.R.; Possingham, H.P.; Laurance, W.F.; Wood, P.; Fekete, B.M.; et al. Sixteen years of change in the global terrestrial human footprint and implications for biodiversity conservation. *Nat. Commun.* **2016**, *7*, 1–11. [CrossRef]
8. Sanderson, E.W.; Jaiteh, M.; Levy, M.; Redford, K.H.; Wannebo, A.V.; Woolmer, G. The human footprint and the last of the wild. *Bioscience* **2002**, *52*, 891–904. [CrossRef]
9. Aplet, G.; Thomson, J.; Wilbert, M. Indicators of wildness: Using attributes of the land to assess the context of wilderness. In Proceedings of the Wilderness Science in a Time of Change, Missoula, MT, USA, 23–27 May 1999; McCool, S.F., Cole, D.N., Borrie, W.T., O'Laughlin, J., Eds.; 2000; RMRS-P-15, pp. 89–98.
10. Watson, J.E.M.; Shanahan, D.F.; Marco, M.D.; Allan, J.; Laurance, W.F.; Sanderson, E.W.; Mackey, B.; Venter, O. Catastrophic declines in wilderness areas undermine global environment targets. *Curr. Biol.* **2016**, *26*, 1–6. [CrossRef]
11. Pimm, S.L.; Jenkins, C.N.; Abell, R.; Brooks, T.M.; Gittleman, J.L.; Joppa, L.N.; Raven, P.H.; Roberts, C.M.; Sexton, J.O. The biodiversity of species and their rates of extinction, distribution, and protection. *Science* **2014**, *344*, 1246752. [CrossRef]
12. Wilcove, D.S.; Rothstein, D.; Dubow, J.; Phillips, A.; Losos, E. Quantifying threats to imperiled species in the United States. *Bioscience* **1998**, *48*, 607–615. [CrossRef]
13. Huston, M. Biological and Diversity, Economics, and Soils. *Science* **1993**, *262*, 1676–1680. [CrossRef] [PubMed]
14. Huston, M.A.; Wolverton, S. The global distribution of net primary production: Resolving the paradox. *Ecol. Monogr.* **2009**, *79*, 343–377. [CrossRef]
15. Belote, R.T.; Prisley, S.; Jones, R.H.; Fitzpatrick, M.; de Beurs, K. Forest productivity and tree diversity relationships depend on ecological context within mid-Atlantic and Appalachian forests (USA). *For. Ecol. Manag.* **2011**, *261*, 1315–1324. [CrossRef]
16. Jenkins, C.N.; Pimm, S.L.; Joppa, L.N. Global patterns of terrestrial vertebrate diversity and conservation. *Proc. Natl. Acad. Sci. USA* **2013**, *110*, E2602–E2610. [CrossRef] [PubMed]
17. Belote, R.T.; Aplet, G.H. Land protection and timber harvesting along productivity and diversity gradients in the Northern Rocky Mountains. *Ecosphere* **2014**, *5*. [CrossRef]
18. Jenkins, C.N.; Van Houtan, K.S.; Pimm, S.L.; Sexton, J.O. US protected lands mismatch biodiversity priorities. *Proc. Natl. Acad. Sci. USA* **2015**, *112*, 5081–5086. [CrossRef] [PubMed]
19. Little, E.L. *Atlas of United States Trees*; U.S. Department of Agriculture Forest Service: Washington, DC, USA, 1971.
20. Theobald, D.M. A general model to quantify ecological integrity for landscape assessments and US application. *Landsc. Ecol.* **2013**, *28*, 1859–1874. [CrossRef]
21. Venter, O.; Sanderson, E.W.; Magrach, A.; Allan, J.R.; Beher, J.; Jones, K.R.; Possingham, H.P.; Laurance, W.F.; Wood, P.; Fekete, B.M.; et al. Global terrestrial Human Footprint maps for 1993 and 2009. *Sci. Data* **2016**, 160067. [CrossRef]

22. Dobrowski, S.Z.; Abatzoglou, J.; Swanson, A.K.; Greenberg, J.A.; Mynsberge, A.R.; Holden, Z.A.; Schwartz, M.K. The climate velocity of the contiguous United States during the 20th century. *Glob. Chang. Biol.* **2013**, *19*, 241–251. [CrossRef]

23. Stephenson, N.L. Climatic control of vegetation distribution: The role of the water balance. *Am. Nat.* **1990**, *135*, 649–670. [CrossRef]

24. Zhao, M.; Heinsch, F.A.; Nemani, R.R.; Running, S.W. Improvements of the MODIS terrestrial gross and net primary production global data set. *Remote Sens. Environ.* **2005**, *95*, 164–176. [CrossRef]

25. Scott, J.M.; Davis, F.W.; McGhie, R.G.; Wright, R.G.; Groves, C.; Estes, J. Nature reserves: Do they capture the full range of America's biological diversity? *Ecol. Appl.* **2001**, *11*, 999–1007. [CrossRef]

26. Aycrigg, J.L.; Davidson, A.; Svancara, L.K.; Gergely, K.J.; McKerrow, A.; Scott, J.M. Representation of ecological systems within the protected areas network of the continental United States. *PLoS ONE* **2013**, *8*, e54689. [CrossRef] [PubMed]

27. The Conservation Biology Institute. *PAD-US (CBI Edition) Version 2.1*; The Conservation Biology Institute: Corvallis, OR, USA, 2015.

28. Belote, R.T.; Jones, R.H.; Hood, S.M.; Wender, B.W. Diversity-invasibility across an experimental disturbance gradient in Appalachian forests. *Ecology* **2008**, *89*. [CrossRef]

29. Grace, J.B. *Structural Equation Modeling and Natural Systems*; Cambridge University Press: Cambridge, UK, 2006.

30. Rosseel, Y. lavaan: An R package for structural equation modeling. *J. Stat. Softw.* **2012**, *48*, 1–36. [CrossRef]

31. Barrett, P. Structural equation modelling: Adjudging model fit. *Pers. Individ. Differ.* **2007**, *42*, 815–824. [CrossRef]

32. Shands, W.E.; Healy, R.G. *The Lands Nobody Wanted: Policy for National Forests in the Eastern United States*; Conservation Foundation: Washington, DC, USA, 1977.

33. Hessburg, P.F.; Agee, J.K. An Environmental Narrative of Inland Northwest United States Forests, 1800–2000. *For. Ecol. Manag.* **2003**, *178*, 23–59. [CrossRef]

34. Lieberman, B.; Gordan, E. *Climate Change in Human History: Prehistory to the present*; Bloomsbury Academic: London, UK, 2018; ISBN 9781472598493.

35. Diamond, J.M. *Guns, Germs and Steel: A Short History of Everybody for the Last 13,000 Years*; Random House: New York, NY, USA, 1998; ISBN 0099302780.

36. Francis, A.P.; Currie, D.J. A globally consistent richness-climate relationship for angiosperms. *Am. Nat.* **2003**, *161*, 523–536. [CrossRef]

37. Dietz, M.S.; Belote, R.T.; Aplet, G.H.; Aycrigg, J.L. The world's largest wilderness protection network after 50 years: An assessment of ecological system representation in the U.S. National Wilderness Preservation System. *Biol. Conserv.* **2015**, *184*, 431–438. [CrossRef]

38. Belote, R.T.; Dietz, M.S.; Jenkins, C.N.; McKinley, P.S.; Irwin, G.H.; Fullman, T.J.; Leppi, J.C.; Aplet, G.H. Wild, connected, and diverse: Building a more resilient system of protected areas. *Ecol. Appl.* **2017**, *27*, 1050–1056. [CrossRef] [PubMed]

39. Noss, R.F.; Platt, W.J.; Sorrie, B.A.; Weakley, A.S.; Means, D.B.; Costanza, J.; Peet, R.K. How global biodiversity hotspots may go unrecognized: Lessons from the North American Coastal Plain. *Divers. Distrib.* **2015**, *21*, 236–244. [CrossRef]

40. Ellison, A.M.; Bank, M.S.; Clinton, B.D.; Colburn, E.A.; Elliott, K.; Ford, C.R.; Foster, D.R.; Kloeppel, B.D.; Knoepp, J.D.; Lovett, G.M.; et al. Loss of foundation species: Consequences for the structure and dynamics of forested ecosystems. *Front. Ecol. Environ.* **2005**, *3*, 479–486. [CrossRef]

41. Lovett, G.M.; Canham, C.D.; Arthur, M.A.; Weathers, K.C.; Fitzhugh, R.D. Forest Ecosystem Responses to Exotic Pests and Pathogens in Eastern North America. *Bioscience* **2006**, *56*, 395–405. [CrossRef]

42. Riitters, K.; Potter, K.; Iannone, B.; Oswalt, C.; Guo, Q.; Fei, S.; Riitters, K.; Potter, K.M.; Iannone, B.V.; Oswalt, C.; et al. Exposure of protected and unprotected forest to plant invasions in the eastern United States. *Forests* **2018**, *9*, 723. [CrossRef]

43. Faurby, S.; Svenning, J.C. Historic and prehistoric human-driven extinctions have reshaped global mammal diversity patterns. *Divers. Distrib.* **2015**, *21*, 1155–1166. [CrossRef]

44. Houghton, R.A.; Hackler, J.L. Changes in terrestrial carbon storage in the United States. 1: The roles of agriculture and forestry. *Glob. Ecol. Biogeogr.* **2000**, *9*, 125–144. [CrossRef]

45. Belote, R.T.; Jones, R.H.; Wieboldt, T.F. Compositional stability and diversity of vascular plant communities following logging disturbance in Appalachian forests. *Ecol. Appl.* **2012**, *22*. [CrossRef]
46. Drummond, M.A.; Loveland, T.R. Land-use pressure and a transition to forest-cover loss in the Eastern United States. *Bioscience* **2010**, *60*, 286–298. [CrossRef]
47. Theobald, D.M.; Zachmann, L.J.; Dickson, B.G.; Gray, M.E.; Albano, C.M.; Landau, V.; Harrison-Atlas, D. *The Disappearing West: Description of the Approach, Data, and Analytical Methods Used to Estimate Natural Land Loss in the Western U.S.*; Conservation Science Partners: Truckee, CA, USA, 2016.
48. Belote, R.T.; Irwin, G.H. Quantifying the national significance of local areas for regional conservation planning: North Carolina's Mountain Treasures. *Land* **2017**, *6*, 35. [CrossRef]
49. Kareiva, P.; Marvier, M. Conserving biodiversity coldspots. *Am. Sci.* **2003**, *91*, 344–351. [CrossRef]
50. Watson, J.E.M.; Venter, O.; Lee, J.; Jones, K.R.; Robinson, J.G.; Possingham, H.P.; Allan, J.R. Protect the last of the wild. *Nature* **2018**, *563*, 27–30. [CrossRef] [PubMed]
51. LaManna, J.A.; Belote, R.T.; Burkle, L.A.; Catano, C.P.; Myers, J.A. Negative density dependence mediates biodiversity-productivity relationships across scales. *Nat. Ecol. Evol.* **2017**, *1*. [CrossRef] [PubMed]

© 2018 by the author. Licensee MDPI, Basel, Switzerland. This article is an open access article distributed under the terms and conditions of the Creative Commons Attribution (CC BY) license (http://creativecommons.org/licenses/by/4.0/).

forests

MDPI

Article

Contrasting Species Diversity and Values in Home Gardens and Traditional Parkland Agroforestry Systems in Ethiopian Sub-Humid Lowlands

Eguale Tadesse [1,2], Abdu Abdulkedir [2], Asia Khamzina [1], Yowhan Son [1,*] and Florent Noulèkoun [1]

[1] Department of Environmental Science and Ecological Engineering, Korea University, 145 Anam-ro, Seongbuk-gu, Seoul 02841, Korea; eguale97@gmail.com (E.T.); asia_khamzina@korea.ac.kr (A.K.); bigflo@korea.ac.kr (F.N.)

[2] Ethiopian Environment and Forest Research Institute (EEFRI), 24536 Addis Ababa, Ethiopia; aabdelkadir@yahoo.com

* Correspondence: yson@korea.ac.kr; Tel.: +82-2-3290-3015

Received: 5 February 2019; Accepted: 11 March 2019; Published: 15 March 2019

Abstract: Understanding the complex diversity of species and their potential uses in traditional agroforestry systems is crucial for enhancing the productivity of tropical systems and ensuring the sustainability of the natural resource base. The aim of this study is the evaluation of the role of home gardens and parklands, which are prominent tropical agroforestry systems, in the conservation and management of biodiversity. Our study quantified and compared the diversity of woody and herbaceous perennial species and their uses in traditional home gardens and parkland agroforestry systems under a sub-humid climate in western Ethiopia. A sociological survey of 130 household respondents revealed 14 different uses of the species, mostly for shade, fuelwood, food, and as traditional medicine. Vegetation inventory showed that the Fisher's α diversity index and species richness were significantly higher in home gardens (Fisher's α = 5.28 ± 0.35) than in parklands (Fisher's α = 1.62 ± 0.18). Both systems were significantly different in species composition (Sørenson's similarity coefficient = 35%). The differences occurred primarily because of the high intensity of management and the cultivation of exotic tree species in the home gardens, whereas parklands harbored mostly native flora owing to the deliberate retention and assisted regeneration by farmers. In home gardens, *Mangifera indica* L. was the most important woody species, followed by *Cordia africana* Lam. and *Coffea arabica* L. On the other hand, *Syzygium guineense* Wall. was the most important species in parklands, followed by *C. africana* and *M. indica*. The species diversity of agroforestry practices must be further augmented with both indigenous and useful, non-invasive exotic woody and herbaceous species, particularly in parklands that showed lower than expected species diversity compared to home-gardens.

Keywords: herbaceous perennial species; household respondents; questionnaire survey; species richness; woody species

1. Introduction

Tropical forests are vital hosts of global biodiversity as they support approximately two-thirds of all known species and contain 65% of the world's endangered species [1]. The removal or destruction of forest cover resulted in significant losses of tropical and global biodiversity, owing to the destruction of forest-based habitats and species [2]. Tree cover continuously decreased in the tropics for the past 17 years [3]. Deforestation is primarily a concern for developing countries in the tropics [4,5], where significant agricultural demand for land is often coupled with a lack of economic incentives for forest conservation [6]. For example, in Ethiopia, the rapid expansion of agricultural land and the degradation of forests are associated with rapid human population growth (2.5% per year), with the population largely depending on extensive agriculture. Ethiopian animal husbandry, in particular, is characterized by the largest number of livestock in Africa, and increasingly claims land and forest resources [7]. The reduction in forest cover and loss of biodiversity, particularly through deforestation, could activate abrupt, irreversible, and harmful changes, including regional climate change, degradation of rainforests to savannas, emergence of new pathogens [1], extinction of native flora and fauna species, and the displacement of indigenous people [8,9].

Agroforestry practices might be able to reconcile the needs for food production and biodiversity conservation via the integration of trees, shrubs, crops, or animals in the same system [10,11] by the provision of habitats for edge species [12], conservation of remnant native species and their gene pools [11,13], provision of corridors and stepping stones for persistence and movement of flora and fauna species by linking fragmented habitats [10,12], erosion control and water recharge, and buffering the logging pressure on the surrounding natural forest. Generally, from the viewpoint of recurrent food shortages, projected climate change, and increasing prices of fossil fuels, agroforestry is attracting increasing interest from the research and development communities as a cost-effective way to enhance food security, while simultaneously contributing to climate change adaptation and mitigation [14]. In Ethiopia, agroforestry was credited as a sustainable farming practice that uses and conserves biodiversity and limits agricultural expansion into natural forests [15]. However, this circa situm (farm-based) conservation of biodiversity was only recently advocated by the Convention on Biological Diversity [10,16,17].

The tropical agroforestry systems including those in Ethiopia are indicative of the complex, multi-layer structure of the natural forest with a rich plant diversity [18] and are shaped by deliberate planting or retention, and assisted regeneration of useful woody species [18,19]. These traditional agroforestry systems represent a valuable source of genetic resources, in addition to the natural and planted forests [20]. Among the tropical agroforestry systems, home gardens in particular exhibit high species diversity and structural complexity [21–23], and are recognized as being essential for the conservation and sustainable management of tropical forest landscapes [24,25]. Parklands are another prominent type of tropical agroforestry, covering relatively large areas of scattered trees and shrubs on cultivated or recently fallowed cropping fields. These many indigenous species of trees are deliberately preserved, and their regeneration is assisted in the agricultural environment because of their specific use [26].

Some characteristic examples of this practice in Ethiopia include *Cordia africana* Lam. intercropping with maize, and *Faidherbia albida*-based agroforestry [26]. The parkland agroforestry systems have significant socio-economic and environmental values [27]. For instance, N_2-fixing woody species in parklands improve soil fertility, enhance crop productivity, and increase soil moisture to facilitate microbial activity such as that of arbuscular mycorrhiza [28]. Home gardens and parklands can also serve as sinks of atmospheric CO_2 [28,29]. Direct benefits from agroforestry systems are in the form of food, medicine, cooking oil, firewood, shelter, tools, and forage [30,31] for domestic use and income generation [32]. The generally rich diversity in structure and composition of tropical agroforestry systems is, however, influenced by climate, elevation, soil moisture, and nutrient availability [33], and farm characteristics such as farm size, cropping pattern, and management [18,34]. Home gardens are reported as having more species than parklands or other agroforestry systems; however, different farming practices influence the potential of agroforestry to accommodate woody plant diversity and uses [35]. Moreover, evidence exists [32,33,36] that the high demand for arable land and unsustainable cropping practices induced degradation of the soil and tree components of agroforestry parklands, particularly in the semi-arid areas of Africa. This increasing anthropogenic pressure requires evaluation of the current status of agroforestry systems and development of adaptive measures such as the domestication of soil-improving tree species [37].

Study of the biological structure of agroforestry systems as indicated by the number and abundance of species provides insights into the relative importance of different plant species, and helps identify important elements of plant diversity, such as threatened and economically important species, to increase their abundance and productivity [38,39] Biodiversity measures (i.e., species diversity and species richness) are widely used as indicators of ecosystem health and human influence on ecological systems [38,40], and are factored in the monitoring of the status of agroforests and in successful conservation management [41–48]. However, vegetation inventories that document biodiversity status are often precluded in tropical developing countries where resources are lacking for extensive field surveys [40].

One of the common approaches for documenting the importance of agroforestry practices for rural livelihoods in developing countries is via study of the indigenous or local knowledge [49]. The current research challenge is to develop user-inspired and user-oriented management approaches [50,51] such as community-based natural resource management, transition management, sustainability, and sustainability education [50]. Acknowledging that success in development is more likely when local knowledge is considered [49,52], there is a need to document the importance of indigenous knowledge for sustainable development of agroforestry [53].

By integrating both local knowledge and ecological assessment, the present study aimed to evaluate the role of home gardens and parklands, the two most prominent tropical agroforestry systems, in the conservation and management of native vegetation in Ethiopia, which covers several agro-climatic zones and is an important spot of tropical biodiversity, yet experiences serious deforestation and land degradation problems. The specific objectives included (i) determining and comparing floristic composition in the agroforestry systems and their diversity and species richness via a field survey in western Ethiopia, and (ii) evaluating the uses, values, and management of woody and herbaceous species by the local population. We hypothesized that home gardens would have higher diversity and, thus, play a greater role in biodiversity conservation than parklands due to the higher intensity of management and use values of home-garden plant species compared to parkland plant species.

2. Materials and Methods

2.1. The Study Area

The present study was carried out in six villages of the Assosa district in western Ethiopia (Figure 1). The area is known for its widespread home-garden and parkland agroforestry practices and rich indigenous knowledge on traditional plant uses [54]. Assosa is one of 21 districts in the Benishangul Gumuz National Regional State of western Ethiopia. The history of Assosa district is marked by significant human settlement authorized by the ex-government of Ethiopia during the major droughts in the 1970s.

The district covers an area of 1991.41 km^2 [54] and is characterized by an elevational range of 1300 to 1470 m above sea level (a.s.l.) and a sub-humid climate with mean minimum and maximum temperatures of 14.4 and 28.5 °C, respectively [55]. The Assosa study area has a mono-modal rainfall pattern from the end of April through October [54]. The average annual rainfall is approximately 1291.2 mm [55].

The dominant soil types are dystric nitisols and orthic acrisols [54,56] with well-drained, reddish-brown clay loam acidic soils [55]. According to Reference [55], the soils in the study area are characterized by very low organic carbon and nitrogen contents, indicative of a low fertility status. The low nutrient status of the soils is constrained by the limited use of both organic and inorganic fertilizers and the loss of nutrients mainly through leaching [55].

Subsistence agriculture is the major economic activity, engaging approximately 80% of the population [55,56]. Major agricultural crops include millet, sorghum, maize, sesame, cotton, soy bean, coffee, and mango. These are produced by rain-fed and, to some extent, irrigated agriculture.

Recurrent crop failures are reported, caused by erratic rainfall in the area, which negatively affect food security [56].

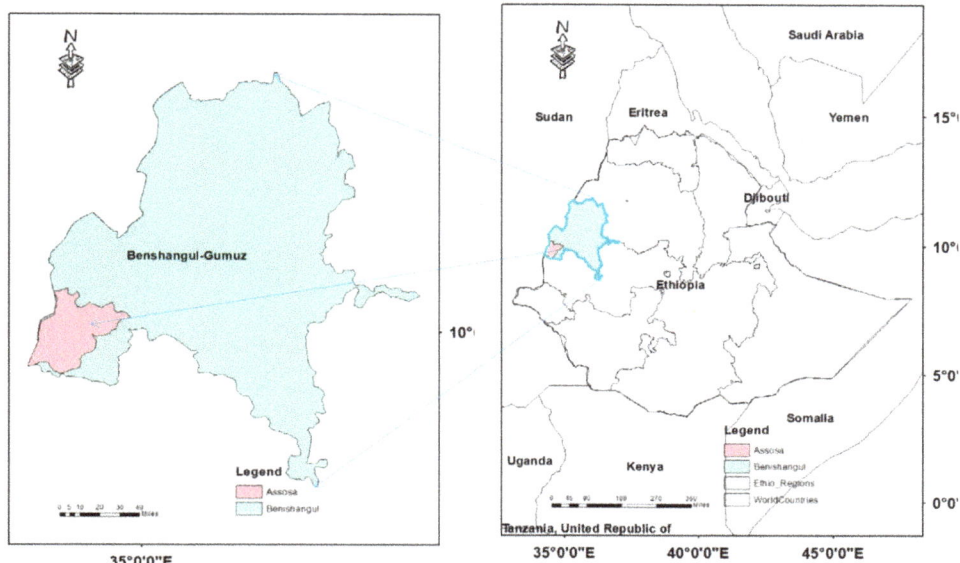

Figure 1. Location of Assosa district in western Ethiopia.

2.2. Household Survey

Based on the presence of agroforestry practices, six out of 74 villages were purposefully selected in the Assosa district. The villages were located between 6 and 21 km away from the Assosa central town. A list of all residents in each village in the Assosa district was collected from the records of the village administration (*kebele*) and development agents. Household respondents (HHs) were chosen using a stratified random sampling approach by adapting the wealth ranking technique of Reference [57], which categorizes farmers in three wealth categories, i.e., poor, moderately endowed, and rich. For each category, a simple random sampling (draw method) was employed to select 8% of the HHs in each village, giving a total of 130 HHs. Of these, 26% were poor, 38% were moderately endowed, and 36% were rich. The HHs were categorized into the three wealth categories by the key informants (KIs), who were farmers and had lived in Assosa for at least 35 years. The KIs were knowledgeable about local situations such as environmental and livelihood changes and local resource management. The information concerning indigenous knowledge on tree species and their uses, tree management practices, and associated constraints was gathered via questionnaire-based interviews with the HHs from 1 February to 15 March 2012.

Each HH had 1-9 family members aged 20-87 years (Table 1). The HHs were engaged in agricultural practices on private, small-land holdings. The 38 female HH heads were widowed or divorced women, relying either on hired labor or who engaged their young children in after-school work on their farmlands. The average land area owned by a farmer was 1.2 ± 0.1 ha and the proportion of land area allocated to parklands (43%) was greater than that allocated to home gardens (15%) (Table 2).

Table 1. Socio-economic characteristics of the 130 household heads interviewed in Assosa district, western Ethiopia.

Socio-Economic Characteristics		Number (%) of Respondents
Sex	Male	92 (70.8)
	Female	38 (29.2)
Age	20–35	17 (13.1)
	36–50	81 (62.3)
	51–65	28 (21.5)
	65–87	4 (3.1)
Literacy status	Literate	76 (58.5)
	Illiterate	54 (41.5)
Marital status	Married	92 (70.8)
	Not married	3 (2.3)
	Widowed	27 (20.7)
	Divorced	8 (6.2)
Wealth category	Poor	34 (26.2)
	Moderately endowed	49 (37.7)
	Rich	47 (36.1)
Family size	1–3 people	29 (22.3)
	4–7 people	85 (65.4)
	8–10 people	16 (12.3)

Table 2. Share of total land area allocated to home gardens, parklands and other land uses per farmer in Assosa district, western Ethiopia.

Village	Average Total Land Area per Farmer (ha)	Percentage of Land Allocated to Home Gardens (%)	Percentage of Land Allocated to Parklands (%)	Percentage of Land Allocated to Other Land Uses (%)
Amba8	0.6 ± 0.04	14	37	49
Megele37	0.9 ± 0.09	18	45	37
Megele39	0.9 ± 0.09	17	50	33
Amba7	0.8 ± 0.03	11	43	46
Nebarkomshga	3.7 ± 0.42	21	51	28
Amba13	1.0 ± 0.03	15	36	49
Total average	1.2 ± 0.09	15	43	42

Note: Other land uses include miscellaneous lands owned by household respondents (HHs) allocated to pasture, cropland with no trees, and shrubs. The sample size was 130 households.

2.3. Vegetation Survey

To obtain an inventory of the woody and herbaceous perennial species, 54 HHs (and, thus, 54 home-garden plots and 54 parkland plots) were selected out of the 130 surveyed HHs, i.e., three HHs from each wealth category, totaling nine HHs per village (six villages in total) and a pair (i.e., managed by the same HH) of home gardens and parklands per HH. We ensured that both land-use types were managed by the same HH to control for variations in management practices within HHs. Moreover, given that the study was carried out in a relatively small area where the site conditions (i.e., climate, topography, altitude) remain relatively homogenous, as described in Section 2.1, we assumed that variations in site conditions have less influence on the vegetation composition compared to differences in land management practices.

The study of the composition of woody and herbaceous species was done in the home gardens and parklands of the selected HHs. In the parklands, 50 m × 50 m quadrants were established as sampling plots because the minimum size of a farmland owned by a local farmer was 2500 m² (based on KI interviews cross-checked with personal observations). In contrast, the sampling in home gardens was performed using 30 m × 30 m quadrants defined by the KIs as the minimum size (~1000 m^2) of a home garden per farmer in the area [35].

Local names of plant species in each sampling plot were identified with the help of members of the local communities participating in the survey. Consequently, the plant species nomenclature was defined following References [58,59]. The species that could not be identified in the field were sampled (mainly for foliage), pressed flat to dry, and transported for identification to the Herbarium Laboratory of Addis Ababa University, Ethiopia.

The total number of woody and herbaceous perennial plants in the sampling quadrants was counted and recorded to determine the relative abundance of each species. The stem diameter at breast height (dbh) of all woody and herbaceous perennial (*Musa* × *paradisiaca* L., *Oxytenanthera abyssinica* (A. Rich.) Munro, and *Carica papaya* L.) plants with dbh ≥ 5 cm was measured in each sampling plot. When branching occurred below 1.3 m of the plant height, the dbh of all branches was measured and the average value was calculated. The dbh value was also used to calculate the basal area of plants with dbh ≥ 5 cm as follows:

$$BA = \frac{\pi (dbh)^2}{4},$$

(1)

where *BA* is the basal area (cm^2).

Species dominance was calculated as the ratio of the total BA of the plants of each species to the total sampled area. The relative abundance (ra), relative dominance (rd), and relative frequency (rf) were calculated as follows:

$$ra = \frac{\text{Number of individuals of species}}{\text{Total number of individuals}} \times 100\%,$$

(2)

$$rd = \frac{\text{Dominance of a species}}{\text{Total dominance of all species}} \times 100\%,$$

(3)

$$rf = \frac{\text{Frequency of species 1}}{\text{Total frequency of all species}} \times 100\%.$$

(4)

Therefore, the importance value index (IVI) indicating the importance of each species in the system was calculated as follows:

$$IVI = ra + rd + rf.$$

(5)

2.4. Species Diversity and Richness

To characterize the species diversity and richness in the studied agroforestry systems, we used Fisher's α index and the species–area relationship (SAR). These indicators were chosen because they are less sensitive to sample size. Fisher's α index is a parametric diversity index, which assumes that species abundance follows a logarithmic distribution [60]. It is a scale-independent indicator of diversity and was computed as follows:

$$S = a * \ln\left(1 + \frac{n}{a}\right),$$

(6)

where *S* is the number of taxa, *n* is the number of individuals, and α is Fisher's α.

The species–area relationship (SAR) is concerned with the number of species in areas of different size, irrespective of the identity of species within the areas [61]. The power function (Equation (7)) is the most commonly used model to describe the form of the species–area curve [62,63].

$$S' = S_0 A^z,$$

(7)

where *S'* is the number of species, *A* is the area, S_0 is the number of species in a unit area (*A* = 1), and *z* is a model parameter (0 < z < 1).

Furthermore, the Sørensen similarity coefficient was chosen to compare the similarity in the species composition of home gardens and parklands because it gives more weight to the species that are common in the samples rather than to those that only occur in either sample [64]. The Sørensen similarity index (Ss) was calculated as follows:

$$Ss = \left(\frac{2a}{2a + b + c} \right) \times 100, \tag{8}$$

where a is the number of species common to both samples, b is the number of species in sample 1, and c is the number of species in sample 2.

2.5. Statistical Analysis

Fisher's α index was compared among the six villages and between the two agroforestry practices. The data distribution could not be normalized by transformations. Therefore, the non-parametric Kruskal–Wallis test was used to check the significance of the differences. The species–area curves were plotted for each land-use type separately, and the SARs were fitted with the power function (Equation (7)). The analyses were performed using Microsoft Excel 2010, SPSS software (v. 24) and R version 3.4.3 [65].

3. Results and Discussion

3.1. Floristic Composition of Home Gardens and Parklands

During the HH survey, four agroforestry practices were identified. The dominant practices were home gardens and parklands, and the less common practices were alley cropping and on-farm boundary planting. The home-garden agroforestry was practiced by all HHs ($n = 130$) and the parklands by only 30 HHs (23% of the total sampled HHs). Alley cropping and on-farm boundary planting were practiced by 10% and 7% of HHs, respectively. All practices were previously reported as representative, particularly in southern Ethiopia [66], as well as in many other tropical regions [19]. In particular, home gardens were stated as being the most common among the smallholder agroforestry practices in the Ethiopian highlands, hosting higher woody species diversity than the nearby natural woodlands or forest lands [54,67].

The Assosa vegetation survey identified 57 woody and herbaceous perennial species (the latter being C. papaya, M. × paradisiaca, and O. abyssinica), with 56 of the species present in home gardens and 22 in parklands. The identified species belonged to 27 plant families. The most dominant family Fabaceae was represented by 11 woody species (19.3% of the total number of species recorded) and was followed by families Euphorbiaceae, Rutaceae, and Myrtaceae, represented by 4–6 species and constituting 10.5%, 8.8%, and 7%, respectively (Table 3).

Overall, 35 species were only found in home gardens but not in parkland agroforestry systems of the sampled HHs. One tree species, Allophylus abyssinicus (Hochst.) was found only in the parklands. A total of 21 species occurred in both agroforestry systems (Table A3, Appendix A). Overall, species composition significantly differed between the systems as judged by the relatively low Sørensen similarity coefficient (35%). Most of the woody species retained by farmers in parklands and home gardens were remnants of the natural vegetation, which covered the area before the settlements appeared in the 1970s and the Ethiopian natural disaster (famine) times. Afterward, planting of both native and exotic species occurred, mostly in home gardens and in some parklands. Planted timber tree species included Albezia gummifera (J.F. Gmel.) C.A. Sm., Melia azedarach L., Cordia africana Lam., Grevillea robusta A. Cunn. ex R.Br., and Eucalyptus camaldulensis Dehnh. For fruit trees, Citrus aurantifolia (Christm.) Swingle, Citrus sinensis (L.), and Mangifera indica L. were identified. Several species such as Catha edulis (Vahl) Endl., Rhamnus prinoides L'Hér., Coffea arabica L., and O. abyssinica were planted as perennial cash crops. These findings (Tables A1 and A2, Appendix A) corroborate with previous studies in the upper Blue Nile Basin and western Ethiopia, which reported the common presence of tree species *Croton macrostachys* Hochst. ex Delile, *Acacia abyssinica* Benth, and *C. africana* Lam. managed by farmers on their agricultural lands [26].

Table 3. Woody and herbaceous perennial species and corresponding families identified in home gardens and parkland agroforestry systems in Assosa district, western Ethiopia.

No.	Family	Number of Species	Percentage	Number of Individuals
1	Fabaceae	11	19.3	113
2	Euphorbiaceae	6	10.5	72
3	Rutaceae	5	8.8	90
4	Myrtaceae	4	7.0	236
5	Bignoniaceae	3	5.3	33
6	Anacardiaceae	2	3.5	240
7	Celastraceae	2	3.5	163
8	Sapindaceae	2	3.5	5
9	Moraceae	2	3.5	23
10	Proteaceae	2	3.5	19
11	Combretaceae	2	3.5	54
12	Acanthaceae	1	1.8	55
13	Annonaceae	1	1.8	1
14	Boraginaceae	1	1.8	172
15	Burseraceae	1	1.8	1
16	Caricaceae	1	1.8	38
17	Casuarinaceae	1	1.8	14
18	Cupressaceae	1	1.8	3
19	Lauraceae	1	1.8	8
20	Meliaceae	1	1.8	31
21	Musaceae	1	1.8	107
22	Poaceae	1	1.8	170
23	Rhamnaceae	1	1.8	131
24	Rosaceae	1	1.8	2
25	Rubiaceae	1	1.8	188
26	Sterculiaceae	1	1.8	13
27	Tiliaceae	1	1.8	1
	Total	57	100	

Based on the IVI ranking, *M. indica* was the most important woody species in home gardens, followed by *C. africana* and *C. arabica* (Tables 4 and A1, Appendix A), whereas, in parkland agroforestry systems, *Syzygium guineense* (Willd.) DC., *M. indica*, *C. africana*, *Terminalia brownii* Fresen., and *O. abyssinica* were the top five most important species (Tables 5 and A2, Appendix A).

Table 4. Importance value index (IVI) ranking of the top 10 woody and herbaceous perennial species in home gardens of Assosa district, western Ethiopia.

Name of the Species	IVI
Mangifera indica	65
Cordia africana	30
Coffea arabica	21
Catha edulis	16
Eucalyptus camaldulensis	15
Oxytenanthera abyssinica	14
Musa x paradisiacal	12
Rhamnus prinoides	12
Syzygium guineense	10
Citrus sinensis	10

Table 5. Importance value index (IVI) ranking of the top 10 woody and herbaceous perennial species in parkland agroforestry systems of Assosa district, western Ethiopia.

Name of the Species	IVI
Syzygium guineense	112
Cordia africana	45
Mangifera indica	29
Terminalia brownii	24
Ficus sur	14
Oxytenanthera abyssinica	13
Calpurnia aurea	8
Dombeya torrida	7
Musa x paradisiacal	5
Eucalyptus camaldulensis	4

3.2. Species Richness and Diversity in Home Gardens and Parklands

A high species diversity is often associated with important ecological services such as nutrient cycling, soil and water conservation, and resilience under anthropogenic pressure [10,12]. In this study, 56 and 22 species were counted in home gardens and parklands, respectively. The SARs predicted that 33 species·ha^{-1} and 4 species·ha^{-1} would be recorded for home gardens and parklands, respectively (Figure 2), suggesting that home gardens are likely to be richer in species compared to parklands. Our findings are reminiscent of the findings by Reference [35], who showed that home-garden agroforestry systems in the sub-humid eco-climatic zone of Ethiopia host higher woody species richness (64 species) than the nearby natural woodlands (32 species) and forest lands (31 species). The accumulation of a greater number of species in home gardens compared to parklands observed in the present study may be attributed to the planting preference of exotic species in home gardens (25 exotic vs. 31 indigenous species) than in parklands (eight exotic vs. 14 indigenous species) (Tables A1 and A2, AAppendix A). The introduced exotic species included perennial cash crops (e.g., *C. edulis*), fruit trees (e.g., *M. indica*), and those used as live fences and windbreaks (e.g., *Jatropha curcas* L.) in home gardens [54]. Although some of the exotic species have the potential to be invasive (e.g., *M. azedarach*, *G. robusta*; Tables A1 and A2, Appendix A), none of them were reported as invasive species in the study area by the HHs.

Comparison of species richness observed in tropical agroforestry globally is complicated by the difference in altitude, amount of rainfall, type of soil, and other factors such as differences in social, environmental, and economic conditions that influence species distribution and provenances [19,54]. For example, relatively low tree species richness (27) was recorded in home gardens of Kandy, Sri Lanka [68]. In Tanzania, East Africa [69,70], both studies counted 53 home-garden tree species and, thus, a higher richness comparable with results from our study (56). Reference [71] encountered 60 tree species in Mexican home gardens and even higher richness was reported from India, where Reference [21] observed 87 home-garden tree species in Assam and 71 tree species in Kerala state. The largest number, 179 woody species in home gardens, was reported from west Java, Indonesia [31].

Higher species richness in parklands than that observed in Assosa (22 species) was reported elsewhere in Ethiopia and in sub-Saharan Africa. For instance, Reference [33] recorded 48 and 41 woody species during fallow periods and crop cultivation, respectively, in parklands of Burkina Faso, West Africa. A study in south central Ethiopia [35] identified 32 woody species in parklands (during the cultivation phase). The lower species richness in Assosa parklands (22 species or an average of 4 species·ha^{-1}) might be associated with the history of human settlement in the 1970s, which increased the demand for agricultural land and wood.

Fisher's α diversity index was significantly higher for home gardens than for parklands in the study area (Table 6). There was no significant difference in Fisher's α index among the villages within each agroforestry system (Table 6). The higher species diversity in home gardens may be attributed to better and intensive management by family labor, in particular women and children [72]. According to the HHs, various silvicultural practices are applied to manage and maintain the plant species for different purposes in the agroforestry systems (Tables 7 and A5, Appendix A). These include manure (e.g., cow dung) application, watering, pruning, trimming, and fumigation-based control of pest and diseases (Tables 7 and A5, Appendix A). Although similar management practices are applied for both parklands and home gardens, the latter is continuously and more intensively managed by family labor (Table 7). The contribution of family labor as an important human capital for the management of home gardens based on the indigenous knowledge, skills, and abilities of the farming community was reported by Reference [73].

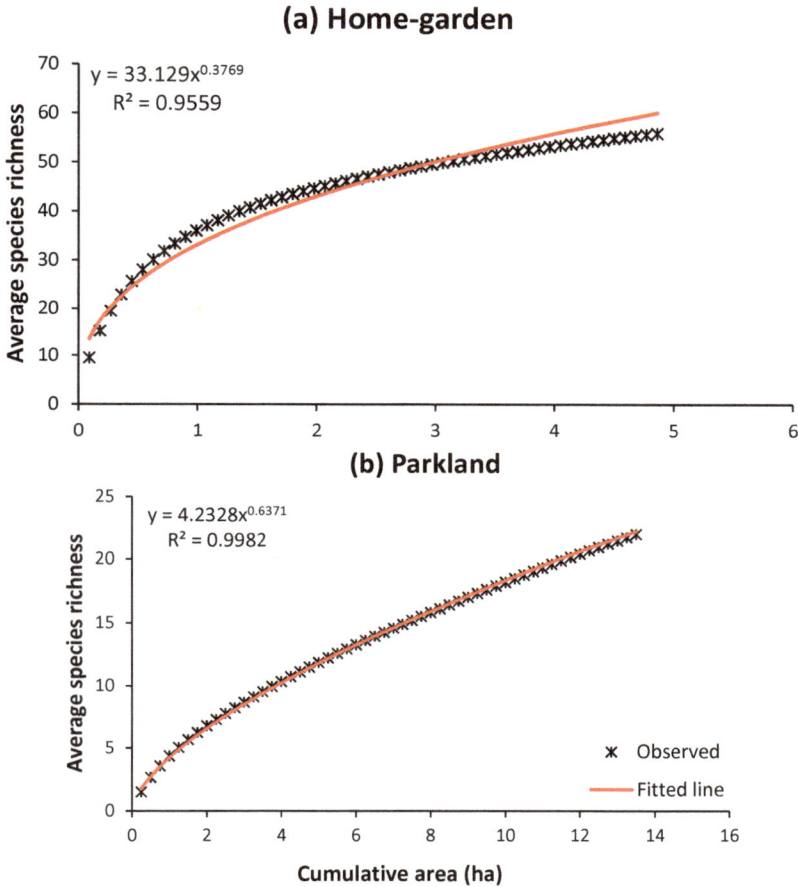

(a) Home-garden

(b) Parkland

Figure 2. Species–area curves for (**a**) home gardens and (**b**) parkland agroforestry systems in Assosa district, western Ethiopia.

Additionally, the higher diversity of home gardens is an indication of secure property rights as testified by the HHs. The farmers unanimously mentioned that they felt more secure about planting commercially important exotic plants in their home gardens, for which they have full ownership rights, than in parklands. Similarly, the importance of secure land tenure for tree planting and biodiversity conservation was reported in several studies [35,74,75]. Altogether, intensive management, access to family labor, and secure property rights led to higher diversity in home gardens than in parklands in the Assosa district.

Table 6. Comparison of Fisher's α diversity index between home gardens and parkland agroforestry systems in Assosa district, western Ethiopia. Values are means ± standard errors.

Villages	Agroforestry Practice	Fisher's α
Amba8	Home gardens	5.91 [a] ± 0.70
	Parklands	1.55 [b] ± 0.41
Megele37	Home gardens	4.31 [a] ± 0.36
	Parklands	1.90 [b] ± 0.57
Megele39	Home gardens	4.36 [a] ± 0.57
	Parklands	0.87 [b] ± 0.17
Amba7	Home gardens	5.63 [a] ± 1.17
	Parklands	1.99 [b] ± 0.33
Nebarkomshga	Home gardens	6.62 [a] ± 1.13
	Parklands	2.18 [b] ± 0.63
Amba13	Home gardens	4.86 [a] ± 0.79
	Parklands	1.24 [b] ± 0.25
All villages	Home gardens	5.28 [a] ± 0.35
	Parklands	1.62 [b] ± 0.18

Note: For each village and for all villages, the mean values with the same superscripts are not significantly different at the $p < 0.05$ significance level.

Table 7. Home-garden and parkland agroforestry management practices as reported by HHs in Assosa district, western Ethiopia.

Management Practices	Home Garden	Parkland
Manure (cow dung) application	++	+
Irrigation	++	−
Soil management activities (harrowing/hoeing/ploughing)	++	+
Tending activities (e.g., trimming, pollarding, pruning, lopping, and coppicing)	++	+
Traditional pest and disease control (fumigation)	++	+
Plant species regeneration activities (seeding planting, assisted natural regeneration)	++	+

Note: (+) and (−) indicate whether the management is applied in the agroforestry systems or not, respectively; (++) indicates a higher intensity of management for a given practice. Further details on the management practices are provided in Table A5 (Appendix A).

3.3. Uses of Agroforestry Species

The woody and herbaceous perennial species in home gardens and parkland agroforestry systems in Assosa were stated by the HHs as sources of primarily food, fiber, fodder, timber, fuelwood, medicine, and other products of commercial value such as fruit from *M. indica* and foliage-derived stimulant from *C. edulis* (Figure 3; Table A4, Appendix A). All agroforestry species were credited with an improved microclimate due to the provision of shade. Next in importance, judged by the number of species named, was the provision of fuelwood (e.g., *E. camaldulensis* and *O. abyssinica*), food (e.g., *M. indica* and *M.* × *paradisiaca*), and traditional medicine (e.g., *C. macrostachyus* and *Justicia schimperiana* (Hochst. ex Nees) T. Anderson) (Table A4, Appendix A). Most significantly, the farmers harvested edible fruits of *M. indica*, *C. cinencis*, and *S. guineense* for domestic consumption and for sale, and the stems of *C. africana*, *A. gummifera*, and *Ficus sur* Forssk for timber (Figure 3). The respondents also mentioned soil fertility maintenance and a range of services associated with live fences, such as protection against soil erosion, microclimate amelioration, and recreational value in addition to the main purpose of border demarcation. Some uses were highly specific, such as the use of foliar juice from *M. azedarach* trees as an insect pest repellant to protect corn seedlings, exemplifying the variety of local knowledge [54,76].

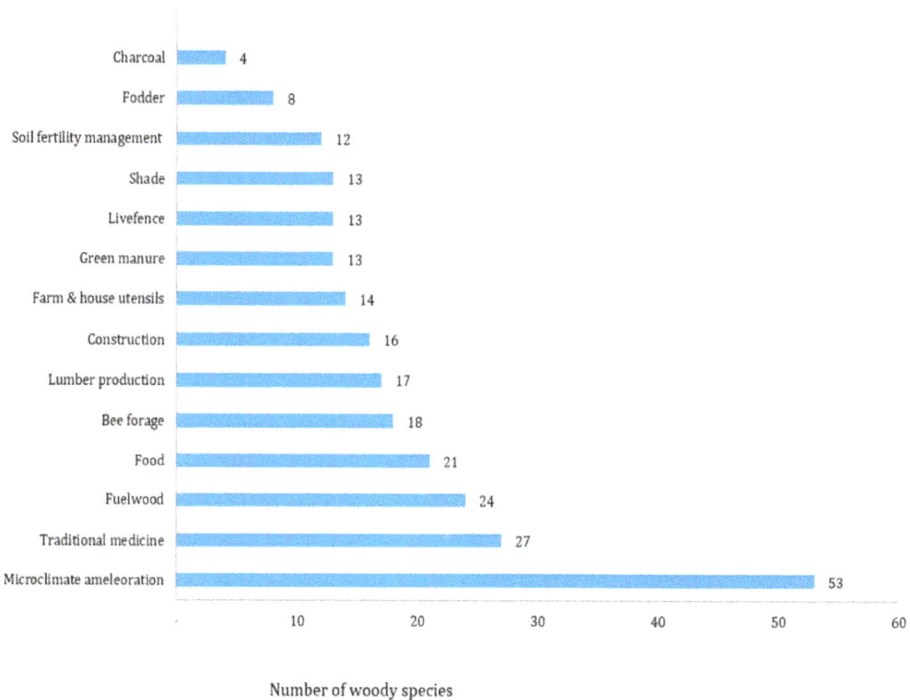

Figure 3. Local uses and values of woody and herbaceous perennial species in home gardens and parkland agroforestry systems in Assosa district, western Ethiopia.

These agroforestry benefits were reported as advantageous for enhancing ecological sustainability and improving the income and livelihood of rural people [19,77]. Thus, both indigenous and exotic tree and shrub species appeared important for farmers in Assosa district. Farmers ranked the indigenous *C. africana* and the exotic fruit species *M. indica* as being most important due to their relative productivity and profitability (Figure 3). However, the respondents stated that they faced insufficient yields from agroforestry species (from both home gardens and parklands), which might be due to the prevalence of management problems such as water shortages [54], as mentioned by the HHs.

The multipurpose use of woody plants is commonly reported in tropical agroforestry systems (e.g., Reference [78]). Similarly, HHs in western Ethiopia mentioned that most of the woody and herbaceous perennial species were used for more than one purpose. For instance, the wild fruit tree *S. guineense* is used for both charcoal making and as a food source, whereas almost all the species in the study area were mentioned to contribute to the improvement of the local microclimate (Figure 3; Table A4, Appendix A). Reference [79] showed that, out of the 2374 plant species identified as being important for smallholders in three tropical regions (Africa, South America, and Southeast Asia), 732, 639, 582, and 421 species were used for fuelwood, human food, animal fodder, and soil improvement, respectively. In Africa alone, 357 tree species were used for fuelwood, 295 for fodder, 295 for food, and 194 species for soil improvement [79]. The relative importance of tree uses is largely consistent with our results, showing the largest number of species being used for fuelwood and food, and a significant number credited with soil fertility management (Figure 3).

As mentioned by the HHs, the benefits of the agroforestry woody and herbaceous perennial species in the amelioration of the local microclimate are multifold, as the plants provide shade for people, grazing animals, and accompanying crops (Table A4, Appendix A). According to Reference [80], shading by tree canopies reduces evaporation from the soil surface and lowers air temperature. Specifically, Reference [81] recorded a 5 °C reduction in ambient temperature in the shaded versus open area during midday in a savanna ecosystem of Senegal.

The study by Reference [79] in tropical agroforestry systems emphasized in situ germplasm conservation of native plant species [82]. The effectiveness of the conservation of native flora is apparent in our study, as suggested by the larger share of indigenous versus exotic species, particularly in the parklands. Results by Reference [79] showed that smallholders are able to use a wide range of both indigenous and exotic trees. Similarly, the HHs in Assosa mentioned various uses of plants both from exotic and indigenous species, emphasizing the economic advantages of the former and traditional uses (such as for medicine) of the latter.

4. Conclusions

This case study in sub-humid western Ethiopia showed that prevailing agroforestry systems of home gardens and parklands that contain nearly 60 woody and herbaceous perennial species supply more than 14 different goods and services to local farmers. These agroforestry practices are, therefore, crucial in conserving the biodiversity on farms.

In spite of the larger area sampled for parklands, the home gardens showed higher species richness and diversity than did parklands owing to better management and protection by family labor, secure land tenure, and also because of the cultivation of exotic species. In contrast, parklands mostly harbored native flora via assisted natural regeneration, but were characterized by lower than expected species richness owing to the agricultural use of land.

The importance of the tree species judged by their useful values as perceived by the HH respondents was in agreement with the overall importance of these species as revealed in the vegetation survey. Therefore, boosting the low diversity of parklands and sustainable management of home gardens is crucial for the conservation of native vegetation and diversification of agroforestry land use. The management of agroforestry species using the local indigenous knowledge of farmers and further augmenting with both indigenous and useful, non-invasive exotic woody and herbaceous species is an important step toward agroforestry development in the study area.

Author Contributions: E.T., conceptualization, data collection, analysis, and writing of the draft manuscript; A.K., commenting, writing, reviewing, and editing; A.A., commenting; Y.S., guidance, commenting, and editing; F.N., commenting and data analysis.

Funding: This study was carried out with the support of 'R & D Program for Forest Science Technology (Project No. 2018110A00-1920-BB01)' provided by Korea Forest Service (Korea Promotion Institute).

Acknowledgments: We acknowledge "Agricultural Technical Vocational Education and Training College" of Assosa, Ethiopia and "World Vission Ethiopia" Assosa-Homosha district office, for allotting transportation for the research work. Korea University, Department of Environmental Science and Ecological Engineering also facilitated the preparation and publication of the manuscript. We are also grateful to the reviewers for their insightful and constructive comments.

Conflicts of Interest: The authors declare no conflicts of interest.

Appendix A

Table A1. Home-garden woody and herbaceous perennial species ordered according to their frequency of occurrence in Assosa district, western Ethiopia.

No.	Species Name	Origin	Potential of Invasiveness *	Basal Area (m²)	Relative Dominance (%)	Relative Frequency (%)	Relative Abundance (%)	Importance Value Index
1	*Mangifera indica*	E	Yes	18.25	42.39	10.1	12.57	65.06
2	*Cordia africana*	I	No	5.5	12.78	9.49	7.97	30.24
3	*Coffea arabica*	I	No	1.06	2.46	7.27	11.08	20.81
4	*Catha edulis*	I	No	0.68	1.58	4.85	9.33	15.76
5	*Citrus sinensis*	E	No	0.28	0.65	4.85	3.05	8.55
6	*Oxytenanthera abyssinica*	I	No	0.51	1.18	4.85	8.4	14.43
7	*Rhamnus prinoides*	I	No		0	4.24	8.15	12.39
8	*Eucalyptus camaldulensis*	E	Yes	3.23	7.5	4.04	3.05	14.59
9	*Carica papaya*	E	No	0.99	2.3	3.23	2.36	7.9
10	*Musa × paradisiaca*	E	No	1.4	3.25	3.23	5.97	12.46
11	*Stereospermum kunthianum*	I	No	0.28	0.65	2.63	0.93	4.21
12	*Syzygium guineense*	I	No	2.39	5.55	2.63	1.68	9.86
13	*Citrus aurantifolia*	E	Yes	0.15	0.35	2.42	1.12	3.89
14	*Melia azedarach*	E	Yes	1.21	2.81	2.42	1.68	6.92
15	*Leucaena leucocephala*	E	Yes	0.09	0.21	2.22	1.8	4.24
16	*Psidium gujava*	E	No	0.11	0.26	2.02	1.18	3.46
17	*Grevillea robusta*	E	Yes	0.16	0.37	1.82	1.06	3.25
18	*Ricinus communis*	I	No	0.1	0.23	1.82	1.24	3.3
19	*Terminalia broxnii*	I	No	0.98	2.28	1.82	1.06	5.15
20	*Croton macrostachyus*	I	No	0.44	1.02	1.62	0.75	3.38
21	*Entada abyssinica*	I	No	0.39	0.91	1.41	2.3	4.62
22	*Justicia schimperiana*	I	No	0.08	0.19	1.41	3.42	5.02
23	*Manihot sculenta*	E	No	0.03	0.07	1.41	0.44	1.92
24	*Persea americana*	E	No	0.17	0.39	1.41	0.5	2.31
25	*Spatodea campanulata*	E	No	0.49	1.14	1.41	0.62	3.17
26	*Albizia gummifera*	I	No	0.51	1.18	1.21	0.5	2.89
27	*Casimiroa edulis*	E	No	0.25	0.58	1.21	0.5	2.29
28	*Ficus sur*	I	No	1.14	2.65	1.21	0.56	4.42
29	*Jatropha curcas*	E	Yes	0.17	0.39	1.21	1.93	3.54
30	*Citrus aurantium*	E	No	0.07	0.16	1.01	0.44	1.61
31	*Jacaranda mimosifolia*	E	Yes	0.14	0.33	1.01	0.5	1.83
32	*Casuarina canninghamiana*	E	No	0.34	0.79	0.81	0.56	2.16
33	*Morus alba*	E	Yes	0.1	0.23	0.81	0.5	1.54

Table A1. Cont.

No.	Species Name	Origin	Potential of Invasiveness *	Basal Area (m²)	Relative Dominance (%)	Relative Frequency (%)	Relative Abundance (%)	Importance Value Index
34	Calpurnia aurea	E	No	0.01	0.02	0.61	0.19	0.82
35	Cupresus lusitanica	E	No	0.1	0.23	0.61	0.19	1.03
36	Sesbania sesban	I	No	0.06	0.14	0.61	0.31	1.06
37	Acacia abyssinica	I	No	0.44	1.02	0.4	0.25	1.68
38	Combretum aculeatum	I	No	0.11	0.26	0.4	0.12	0.78
39	Maytenus senegalensis	I	No	0.01	0.02	0.4	0.12	0.55
40	Prunus persica	E	No	0.02	0.05	0.4	0.12	0.57
41	Tamarindus indica	I	No	0.03	0.07	0.4	0.12	0.6
42	Annona senegalensis	I	No	0.01	0.02	0.2	0.06	0.29
43	Boswellia papyrifera	I	No	0.02	0.05	0.2	0.06	0.31
44	Cajanus cajan	E	No	0	0	0.2	0.06	0.26
45	Citrus medica	E	No	0.01	0.02	0.2	0.06	0.29
46	Dodonaea angustifolia	I	No	0	0	0.2	0.06	0.26
47	Dombeya torrida	I	No	0.03	0.07	0.2	0.06	0.33
48	Erythrina abyssinica	I	No	0.14	0.33	0.2	0.12	0.65
49	Euphorbia abyssinica	I	No	0.01	0.02	0.2	0.06	0.29
50	Euphorbia trucalii	I	No	0	0	0.2	0.06	0.26
51	Faurea speciose	E	No	0.15	0.35	0.2	0.06	0.61
52	Grewia mollis	I	No	0.02	0.05	0.2	0.06	0.31
53	Lannea fruticose	I	No	0.14	0.33	0.2	0.06	0.59
54	Millettia ferruginea	I	No	0.01	0.02	0.2	0.12	0.35
55	Myrtus communis	I	No	0.01	0.02	0.2	0.37	0.6
56	Pilostigma thonningii	I	No	0.03	0.07	0.2	0.06	0.33
					100	100	100	

Note: I = indigenous, E = exotic. * The potential of invasiveness of exotic species was derived from References [83,84].

Table A2. Parkland woody and herbaceous perennial species ordered according to the frequency of occurrence in Assosa district, western Ethiopia.

No.	Species Name	Origin	Potential of Invasiveness *	Basal Area (m²)	Relative Dominance (%)	Relative Frequency (%)	Relative Abundance (%)	Importance Value Index
1	Syzygium guineense	I	No	6.89	45	32.5	34.04	111.55
2	Cordia africana	I	No	2.04	13.32	20	11.7	45.03
3	Mangifera indica	E	Yes	1.35	8.82	11.25	8.51	28.58
4	Terminalia brownii	I	No	1.29	8.43	7.5	7.98	23.9

Table A2. Cont.

No.	Species Name	Origin	Potential of Invasiveness *	Basal Area (m²)	Relative Dominance (%)	Relative Frequency (%)	Relative Abundance (%)	Importance Value Index
5	Ficus sur	I	No	1.32	8.62	3.75	1.6	13.97
7	Calpurnia aurea	I	No	0.18	1.18	3.75	3.46	8.38
6	Oxytenanther abyssinica	I	No	0.22	1.44	2.5	9.31	13.25
8	Dombeya torrida	I	No	0.41	2.68	1.25	3.19	7.12
9	Musa × paradisiaca	E	No	0.2	1.31	1.25	2.93	5.48
10	Catha edulis	I	No	0.01	0.07	1.25	2.93	4.24
11	Combretum aculeatum	I	No	0.24	1.57	1.25	1.33	4.15
12	Coffea arabica	I	No	0.01	0.07	1.25	2.66	3.97
13	Allophylus abyssinicus	I	No	0.2	1.31	1.25	1.06	3.62
14	Eucalyptus camaldulensis	E	Yes	0.07	0.46	1.25	1.86	3.57
15	Lannea fruticose	I	No	0.15	0.98	1.25	1.33	3.56
16	Casuarina canninghamiana	E	No	0.14	0.91	1.25	1.33	3.49
17	Acacia abyssinica	I	No	0.2	1.31	1.25	0.8	3.35
18	Melia azedarach	E	Yes	0.15	0.98	1.25	1.06	3.29
19	Albezia gumifera	I	No	0.13	0.85	1.25	0.8	2.9
20	Citrus sinensis	E	No	0.02	0.13	1.25	1.06	2.44
21	Citrus aurantifolia	E	Yes	0.02	0.13	1.25	0.8	2.18
22	Grevillea robusta	E	Yes	0.07	0.46	1.25	0.27	1.97
	Sum				100	100	100	100

Note: I = indigenous, E = exotic. * The potential of invasiveness of exotic species was derived from References [83,84].

Table A3. The importance value index (IVI) of woody and herbaceous perennial species common to parklands and home gardens of Assosa district, western Ethiopia.

No.	Species and the Respective Family Name	Importance Value Index	
		Home Garden	Parkland
1	Mangifera indica (Anacardiaceae)	65.06	28.58
2	Cordia africana (Boraginaceae)	30.24	45.03
3	Coffea arabica (Rubiaceae)	20.81	3.97
4	Catha edulis (Celasteraceae)	15.76	4.24
5	Eucalyptus camaldulensis (Myrtaceae)	14.59	3.57
6	Oxytenanther abyssinica (Poaceae)	14.43	13.25
7	Musa × paradisiaca (Musaceae)	12.46	5.48

Table A3. *Cont.*

No.	Species and the Respective Family Name	Importance Value Index	
		Home Garden	Parkland
8	*Syzygium guineense* (Myrtaceae)	9.86	111.55
9	*Citrus sinensis* (Rutaceae)	8.55	2.44
10	*Melia azedarach* (Meliaceae)	6.92	3.29
11	*Terminalia brownii* (Combretaceae)	5.15	23.9
12	*Ficus sur* (Moraceae)	4.42	13.97
13	*Citrus aurantifolia* (Rutaceae)	3.89	2.18
14	*Grevillea robusta* (Proteaceae)	3.25	1.97
15	*Albezia gumifera* (Fabaceae)	2.89	2.9
16	*Casuarina canninghamiana* (Casuarinaceae)	2.16	3.49
17	*Acacia abyssinica* (Fabaceae)	1.68	3.35
18	*Calpurnia aurea* (Fabaceae)	0.82	8.38
19	*Combretum aculeatum* (Combretaceae)	0.78	4.15
20	*Lannea fruticose* (Anacardiaceae)	0.59	3.56
21	*Dombeya torrida* (Sterculiaceae)	0.33	7.12

Table A4. Major woody and herbaceous perennial species and their uses in Assosa district, western Ethiopia as stated by the households. "?" stands for other purposes not mentioned by the HHs.

Species and the Respective Family Name	Proposed Uses			
Catha edulis (Celastraceae)	Stimulant	Microclimate amelioration	?	?
Coffea arabica (Rubiaceae)	Stimulant	Microclimate amelioration	?	?
Terminalia brownii (Combretaceae)	Fumigation	Microclimate amelioration	?	?
Faurea speciose (Proteaceae)	Fumigation	Microclimate amelioration	?	?
Acacia abyssinca (Fabaceae)	Charcoal making	Microclimate amelioration	?	?
Combretum aculeatum (Combretaceae)	Charcoal making	Microclimate amelioration	?	?
Eucalyptus camaldulensis (Myrtaceae)	Construction	Fuelwood	Microclimate amelioration	?
Syzygium guineense (Myrtaceae)	Construction	Food	Charcoal making	Microclimate amelioration
Sesbania sesban (Fabaceae)	Soil fertility enhancement	Fodder	Microclimate amelioration	?
Albezia gumifera (Fabaceae)	Soil fertility enhancement	Shade	Microclimate amelioration	?
Piliostigma thonningii (Fabaceae)	Fodder	Bee forage	Traditional medicine	Microclimate amelioration
Cordia africana (Boraginaceae)	Fodder	Food	Shade & Microclimate amelioration	House utensils
Leucaena leucocephala (Fabaceae)	Fodder		Microclimate amelioration	?
Morus alba (Moraceae)	Fodder	Live fence	Microclimate amelioration	?
Jatropha curcas (Euphorbiaceae)	Live fence	Fuelwood	Microclimate amelioration	?

Table A4. Cont.

Species and the Respective Family Name	Proposed Uses			
Entada abyssinica (Fabaceae)	Live fence	Fuelwood	Microclimate amelioration	?
Grewia mollis (Tiliaceae)	Farm implements	House utensils	Microclimate amelioration	?
Ficus sur (Moraceae)	Farm implements	House utensils	Food	Microclimate amelioration
Justicia schimperiana (Acanthaceae)	Bee forage	Traditional medicine	Live fence	Microclimate amelioration
Millettia ferruginea (Fabaceae)	Shade	Soil fertility enhancement	Microclimate amelioration	?
Mangifera indica (Anacardiaceae)	Food	Shade	Microclimate amelioration	?
Croton macrostachyus (Euphorbiaceae)	Traditional medicine	Shade	Microclimate amelioration	?
Psidium guajava (Myrtaceae)	Traditional medicine	Microclimate amelioration	?	?
Manihot esculenta (Euphorbiaceae)	Food	?	?	?
Musa × paradisiaca (Musaceae)	Food	?	?	?
Melia azedarach (Meliaceae)	Traditional medicine	Fuelwood	Microclimate amelioration	?

Table A5. Management practices in home gardens and parkland agroforestry systems, as stated by the HHs in Assosa district, western Ethiopia.

Woody Species Regeneration Methods		Management of Woody Species				Agroforestry Management Problems		Impacts of Pests and Diseases on Important Agroforestry Woody Species		
Sources of Seedlings	Number of HHs	Management Activities	Species Mostly Receiving the Management	Purpose of Management	Number of HHs	Management Problem	Number of HHs	Species Mostly Affected	Causative Agents	Number of HHs
Government nurseries	99	Pollarding	Cordia africana Lam.	To reduce shade effects on crops	120	Land tenure	116	Mangifera indica L.	Aphids and ants	23
Private and communal nurseries	27	Coppicing	C. africana	To obtain planting material	120	Termite attack	93	C. africana	Wood borers	20
Neighboring village nurseries	22	Pollarding	Eucalyptus camaldulensis Dehnh.	To obtain construction wood	85	Low survival of seedlings	41	Coffea arabica L.	Insects	22
Naturally regenerating seedlings	10	Coppicing	E. camaldulensis	To obtain planting material	85	Scarcity of water	35	Citrus sinensis (L.) Osbeck, C. aurantium L., C. medica L.	Black spots on fruits and leaves	23
-	-	Pollarding	Jatropha curcas L.	To optimize the height of live fences and to harvest fuelwood	25	Disease and pest attacks (other than termite attack)	23	Catha edulis (Vahl) Forssk. ex Endl.	Black spots on stems and leaves	15
-	-	-	-	-	-	Decline of soil fertility	22	-	-	-
-	-	-	-	-	-	Scarcity of seedlings	20	-	-	-

References

1. Myers, N.; Mittermeier, R.A.; Mittermeier, C.G.; da Fonseca, G.A.B.; Kent, J. Biodiversity hotspots for conservation priorities. *Nature* **2000**, *403*, 853–858. [CrossRef]
2. Herkenrath, P.; Harrison, J. The 10th meeting of the Conference of the Parties to the Convention on Biological Diversity-a breakthrough for biodiversity? *Oryx* **2011**, *45*, 1–2. [CrossRef]
3. Córdova, R.; Hogarth, N.; Kanninen, M. Sustainability of smallholder livelihoods in the ecuadorian highlands: A comparison of agroforestry and conventional agriculture systems in the indigenous territory of Kayambi People. *Land* **2018**, *7*, 45. [CrossRef]
4. Chakravarty, S.; Ghosh, S.; Suresh, C. Deforestation: Causes, effects and control strategies. In *Global Perspectives on Sustainable Forest Management*; IntechOpen: London, UK, 2011; pp. 3–29. [CrossRef]
5. Scrieciu, S.S. Economic causes of tropical deforestation: A global empirical application. In *Development Economics and Public Policy Working Paper 4*; Institute for Development Policy and Management, University of Manchester: Manchester, UK, 2003; Available online: http://hummedia.manchester.ac.uk/institutes/gdi/publications/workingpapers/depp/depp_wp04.pdf (accessed on 15 March 2018).
6. Denboba, M.A. *Forest Conversion-Soil Degradation-Farmers' Perception Nexus: Implications for Sustainable Land Use in the Southwest of Ethiopia*; Cuvillier Verlag: Göttingen, Germany, 2005; Volume 26, pp. 1–9. Available online: https://cuvillier.de/uploads/preview/public_file/5336/3865374441.pdf (accessed on 28 April 2018).
7. CSA (Central Statistic Agency). *2007 Population and Housing Census of Ethiopia: Administrative Report*; Central Statistical Agency: Addis Ababa, Ethiopia, 2012; pp. 1–117. [CrossRef]
8. Schulze, C.H.; Waltert, M.; Kessler, P.J.A.; Pitopang, R.; Shahabuddin; Veddeler, D.; Mühlenberg, M.; Gradstein, S.R.; Leuschner, C.; Steffan-Dewenter, I.; et al. Biodiversity indicator groups of tropical land-use systems: Comparing plants, birds, and insects. *Ecol. Appl.* **2004**, *14*, 1321–1333. [CrossRef]
9. Rojahn, D.A. Incentive mechanisms for a sustainable use system of the montane rain forest in Ethiopia. Inaugural-Dissertation. Berlin. 2006. Available online: https://www.zef.de/uploads/tx_zefportal/Publications/32e7_Thesis_Rojahn.pdf (accessed on 23 February 2018).
10. McNeely, J.A.; Schroth, G. Agroforestry and biodiversity conservation-traditional practices, present dynamics, and lessons for the future. *Biodivers. Conserv.* **2006**, *15*, 549–554. [CrossRef]
11. Harvey, C.A.; González Villalobos, J.A. Agroforestry systems conserve species-rich but modified assemblages of tropical birds and bats. *Biodivers. Conserv.* **2007**, *16*, 2257–2292.
12. Jose, S. Agroforestry for ecosystem services and environmental benefits: An overview. *Agrofor. Syst.* **2009**, *76*, 1–10. [CrossRef]
13. Dawson, I.K.; Guariguata, M.R.; Loo, J.; Weber, J.C.; Lengkeek, A.; Bush, D.; Cornelius, J.; Guarino, L.; Kindt, R.; Orwa, C.; et al. What is the relevance of smallholders' agroforestry systems for conserving tropical tree species and genetic diversity in circa situm, in situ and ex situ settings? A review. *Biodivers. Conserv.* **2013**, *22*, 301–324. [CrossRef]
14. Garrity, D.P.; Akinnifesi, F.K.; Ajayi, O.C.; Weldesemayat, S.G.; Mowo, J.G.; Kalinganire, A.; Larwanou, M.; Bayala, J. Evergreen Agriculture: A robust approach to sustainable food security in Africa. *Food Secur.* **2010**, *2*, 197–214. [CrossRef]
15. Khumalo, S.; Chirwa, P.W.; Moyo, B.H.; Syampungani, S. The status of agrobiodiversity management and conservation in major agroecosystems of Southern Africa. *Agric. Ecosyst. Environ.* **2012**, *157*, 17–23. [CrossRef]
16. Boshier, D.H.; Gordon, J.E.; Barrance, A.J. Prospects for Agro-Ecosystems: Mesoamerican Dry-Forest. Available online: http://forest-genetic-resources-training-guide.bioversityinternational.org/fileadmin/bioversityDocs/Training/FGR_TG/additional_materials/BoshierGordonBarrance2004.pdf (accessed on 22 September 2017).
17. Balmford, A.; Bennun, L.; Ten Brink, B.; Cooper, D.; Côté, I.M.; Crane, P.; Dobson, A.; Dudley, N.; Dutton, I.; Green, R.E.; et al. The convention on biological diversity's 2010 target. *Science* **2005**, *307*, 212–213. [CrossRef]
18. Kumar, B.M.; Nair, P.K.R. The enigma of tropical home-gardens. *Agrofor. Syst.* **2004**, *61–62*, 135–152. [CrossRef]
19. Nair, P.K. *Classification of agroforesty systems*; Kluwer Academic Publishers: Dordrecht, the Netherlands; Boston, MA, USA; London, UK, 1993; Volume 73, ISBN 0-7923-2134-0.

20. Sistla, S.A.; Roddy, A.B.; Williams, N.E.; Kramer, D.B.; Stevens, K.; Allison, S.D. Agroforestry practices promote biodiversity and natural resource diversity in atlantic Nicaragua. *PLoS ONE* **2016**, *11*, 1–20. [CrossRef]

21. Das, T.; Das, A.K. Inventorying plant biodiversity in home-gardens: A case study in Barak Valley, Assam, North East India. *Curr. Sci.* **2005**, *89*, 155–163. Available online: https://pdfs.semanticscholar.org/f942/1ffd047fb538187f4c7404614bc6b7ebf1b4.pdf (accessed on 27 September 2018).

22. Pandey, C.B.; Rai, R.B.; Singh, L.; Singh, A.K. Home-gardens of Andaman and Nicobar, India. *Agric. Syst.* **2007**, *92*, 1–22. [CrossRef]

23. Asfaw, B.; Lemenih, M. Traditional agroforestry systems as a safe haven for woody plant species: A case study from a topo-climatic gradient in south central Ethiopia. *For. Trees Livelihoods* **2012**, *8028*. [CrossRef]

24. Schroth, G.; Harvey, C.A.; da Fonseca, G.A.; Gascon, C.; Vasconcelos, H.L.; Izac, A.M.N. (Eds.) *Agroforestry and Biodiversity Conservation in Tropical Landscapes*; Island Press: Washington, DC, USA, 2004; ISBN 1-55963-357-3.

25. Kang, B.T.; Akinnifesi, F.K. Agroforestry as alternative land-use production systems for the tropics. *Nat. Resour. Forum* **2000**, *24*, 137–151. [CrossRef]

26. Bishaw, B.; Abdelkadir, A. Agroforestry and community forestry for rehabilitation of degraded watersheds on the Ethiopian highlands. *Combat. Famine Ethiop.* **2003**, *7*, 1–22. Available online: https://scholarworks.wmich.edu/cgi/viewcontent.cgi?referer=https://scholar.google.co.kr/&httpsredir=1&article=1056&context=africancenter_icad_archive (accessed on 17 March 2017).

27. Nguyen, Q.; Hoang, M.H.; Öborn, I.; van Noordwijk, M. Multipurpose agroforestry as a climate change resiliency option for farmers: An example of local adaptation in Vietnam. *Clim. Chang.* **2013**, *117*, 241–257. [CrossRef]

28. Nair, P.K.R.; Kumar, B.M.; Nair, V.D. Agroforestry as a strategy for carbon sequestration. *J. Plant Nutr. Soil Sci.* **2009**, *172*, 10–23. [CrossRef]

29. Nair, P.K.R.; Nair, V.D.; Kumar, B.M.; Haile, S.G. Soil carbon sequestration in tropical agroforestry systems: A feasibility appraisal. *Environ. Sci. Policy* **2009**, *12*, 1099–1111. [CrossRef]

30. Awad, B.; Tahir, E.; Gebauer, J. *Non-timber Forest Products: Opportunities and Constraints for Poverty Reduction in the Nuba Mountains, South Kordofan, Sudan*; Deutscher Tropentag: Berlin, Germany, 2004; Available online: http://www.tropentag.de/2004/abstracts/full/93.pdf (accessed on 26 April 2018).

31. Steppler, H.A.; Nair, P.K.R. Agroforestry-a decade of development. *Exp. Agric.* **1987**, *393*. [CrossRef]

32. Gijsbers, H.J.M.; Kessler, J.J.; Knevel, M.K. Dynamics and natural regeneration of woody species in farmed parklands in the Sahel region (Province of Passore, Burkina Faso). *For. Ecol. Manag.* **1994**, *64*, 1–12. [CrossRef]

33. Nikiema, A. Agroforestry parkland species diversity: Uses and management in semi-arid west Africa (Burkina Faso). Ph.D. Thesis, Wageningen University, Wageningen, the Netherlands, 2005.

34. Brown, R.L.; Reilly, L.A.J.; Peet, R.K. Species richness: Small scale. *eLS* **2016**, 1–9. [CrossRef]

35. Tolera, M.; Asfaw, Z.; Lemenih, M.; Karltun, E. Woody species diversity in a changing landscape in the south-central highlands of Ethiopia. *Agric. Ecosyst. Environ.* **2008**, *128*, 52–58. [CrossRef]

36. Kassa, H.; Gebrehiwet, K.; Yamoah, C. Balanites aegyptiaca, a potential tree for parkland agroforestry systems with sorghum in northern Ethiopia. *J. Soil Sci. Environ. Manag.* **2010**, *1*, 107–114.

37. Irisha, J.; Ilham, J.; Esa, N. Contribution of local knowledge towards urban agroforestry as a sustainable approach on climate change adaptation. In *SHS Web of Conferences*; EDP Sciences: Les Ulis, France, 2018; pp. 1–5. [CrossRef]

38. Hamilton, A.J. Species diversity or biodiversity? *J. Environ. Manag.* **2005**, *75*, 89–92. [CrossRef]

39. Davari, N.; Jouri, M.H.; Ariapour, A. Comparison of measurement indices of diversity, richness, dominance, and even-ness in rangeland ecosystem (case study: Jvaherdeh-Ramesar). *J. Rangel. Sci.* **2011**, *2*, 389–398. Available online: http://www.rangeland.ir/article_513066_9dd37707d60c8765a5009f1700761eab.pdf (accessed on 28 March 2018).

40. Ssegawa, P.; Nkuutu, D.N. Diversity of vascular plants on Ssese islands in lake Victoria, central Uganda. *Afr. J. Ecol.* **2006**, *44*, 22–29. [CrossRef]

41. Aigbe, H.I.; Akindele, S.O.; Onyekwelu, J. Tree species diversity and density pattern in Afi river forest reserve, Nigeria. *Int. J. Sci. Technol. Res.* **2014**, *3*, 178–185.

42. Duelli, P.; Obrist, M.K. Biodiversity indicators: The choice of values and measures. *Agric. Ecosyst. Environ.* **2003**, *98*, 87–98. [CrossRef]

43. Sarkar, S. *Defining Biodiversity; Assessing Biodiversity*; University of Texas at Austin: Austin, TX, USA, 2002; Volume 85, ISBN 0026-9662.

44. Gotelli, N.J.; Colwell, R.K. Quantifying biodiversity: Procedures and pitfalls in the measurement and comparison of species richness. *Ecol. Lett.* **2001**, *4*, 379–391. [CrossRef]

45. Solow, A.; Polasky, S. Measuring biological diversity: Rejoinder. *Environ. Ecol. Stat.* **1994**, *1*, 107. Available online: https://ricottalab.files.wordpress.com/2015/10/solow-polasky-1994.pdf (accessed on 10 July 2017).

46. Liu, Z.; Liu, G.; Fu, B.; Zheng, X. Relationship between plant species diversity and soil microbial functional diversity along a longitudinal gradient in temperate grasslands of Hulunbeir, inner Mongolia, China. *Ecol. Res.* **2008**, *23*, 511–518. [CrossRef]

47. Omernik, J.M. The misuse of hydrologic unit maps for extrapolation, reporting and ecosystem management. *J. Am. Water Resour. Assoc.* **2003**, *39*, 563–573. Available online: http://www.ecologicalregions.info/data/pubs/Omernik2003JAWRA_Misuse_of_HydroUnits.pdf (accessed on 22 March 2018).

48. Salarian, T.; Jouri, M.H.; Askarizadeh, D.; Mahmoudi, M. The study of diversity indices of plants species using SHE Method (Case study: Javaherdeh rangelands, Ramsar, Iran). *J. Rangel. Sci.* **2015**, *5*, 27–34. Available online: http://www.rangeland.ir/article_512665_202f2271f9eb3bc640bc7d5442dc19fd.pdf (accessed on 18 December 2017).

49. Mbuya, B.L.; Msanga, H.; Ruffo, C.; Birnie, A.; Joseph Banda, B.; Banda, P. *Agroforestry extension manual for eastern Zambia*; The Technical Handbook Series of the Regional Land Management Unit (RELMA/Sida); Regional Land Management Unit (RELMA/Sida): Nairobi, Kenya, 1992; Volume 14, ISBN 9966-896-03-1.

50. Raymond, C.M.; Fazey, I.; Reed, M.S.; Stringer, L.C.; Robinson, G.M.; Evely, A.C. Integrating local and scientific knowledge for environmental management. *J. Environ. Manag.* **2010**, *91*, 1766–1777. [CrossRef]

51. Asfaw, Z.; AAgren, G.I. Farmers' local knowledge and topsoil properties of agroforestry practices in Sidama, Southern Ethiopia. *Agrofor. Syst.* **2007**, *71*, 35–48. [CrossRef]

52. Salick, J.; Anderson, D.; Woo, J.; Sherman, R.; Cili, N.; Dorje, S. Bridging scales and epistemologies: Linking local knowledge and global science in multi-scale assessments. *Millenn. Ecosyst. Assess.* **2004**, 1–12. [CrossRef]

53. Kumar, B.M.; Takeuchi, K. Agroforestry in the western Ghats of peninsular India and the satoyama landscapes of Japan: A comparison of two sustainable land use systems. *Sustain. Sci.* **2009**, *4*, 215–232. [CrossRef]

54. Tadesse, E.; Asfaw, Z. Woody species richness use diversity and management in agroforestry practices: The case of Assosa District Benishangul Gumuz Region. *J. Biodivers. Manag.* **2017**. [CrossRef]

55. Muleta, D. *Plant Nutrient Soil Fertility and Plant Nutrient Management*; Ethiopian Institute of Agricultural Research (EIAR): Addis Ababa, Ethiopia, 2018; pp. 164–166. ISBN 978-99944-66-52-8. Available online: http://www.eiar.gov.et (accessed on 15 October 2018).

56. Teshome, M.; Eshete, A.; Bongers, F. Uniquely regenerating frankincense tree populations in western Ethiopia. *For. Ecol. Manag.* **2017**, *389*, 127–135. [CrossRef]

57. Crowley, E.L. *Rapid Data Collection Using Wealth Ranking and Other Techniques*; Working Paper; International Centre for Research in Agroforestry/Tropical Soils Biology and Fertility Program: Nairobi, Kenya, 1997; Available online: https://scholar.google.co.kr/scholar?hl=ko&as_sdt=0%2C5&q (accessed on 20 November 2017).

58. Azene, B.-T.; Tengnäs, B. *Useful Trees and Shrubs of Ethiopia: Identification, Propagation, and Management for 17 Agroclimatic Zones*; Relma in ICRAF Project World Agroforestry Centre: Nairobi, Kenya, 2007; 522p, ISBN 92-9059-212-5.

59. Awas, T. Plant Diversity in Western Ethiopia: Ecology, Ethnobotany and Conservation. Ph.D. Thesis, University of Oslo, Oslo, Norway, 2007.

60. Magurran, A.E. *Ecological diversity and its measurement*; Springer: Dordrecht, the Netherlands, 1988; ISBN 978-94-015-7360-3.

61. Ugland, K.I.; Gray, J.S.; Ellingsen, K.E. The species–accumulation curve and estimation of species richness. *J. Anim. Ecol.* **2003**, *72*, 888–897.

62. Arrhenius, O. Species and Area. *J. Ecol.* **1921**, *9*, 95–99.

63. Proença, V.; Pereira, H.M. Species–area models to assess biodiversity change in multi-habitat landscapes: The importance of species habitat affinity. *Basic Appl. Ecol.* **2013**, *14*, 102–114.

64. Wang, X.H.; Kent, M.; Fang, X.F. Evergreen broad-leaved forest in eastern China: Its ecology and conservation and the importance of resprouting in forest restoration. *For. Ecol. Manag.* **2007**, *245*, 76–87. [CrossRef]

65. R Core Team. A Language and Environment for Statistical Computing. Reference Index. 2017. Available online: http://softlibre.unizar.es/manuales/aplicaciones/r/fullrefman.pdf (accessed on 3 February 2019).

66. Woldeyes, F.; Asfaw, Z.; Demissew, S.; Roussel, B. Home-gardens (Aal-oos-gad) of the basketo people of southwestern Ethiopia: Sustainable agro-ecosystems characterizing a traditional landscape. *Ethnobot. Res. Appl.* **2016**, *14*, 549–563. [CrossRef]

67. Mengistu, B.; Asfaw, Z. Woody species diversity and structure of agroforestry and adjacent land uses in Dallo Mena district, south-east Ethiopia. *Nat. Resour.* **2016**, *7*, 515–534. [CrossRef]

68. Jacob, V.J.; Alles, W.S. Kandyan gardens of Sri Lanka. *Agrofor. Syst.* **1987**, *5*, 123–137. [CrossRef]

69. Fernandes, E.C.M.; Oktingati, A.; Maghembe, J. The Chagga home-gardens: A multistoried agroforestry cropping system on Mt. Kilimanjaro (northern Tanzania). *Agrofor. Syst.* **1985**, *2*, 73–86. [CrossRef]

70. Subitha, P.; Sukumaran, S.; Jeeva, S. Inventorying Plant Diversity in the Home-Gardens of Kuzhicodu Village, Kanyakumari District, Tamilnadu, India. 2016, Volume 6, pp. 28–43. Available online: http://www.jsrr.net/Vol%206%20No%201%20April%2016/P%20Subitha28-43.pdf (accessed on 12 August 2017).

71. Chandrashekara, U.M. Tree species yielding edible fruit in the coffee-based home-gardens of Kerala, India: Their diversity, uses and management. *Food Secur.* **2009**, *1*, 361–370. [CrossRef]

72. Rugalema, G.H.; Johnsen, F.H.; Rugambisa, J. The home-garden agroforestry system of Bukoba district, north-western Tanzania. 2. Constrainsts to farm productivity. *Agrofor. Syst.* **1994**, *26*, 205–214. [CrossRef]

73. Gebrehiwot, M. Recent Transitions in Ethiopian Home-Garden Agroforestry: Driving Forces and Changing Gender Relations. Ph.D. Thesis, Swedish University of Agricultural Sciences, Umeå, Sweden, 2013.

74. Pattanayak, S.K.; Mercer, D.E.; Sills, E.; Yang, J.C. Taking stock of agroforestry adoption studies. *Agrofor. Syst.* **2003**, *57*, 173–186. Available online: https://link.springer.com/content/pdf/10.1023%2FA%3A1024809108210.pdf (accessed on 1 February 2019).

75. Walters, B.B. Do property rights matter for conservation? Family land, forests and trees in Saint Lucia, west Indies. *Hum. Ecol.* **2012**, *40*, 863–878. [CrossRef]

76. Biggelaar, C.D. *A Synthesis of Results of the FTPP Farmer-Initiated Research and Extension Practices Initiative in East Africa: Towards a Development-Oriented Approach to Natural Resource Mana*; Network Paper-Rural Development Forestry Network 21f; ODI: London, UK, 1997; Available online: https://www.odi.org/sites/odi.org.uk/files/odi-assets/publications-opinion-files/1160.pdf (accessed on 18 April 2017).

77. Dury, S.J. Agroforestry for soil conservation. *Agric. Syst.* **1991**, *35*, 472–473. [CrossRef]

78. Perera, A.H.; Rajapakse, R.M.N. A baseline study of Kandyan forest gardens of Sri Lanka: Structure, composition and utilization. *For. Ecol. Manag.* **1991**, *45*, 269–280. [CrossRef]

79. Emmanuel, T. *Agroforestry, Food and Nutritional Security*; ICRAF Working Paper No. 170; World Agroforestry Centre: Nairobi, Kenya, 2013; pp. 28–30. [CrossRef]

80. Dupriez, H.; Leener, P.D. *Trees and multistorey Agriculture in Africa*; Terres et Vie: Nivelles, Belgium, 1998; p. 280. ISBN 287105010x.

81. Akpo, L.E.; Goudiaby, V.A.; Grouzis, M.; Le Houerou, H.N. Tree shade effects on soils and environmental factors in a savanna of Senegal. *West Afr. J. Appl. Ecol.* **2005**, *7*, 41–52.

82. Baig, M.B.; Ahmad, S.; Khan, N.; Khurshid, M. Germplasm conservation of multipurpose trees and their role in agroforestry for sustainable agricultural production in Pakistan. *Int. J. Agric. Biol.* **2008**, *10*, 340–348. Available online: http://www.fspublishers.org (accessed on 7 January 2018).

83. CABI ISC (Centre for Agriculture and Biosciences International Invasive Species Compendium). Invasive Species Compendia Database. 2016. Available online: http://www.cabi.org/isc/ (accessed on 2 February 2019).

84. Invasive Species Specialist Group. *The Global Invasive Species Database*; Version 2015.1.; Invasive Species Specialist Group: Rome, Italy; IUCN: Gland, Switzerland, 2015.

© 2019 by the authors. Licensee MDPI, Basel, Switzerland. This article is an open access article distributed under the terms and conditions of the Creative Commons Attribution (CC BY) license (http://creativecommons.org/licenses/by/4.0/).

forests

MDPI

Article

Landscape-Scale Mixtures of Tree Species are More Effective than Stand-Scale Mixtures for Biodiversity of Vascular Plants, Bryophytes and Lichens

Steffi Heinrichs [1,*], Christian Ammer [1], Martina Mund [1], Steffen Boch [2], Sabine Budde [1], Markus Fischer [3], Jörg Müller [4,5], Ingo Schöning [6], Ernst-Detlef Schulze [6], Wolfgang Schmidt [1], Martin Weckesser [1] and Peter Schall [1]

[1] Department Silviculture and Forest Ecology of the Temperate Zones, University of Goettingen, Büsgenweg 1, D-37077 Göttingen, Germany; christian.ammer@forst.uni-goettingen.de (C.A.); mmund@gwdg.de (M.M.); sabine@budde-forst.de (S.B.); wschmid1@gwdg.de (W.S.); martinweckesser@web.de (M.W.); peter.schall@forst.uni-goettingen.de (P.S.)
[2] Swiss Federal Research Institute WSL, Zürcherstrasse 111, CH-8903 Birmensdorf, Switzerland; steffen.boch@wsl.ch
[3] Institute of Plant Sciences, University of Bern, CH-3013 Bern, Switzerland; Markus.Fischer@ips.unibe.ch
[4] Institute for Biochemistry and Biology, University of Potsdam, Maulbeerallee 1, 14469 Potsdam, Germany; joerg.mueller@sielmann-stiftung.de
[5] Department of Nature Conservation, Heinz Sielmann Foundation, Unter den Kiefern 9, D-14641 Wustermark, Germany
[6] Max-Planck Institute for Biogeochemistry, D-07745 Jena, Germany; ingo.schoening@bgc-jena.mpg.de (I.S.); dschulze@bgc-jena.mpg.de (E.-D.S.)
*   Correspondence: sheinri@gwdg.de; Tel.: +49-551-39-5974

Received: 27 December 2018; Accepted: 17 January 2019; Published: 19 January 2019

Abstract: Tree species diversity can positively affect the multifunctionality of forests. This is why conifer monocultures of Scots pine and Norway spruce, widely promoted in Central Europe since the 18th and 19th century, are currently converted into mixed stands with naturally dominant European beech. Biodiversity is expected to benefit from these mixtures compared to pure conifer stands due to increased abiotic and biotic resource heterogeneity. Evidence for this assumption is, however, largely lacking. Here, we investigated the diversity of vascular plants, bryophytes and lichens at the plot (alpha diversity) and at the landscape (gamma diversity) level in pure and mixed stands of European beech and conifer species (Scots pine, Norway spruce, Douglas fir) in four regions in Germany. We aimed to identify compositions of pure and mixed stands in a hypothetical forest landscape that can optimize gamma diversity of vascular plants, bryophytes and lichens within regions. Results show that gamma diversity of the investigated groups is highest when a landscape comprises different pure stands rather than tree species mixtures at the stand scale. Species mainly associated with conifers rely on light regimes that are only provided in pure conifer forests, whereas mixtures of beech and conifers are more similar to beech stands. Combining pure beech and pure conifer stands at the landscape scale can increase landscape level biodiversity and conserve species assemblages of both stand types, while landscapes solely composed of stand scale tree species mixtures could lead to a biodiversity reduction of a combination of investigated groups of 7 up to 20%.

Keywords: *Fagus sylvatica*; *Pinus sylvestris*; *Picea abies*; *Pseudotsuga menziesii*; forest management; tree species diversity; forest conversion; gamma diversity; landscape scale; Biodiversity Exploratories

1. Introduction

In recent years, the effect of tree species diversity on forest ecosystem functions has been intensively investigated from local to continental scale (e.g., [1–4]). Research results generally hint towards a positive effect of tree species diversity on primary productivity [5] and stability [6] compared to respective monocultures, though depending on environmental conditions and species combinations [7] and varying in space and time [8]. Neighborhood interactions among different tree species in terms of stress release and facilitation have been shown to drive both diversity-productivity (e.g., [9,10]) and diversity-stability relationships [11,12]. This indicates the importance of a distinct intermingling of different tree species at the stand scale to support some key forest ecosystem functions.

Tree species mixtures within forest stands are also assumed to conserve and promote associated understory biodiversity better than monocultures (e.g., [13–15]). The understory is a key component of the diversity of primary producers in temperate forests and contributes to element cycling and functioning of above and belowground food webs [16–18]. Thus, understanding its responses to canopy changes is essential for implementing biodiversity-orientated forest management concepts including the broad promotion of mixed stands instead of monocultures.

Within mixtures, understory species (including vascular plants and soil dwelling and epiphytic cryptogams) may either respond positively to the presence of a specific host tree species (e.g., [19]) or may benefit from the small-scale heterogeneity within stands provided by different tree species in terms of light transmittance [20], nutrient and water availability [21] or litter accumulation [22]. These findings are in accordance with the positive heterogeneity-diversity relationship (e.g., [23,24]). However, pure additive effects of different tree species within a stand may lead to an accumulation of species associated to either tree species but they may not contribute to increased landscape-level diversity [25]. Interactive effects among tree species, on the other hand, may result in an increased species pool in mixtures compared to a combination of pure stands and thus increase diversity at the landscape level [13,26]. For example, shade-tolerant understory species can be impeded by a dense and persistent litter layer (e.g., due to a negative effect on seedling establishment or due to different morphological abilities of plants to emerge through layers of tree litter) in a closed-canopy monoculture of a given tree species [22], and the same species could be outcompeted by dominant light-demanding understory species in a more open stand of another tree species. Instead, within mixtures of both tree species where litter is reduced but the canopy is still closed, these understory species might find suitable conditions. In addition, a more efficient resource use of neighboring and complementary tree species in mixtures (e.g., by a more effective fine root system, [27]) might limit these resources for understory species in the direct vicinity creating wider resource gradients within mixed stands [13].

While additive [19] and interactive [26] effects among tree species have been verified for epiphytic communities such as bryophytes and lichens, there is not much scientific evidence for a positive effect of mixtures regarding understory vascular plant diversity. In their review studies, Barbier et al. [28] and Cavard et al. [25] showed that maximum diversity (mainly expressed as species richness or species diversity (Shannon-diversity) at the stand scale) was mainly observed in pure stands of different tree species and not in mixed stands composed of these species. So far, previous studies have generally focused on alpha diversity at a particular plot or stand level. This focus may have masked the heterogeneity among mixed stands, e.g. in terms of mixture ratios of different tree species, and potentially underestimated the supported species pool. Landscape-level comparisons are rare (see [29]) and have mainly contrasted species composition of mixtures to only one of the respective pure stand types [25].

Conifer monocultures of Scots pine (*Pinus sylvestris* L.) and Norway spruce (*Picea abies* (L.) H. Karst.) have been widely promoted outside their natural range on harvested sites across Central Europe since the 18th and 19th century. Their fast growth rate, undemanding regeneration and management, and manifold usability of their wood made them the economically most important tree species in temperate forests of Europe [30,31]. Decreasing site quality and growth reductions on soils with limited cation availability, the susceptibility to wind throw and pathogens [32,33], and the call for other ecosystem services apart from timber production have led to a large-scale conversion of pure conifer stands into mixed stands with site-adapted broadleaved tree species in recent decades [30,34,35]. As European beech (*Fagus sylvatica* L.) would naturally dominate large parts of Central Europe [36], beech is the main tree species used in this conversion process. A positive effect of forest conversion on forest biodiversity is widely assumed (e.g., [37,38]), but studies confirming this assumption, particularly with respect to understory biodiversity at the landscape scale, are missing.

Here, we assessed the effect of tree species mixtures of European beech (in the following beech) with a conifer species on plot level (alpha diversity) and landscape level (gamma diversity) diversity of vascular plants, bryophytes and lichens (including species growing on soil, deadwood or epiphytically on the bark of trees and shrubs) compared to the respective pure stands. We used data from four regions in Germany differing in climate, geology, tree species composition and forest management. While the western and eastern lowlands of northern Germany were largely forested with Scots pine (in the following pine) on exploited, nutrient poor sandy soils, Norway spruce (in the following spruce) was largely planted in mountain areas in central and southern Germany on different substrates. According to current management plans, mixtures with beech will largely replace these conifer monocultures in the future (e.g., [38,39]). The non-native Douglas fir (*Pseudotsuga menziesii* (Mirb.) Franco) shows higher growth rates and higher stability e.g. in terms of drought [40] than spruce in low mountain ranges, so that it is regarded as an important conifer species in mixture with beech under the expected climate change.

The objectives of our study were: (1) to quantify alpha and gamma diversity of understory vascular plants, bryophytes and lichens in pure and mixed stands by investigating tree species combinations of beech and pine, beech and spruce and beech and Douglas fir (no data for bryophytes and lichens); and (2) to identify an optimized composition of pure and mixed stands within hypothetical forest landscapes in favor of the three groups and in favor of a regional biodiversity when combining these groups.

We generally expected a higher gamma diversity in mixed compared to pure stands as mixtures should contain species that are associated to beech as well as species associated to conifer forests. By this, hypothetical forest landscapes comprising mixed stands only should be equally (only additive effects between tree species) or more (additive and interactive effects between tree species) diverse than a combination of monocultures of the given tree species. If the presence of pure and mixed stand types at the landscape level supports maximum gamma diversity (e.g., [26]), the different stand types should harbor characteristic species that either benefit from a gradient of resources in mixtures or from a higher resource quantity within pure stands [41].

2. Materials and Methods

2.1. Study Regions and Data Sampling

We analyzed data sampled in four regions in Germany (Table 1) ranging from the northern lowlands in western (Northwestern (NW-)Germany) and eastern Germany (Schorfheide-Chorin) to mountain areas of central (Solling Hills) and southern Germany (Schwäbische Alb). All four regions would be naturally dominated by beech (with a small contribution of sessile oak in eastern Germany) but are currently (co-) dominated by conifers, mainly by pine and spruce (Table 1).

In the four regions, vegetation surveys in 400 m^2 plots (20 m × 20 m) were conducted as part of different research projects covering a range of tree species compositions (for details, see, [42–44]). In all regions, plant species composition was assessed per plot for the tree layer ($\geq$5 m height) and for the understory (woody species <5 m height and all herbaceous species) by recording the species presence and their cover value in %. In the study regions Schorfheide-Chorin and Schwäbische Alb, as part of the Biodiversity Exploratories, the presence of bryophyte and lichen species was additionally recorded on the ground, on deadwood and on the bark of shrubs and trees up to a height of about 2 m of the stem on a subset of the vegetation survey plots (for details see [45,46]). In this study, we excluded species growing on rocks to minimize potential impacts of local to regional differences in geology and geomorphology. By focusing on the understory, we may have underestimated the overall cryptogam species richness by not including species restricted to tree crowns. This bias can be particularly high for lichen diversity in beech forests [47], but can be neglected for bryophytes [47,48].

Table 1. Characteristics of the study regions. Data are based on information on forest growth regions and districts provided by Gauer and Aldinger [49].

	Northwestern Germany	Schorfheide-Chorin	Solling Hills	Schwäbische Alb
Area	~25,000 km²	~1300 km²	~350 km²	~420 km²
Coordinates	53°18′ N–53°39′ N; 8°30′ E–10°40′ E	52°52′ N–53°12′ N; 13°37′ E–14°1′ E	51°40′ N–51°50′ N; 9°26′ E–9°44′ E	48°21′ N–48°31′ N; 9°13′ E–9°31′ E
Elevation	0–150 m a.s.l.	3–140 m a.s.l.	300–450 m a.s.l.	460–860 m a.s.l.
Bedrock	Glacial sedimentary deposits (partly with loess cover); old Pleistocene	Glacial series; young Pleistocene	Red sandstone with loess cover	Jurassic limestone
Predominant soil type	Sandy-podsol-cambisol	Cambisol	Acid silty loam cambisol	Cambisol and leptosol
Mean annual temperature	8.1–9.3 °C	8.2–8.6 °C	7.3–7.8 °C	5.7–7.6 °C
Annual Precipitation	560–840 mm	555–590 mm	915–1030 mm	843–1096 mm
Potential natural vegetation	Acidic beech forest	Acidic to mesic beech forests (partly with sessile oak)	Acidic beech forest	Beech forests on limestone
Tree species composition	Forest area ~ 27% Share pine ~ 37% Spruce ~ 25% Beech ~ 5% Other conifers ~ 7%	Forest area ~ 48% Share pine ~ 69% Beech ~ 7% Oak ~ 6%	Forest area ~ 94% Share spruce ~ 62% Beech ~ 30% Oak ~ 8%	Forest area ~ 47% Share beech ~ 39% Spruce ~ 38% Other broadleaves ~ 12%
Investigated forest types (stand age, establish-ment)	Pure beech (76–120 yrs, natural regeneration) Pure pine (51–82 yrs, planted) Pure Douglas fir (53–64 yrs, planted) Beech/pine mixture (79–161 yrs, beech planted under pine) Beech/Douglas fir mixture (61–98 yrs, Douglas fir planted in beech regeneration)	Pure beech (~70–160 yrs, natural regeneration) Pure pine (~30–70 yrs, planted) Beech/pine mixture (46–130 yrs, beech planted under pine)	Pure beech (52–150 yrs, natural regeneration) Pure spruce (54–132 yrs, planted) Beech/spruce mixture (48–149 yrs, spruce planted in beech regeneration gaps)	Pure beech (~60–100 yrs, natural regeneration) Pure spruce (~40–60 yrs, planted) Beech/spruce mixture (~60–100 yrs, spruce planted in beech regeneration gaps)
No. of plots	100	278	167	196

For the present analyses, we selected those surveys conducted in pure forest stands of beech or one conifer tree species and in forest stands representing mixtures of both tree species. In summary, three tree species combinations were studied: (1) beech and pine and respective mixtures in NW-Germany and Schorfheide-Chorin); (2) beech and Douglas fir and respective mixtures in NW-Germany; and (3) beech and spruce with respective mixtures in the regions Solling and Schwäbische Alb (Table 1). Thereby, plots were selected and classified as pure or mixed based on the species composition of the tree layer. When either beech or the conifer species had a share of >90% on the accumulated tree layer cover (=sum of cover values of all species of the tree layer), plots represented pure forest stands. In mixed stands, beech and conifers had a combined share of >70% on accumulated tree layer cover with no other tree species exceeding a 10% share. Thus, the mixed forests comprised a large gradient ranging from beech to conifer dominance (Table A1 in Appendix A). Since we were interested in the differences between the stand types independent of large-scale anthropogenic or natural disturbances, only survey plots with a minimum accumulated tree layer cover of 30% were included in this study. The third region of the Biodiversity Exploratories, the Hainich-Dün (see, [44–46]), was not considered, as studied plots mainly covered pure beech stands and mixtures of beech with other broadleaved tree species, while mixtures with conifers were missing.

In the stands of the regions Solling and NW-Germany, pH values of the upper mineral soil were measured in 1 M KCl and a soil/solution ratio of 1:2 [42,43]. In Schorfheide-Chorin and Schwäbische Alb, pH values of the upper mineral soil were determined across the investigated forest stands in 0.01 M CaCl$_2$ using a soil/solution ratio of 1:2.5 [46].

All investigated stands represent regularly thinned (every 5 to 10 years) age class forests. Beech stands regenerated naturally after repeated shelterwood cuttings mainly on ancient forest sites, while conifer stands were planted after clear cutting in the past ([42,43,50,51]; Table 1). The mature mixtures of beech and pine date back to the beginning and middle of the 20th century when beech was planted into young pure pine stands to improve site conditions [51]. Beech/Douglas fir mixtures resulted from plantings of Douglas fir saplings into natural beech regeneration of similar age. Due to the fast growth rate of Douglas fir, these stands developed into two-layered stands. Spruce within mixtures is generally 15 to 25 years younger than beech and was introduced into beech forest gaps by planting [52]. Mixtures generally represent single tree to group mixtures of beech with conifers.

Stands of Schwäbische Alb were predominantly in the mature timber stage with beech stands and beech/spruce mixtures being 60 to 100 years old and pure spruce stands being 40 to 60 years old. In Solling, stand age ranged from 48 to 150 years with 40% (pure beech) to 60% (pure spruce) of the plots being younger than 90 years. In NW-Germany, stand age varied between 51 and 161 years with pure beech stands, beech/pine and beech/Douglas fir mixtures being mainly older than 90 years. Pure pine and pure Douglas fir stands were younger than 85 and 65 years, respectively. Pure pine stands in Schorfheide-Chorin were equally distributed between mature and immature timber with a stand age of ca. 30 to 70 years, while pure beech stands and beech/pine mixtures were almost exclusively represented by stands in the mature timber stage. While mixed stands were approximately 50 to 70 years (up to 130 years) old, beech stands were mainly older than 70 years up to 160 years (Tables 1 and A1).

Selected plots were representative for the surrounding forest stands of ca. 0.5 to 4 ha size.

2.2. Data Analysis

Data analyses focused on two main points: (1) the characterization of pure and mixed stands of each region regarding mean tree layer cover, soil pH, alpha and gamma diversity, the average number of exclusive species per stand type and an identification of characteristic species per stand type. (2) The identification of a composition of pure and mixed stands in hypothetical forest landscapes that can maximize the gamma diversity of vascular plants for the four study regions and additionally for bryophytes and lichens for the regions Schorfheide-Chorin and Schwäbische Alb. For the latter regions, we further determined a stand type composition for a maximum combined regional diversity of vascular plants, bryophytes and lichens.

All analyses were conducted using the *R* software version 3.5.0 [53]. Nomenclature of vascular plant species follows Wisskirchen and Haeupler [54], of bryophyte species Koperski et al. [55], and of lichen species Wirth [56].

2.2.1. Analyzing Differences among Stand Types

Mean tree layer cover and soil pH were compared between pure and mixed stands across all available plots and soil sampling points within regions using the Kruskal-Wallis-H-test followed by the Mann-Whitney-U test. All other comparisons for vascular plants were based on 1000 resamplings of 17 plots per stand type to avoid effects of unequal sample sizes across stand types and regions on species richness [57]. Thus, from the number of available plots (Table A1) we randomly drew 17 plots per stand type and repeated this 1000 times. We focused on 17 plots to allow for at least 1000 unique plot combinations per region (e.g., for NW-Germany drawing 17 out of 20 plots results in 1140 unique plot combinations). Alpha diversity was quantified as mean species richness per plot per resampling. Gamma diversity represented the accumulated species richness across the 17 plots per resampling (*R* package iNEXT version 2.0.12, function ChaoRichness; [58]). For bryophytes and lichens, we increased the number of randomly drawn plots to the maximum number of plots allowing for at least 1000 unique plot combinations (i.e., 36 plots for bryophytes (minimum of 9139 unique combinations) and 22 plots for lichens in Schorfheide-Chorin (minimum of 2300 unique combinations); 26 plots for bryophytes in Schwäbische Alb (minimum of 3654 unique combinations)). For lichens in Schwäbische Alb, only 12 plots were available (Table A1). We therefore resampled 10 out of 12 plots up to 66 unique combinations only.

2.2.2. Finding an Optimized Composition of Stand Types for Gamma Diversity of Vascular Plants, Bryophytes and Lichens

For identifying a composition of pure and mixed stands that can maximize gamma diversity of vascular plants, bryophytes and lichens, we resampled the same number of pure and mixed stands (see Section 2.2.1) in a way that all compositional combinations were represented in steps of 10% with 1000 replications (=hypothetical forest landscapes). In total, we built 66 compositions of stand types (=66 points in Figure 1) comprising only pure beech, pure conifer or mixed stands (cyan points; Figure 1) as well as compositions with almost equal proportions of the three stand types (orange point in Figure 1 with 40% mixed stands (=e.g., 7 of 17 resampled plots) and 30% pure beech and conifer stands (=e.g., 5 plots each of 17 resampled)). For each composition, gamma diversity was quantified per resampling using the accumulated species richness across resampled plots. We analyzed the effect of stand type composition on gamma diversity using general additive models with two factorial full tensor product spline smoothers with function te (*R* package mgcv version 1.8-23 [59] and function gam (gamma diversity ~ te(conifer, beech))). We report R^2 and the degrees of freedom based on the 66,000 replications (66 compositions × 1000 resamplings) or the 4356 replications for lichens in Schwäbische Alb (66 compositions × 66 resamplings). Ternary diagrams were used to visualize response surfaces (*R* package ggtern version 2.2.1 [60]). We inferred for significant differences of gamma diversity between compositions by pairwise comparisons of resamplings between stand type compositions with average maximum and average minimum gamma diversity using the two-sided $p < 0.05$ (e.g., for the stand type composition with the on average highest gamma diversity at least 975 resamplings resulted in higher gamma diversity compared to the stand type composition with on average minimum gamma diversity). In the same way we checked for significant differences between the maximum diversity stand type composition and gamma diversity of 100% pure (pure beech and pure conifer) and 100% mixed stands.

2.2.3. Quantification of Exclusive Species Numbers per Stand Type

Stand types would contribute to gamma diversity at the landscape level, when they support exclusive species only occurring in a specific stand type. For mixed stands, if they contain both, species associated with beech and species associated with conifer stands, we hypothesized that the number of exclusive species should be low for pure and mixed stand types, when directly contrasted. To test this, we quantified the number of exclusive species per stand type for each tree species combination and taxonomic group investigated. Quantification of exclusive species was again based on 1000 resamplings.

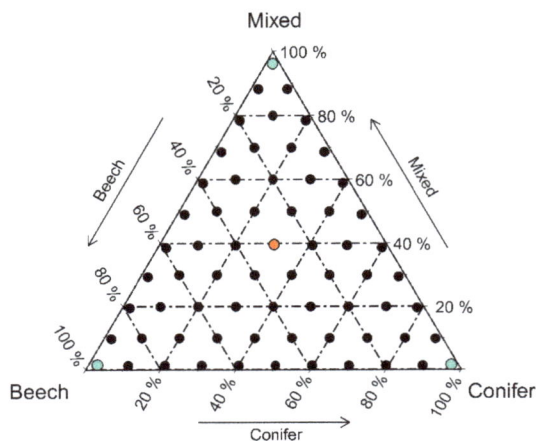

Figure 1. Stand type compositions (66 points) representing all combinations of pure beech, pure conifer and mixed stands in steps of 10% and arranged within a schematic triangle. The corners represent hypothetical forest landscapes of 100% pure beech, pure conifer and mixed stands, respectively (cyan points). Towards the center of the triangle, the three stand types are mixed with the center point (orange point) representing a composition with 40% mixed, 30% pure beech and 30% pure conifer stands.

2.2.4. Identification of Characteristic Species per Stand Type

The quantification of exclusive species only returns an absolute number of species per stand type and resampling independent of a species' general specifity to or frequency in a specific stand type. To identify vascular plant and cryptogam species that are significantly associated with the investigated stand types (=characteristic species), we additionally conducted indicator species analyses according to Dufrêne and Legendre [61] using the *R* package indicspecies version 1.7.6 and the function strassoc. This analysis calculates the specifity of a species as the number of occurrences of a species within a stand type relative to the number of occurrences across all stand types (S = specifity) and the frequency of the species within the stand type (F = frequency). Multiplying S and F results in an indicator value between 0 (species not occurring in a specific stand type) and 1 (species occurring exclusively and always in a specific stand type). Based on resamplings of 17 plots per stand type, we calculated the values S and F per species and stand type for 1000 times (vascular plants in NW-Germany: 10 plots out of 20 plots for 1000 times; lichens in Schorfheide-Chorin: 12 plots out of a minimum of 25 plots for 1000 times; lichens in Schwäbische Alb: 6 plots out of a minimum of 12 for 500 times). Multiplying S and F then calculated the indicator value per species and stand type for all species with $S > 0.4$ and $F > 0.2$ in at least one stand type. By this, we excluded species with an average frequency lower than 20% across resamplings. Specifity was set to 40% to allow for species with preferences for two stand types (e.g., when $S = 0.4$ in a pure and in the mixed stand, both stand types cover 80% of the occurrences of a particular species). Characteristic species per stand type were determined by pairwise comparisons of calculated indicator values between the pure and the mixed stands (i.e., for two-sided $p < 0.05$ at least 975 of 1000 pairwise comparisons showed a larger indicator value for one stand type compared to the others). Species were characteristic for two stand types when indicator values did not differ between them but were larger compared to the third stand type. Identified species were categorized by the environmental Ellenberg indicator values (EIV) for light (L), moisture (M), nutrients (N) and acidity/reaction (R; [62]). EIV assign vascular plant, bryophyte and lichen species along 9-point scales with the value 1 representing species indicating deep shade, dry, nutrient poor or acidic conditions [62]. Species were further classified into forest affinity categories according to the forest species list for vascular plants, bryophytes and lichens ranging from closed forest species to those preferring open site conditions [63].

2.2.5. Finding an Optimized Composition of Stand Types for Regional Diversity

To summarize the effect of stand type composition across vascular plants, bryophytes and lichens for the regions Schorfheide-Chorin and Schwäbische Alb, we quantified multidiversity [64] as a measure for regional diversity combining the three investigated groups. This regional diversity was calculated as the average relative diversity of the taxonomic groups weighted by the species number of the groups (log weighting) to account for general differences in the number of vascular plants, bryophytes and lichens. When a composition of stand types shows a regional diversity of nearly 100%, then all three groups are equally supported close to the optimum. We analyzed the effect of stand type composition on regional diversity using general additive models with two factorial spline smoothers (as we did for single taxa), taking the relative diversity of the taxonomic groups as response variable and their species number as weight. For regional diversity, we additionally considered Shannon diversity (ChaoShannon function of the iNEXT *R* package [58]) which down weights infrequent species.

3. Results

3.1. Environmental Conditions

Pure conifer stands were characterized by lowest tree layer cover compared to the other stand types (except for beech/spruce mixtures in Schwäbische Alb), while pure beech stands showed highest values with no significant difference to beech/pine and beech/Douglas fir mixtures (Figure 2a).

There were no significant differences in soil pH among stand types in NW-Germany (for neither tree species combination, Figure 2b). For the other regions, pure beech stands showed higher soil pH values compared to pure pine and beech/pine mixtures in Schorfheide-Chorin and compared to pure spruce stands in Solling and Schwäbische Alb. Pure beech and beech/spruce mixtures showed no significant differences.

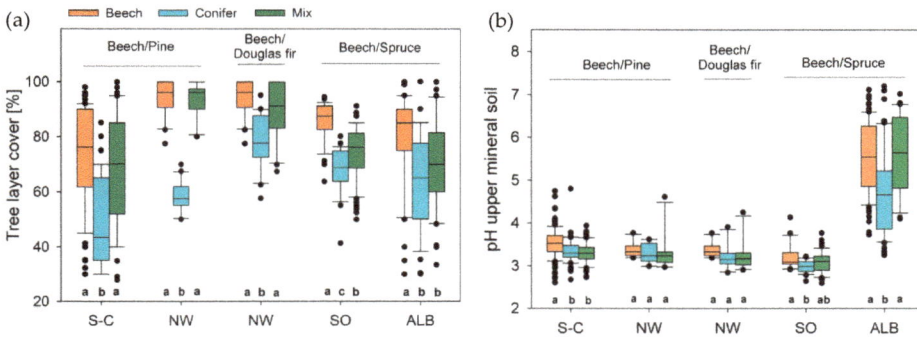

Figure 2. (**a**) Tree layer cover and (**b**) soil pH for pure and mixed stands sampled in four regions across Germany. Diagrams are arranged according to the considered tree species combination. Tree layer cover is based on the number of vegetation survey plots listed in Table A1. Soil pH is based on the following sample sizes: S-C: beech *n* = 139, pine *n* = 60, mix *n* = 76; NW (both combinations): *n* = 10 per forest type; SO: beech *n* = 23, spruce *n* = 24, mix *n* = 47; ALB: beech *n* = 87, spruce *n* = 53; mix *n* = 32. Different lowercase letters mark significant differences between stand types within regions (S-C: Schorfheide-Chorin, NW: Northwestern Germany, SO: Solling, ALB: Schwäbische Alb).

3.2. Diversity Patterns of Vascular Plants

Alpha diversity of vascular plants was highest for all conifer stands with no significant difference to pure beech stands in Schorfheide-Chorin and to mixed stands in Schwäbische Alb. Beech/pine mixtures showed lowest values in both study regions, whereas diversity values of beech/Douglas fir and beech/spruce mixtures were intermediate (Table 2). Within mixtures, different proportions of the two tree species had only small effects on alpha diversity (Figure A1). Species richness slightly decreased with an increasing proportion of beech compared to pine in Schorfheide-Chorin and compared to spruce in Solling.

Gamma diversity patterns were similar with highest values for pure conifer stands across regions and tree species combinations except for Schorfheide-Chorin. For all regions, mixtures showed lower gamma diversity than at least one of the pure stand type (Table 2). Mixtures of beech and pine showed lowest gamma diversity values, whereas mixtures of beech and Douglas fir/spruce were intermediate between the respective pure stands.

Table 2. Alpha and gamma diversity (species richness) of vascular plants in pure beech, pure conifer and mixed stands (=corners of the triangle in Figure 1), and the composition of pure and mixed stands within hypothetical forest landscapes that supports maximum and minimum gamma diversity of vascular plants. Given are mean values of the 1000 resamplings based on 17 plots per resampling. Minimum and maximum resampling values are given in parenthesis [a].

	Beech/Pine (Be/Pi)	Beech/Douglas fir (Be/Dgl)		Beech/Spruce (Be/Spr)	
Region	Schorfheide-Chorin	Northwestern Germany		Solling	Schwäbische Alb
n	17	17	17	17	17
	Alpha Diversity				
Beech	14.0a (7.2–20.8)	9.0b (6.7–10.2)	8.9c (6.7–10.2)	10.4c (6.1–14.8)	22.5b (16.8–28.8)
Conifer	15.9a (13.1–18.7)	18.8a (17.8–19.8)	21.4a (18.6–23.4)	26.1a (20.7–31.8)	36.9a (25.8–48.7)
Mixed	8.6b (4.9–13.1)	7.1c (6.1–8.0)	14.7b (11.9–16.7)	17.7b (11.6–23.6)	33.1a (25.7–40.6)
	Gamma Diversity				
Beech	71.2a (49–94)	53.3c * (41–56)	53.2c * (41–56)	40.9c * (22–56)	87.7c * (60–114)
Conifer	64.5ab (48–83)	60.9a (54–64)	78.9a (67–83)	77.4a (62–88)	153.5a (117–187)
Mixed	49.0b * (29–64)	39.0c * (30–42)	66.2b * (44–70)	61.6b* (43–78)	123.7b * (101–148)
	Stand Type Composition				
Maximum gamma diversity	60%Be-40%Pi-0%Mix	40%Be-60%Pi-0%Mix	0%Be-100%Dgl-0%Mix	0%Be-100%Spr-0%Mix	0%Be-100%Spr-0%Mix
	74.8 (49–97)	61.8 (47–76)	78.9 (67–83)	77.4 (62–88)	153.5 (117–187)
Minimum gamma diversity	0%Be-0%Pi-100%Mix	0%Be-0%Pi-100%Mix	100%Be-0%Dgl-0%Mix	100%Be-0%Spr-0%Mix	100%Be-0%Spr-0%Mix
	49.0 * (29–64)	39.0 * (30–42)	53.2 * (41–56)	40.9 * (22–56)	87.7 * (60–114)

[a] Significant difference among stand types and among stand type compositions was inferred by pairwise comparisons of resamplings (i.e., for two-sided $p < 0.05$ at least 975 of 1000 comparisons showed larger values for one stand type compared to the other). Different lowercase letters mark significant differences in alpha and gamma diversity among the three stand types. * marks a significant difference to the landscape composition supporting maximum gamma diversity.

The composition approach showed that pure stand types supported a maximum gamma diversity of vascular plants (Figure 3a–e, Table 2). This accounted either for a combination of pure beech and pure pine stands (Figure 3a,b) or for pure Douglas fir or spruce stands only (Figure 3c–e). Maximum gamma diversity was significantly higher compared to the gamma diversity of a hypothetical landscape composed of mixed stands only (Table 2). On average beech/pine mixtures reduced the diversity of vascular plants by 34.5% (Schorfheide-Chorin) to 36.9% (NW-Germany) compared to the maximum, whereas the beech/Douglas fir mixture showed a reduction by 16.1% and the beech/spruce mixtures by 19.4% (Schwäbische Alb) to 20.4% (Solling). In both regions with pure pine stands, beech/pine mixtures supported minimum gamma diversity. For the Douglas fir and spruce combinations, 100% pure beech stands were least diverse. Maximum and minimum values were comparable among regions with acidic soil conditions (see Figure 2b), but were twice as much for the Schwäbische Alb on limestone. Nevertheless, the latter region showed a similar diversity pattern compared to Solling with the same tree species combination (Figure 3d,e).

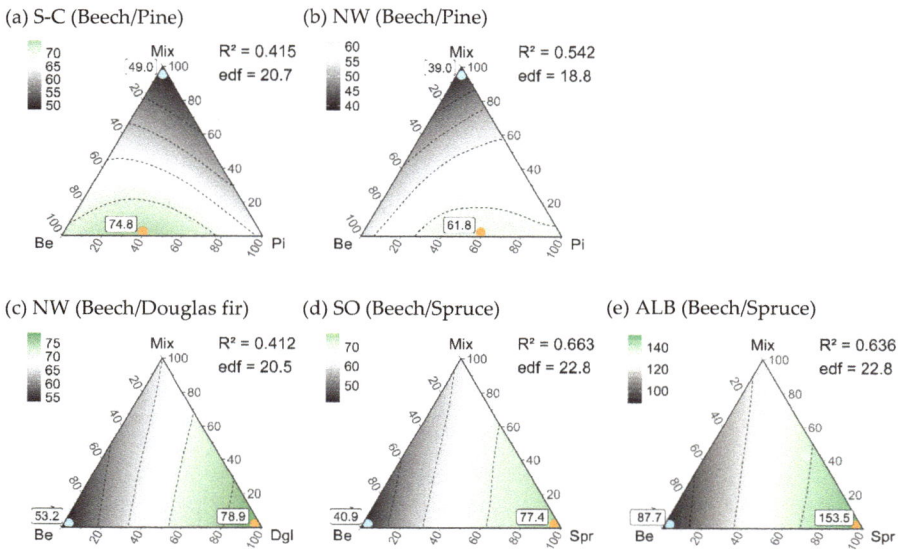

Figure 3. Gamma diversity (species richness) of vascular plants along compositional gradients of pure beech (Be), pure conifer (Pi = Pine, Dgl = Douglas fir, Spr = Spruce) and mixed stands (Mix) for (**a**) Schorfheide-Chorin (S-C), (**b,c**) Northwestern Germany (NW), (**d**) Solling (SO), (**e**) Schwäbische Alb (ALB). The stand type composition was varied in steps of 10% using 1000 resamplings of 17 plots per step (66 unique stand type compositions). The diversity response to stand type composition is characterized by R^2 and estimated degrees of freedom (edf). Labelled dots mark the maximum (orange) and minimum (light blue) gamma diversity.

Stand type composition patterns for maximum diversity are in line with detected exclusive species numbers (Figure 4). While beech/pine mixtures had on average 6.5 (Schorfheide-Chorin) and 7.6 (NW-Germany) exclusive species per resampling when compared to the respective pure stands, pure pine stands supported on average 19.5 and 19.3 exclusive species. Mean exclusive species numbers of pure beech stands (26.5 species in Schorfheide-Chorin, 12.1 species in NW-Germany) also exceeded numbers of beech/pine mixtures.

When compared with pure spruce stands and beech/spruce mixtures, pure beech stands supported the lowest number of exclusive species (with no significant difference to mixtures). In general, the conifer stands on acidic sites (see Figure 2b) showed a remarkable constancy in exclusive species numbers across tree species ranging from on average 18.3 (Douglas fir) to 21.3 (spruce in Solling) exclusive species per resampling. On calcareous soils of Schwäbische Alb, stand types had twice as much exclusive species compared to the acidic Solling region. As the number of exclusive species characterizes resamplings ($n = 17$), numbers may change with increasing sampling completeness [65] indicating instable gamma diversity patterns. However, differences in exclusive species numbers remained relatively stable among stand types with an increasing number of resampled plots (Figure A2). Pure spruce stands of Schwäbische Alb and beech/Douglas fir mixtures showed a steady increase in exclusive species numbers with sampling effort. Thus, the detected stand type composition pattern for Schwäbische Alb even intensified with a higher number of resampled plots (Figure A3). A gamma diversity comparison of stand type combinations with pure Douglas fir stands and beech/Douglas fir mixtures, though, seems to require a higher sampling effort than covered by the present study.

Threatened or protected vascular plant species mainly occurred in the Schwäbische Alb and showed no significant response to stand type composition (Figure A4a,d; Table A2).

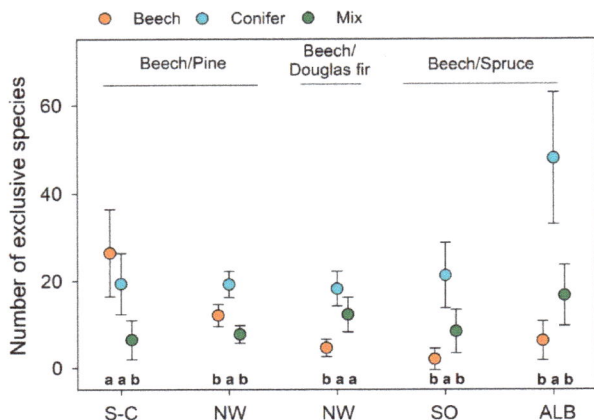

Figure 4. Mean number of exclusive species in pure beech, pure conifer and mixed stands of the four study regions based on pairwise comparisons of 1000 resamplings of 17 plots per stand type and region. Error bars indicate the 95% quantile of resamplings. Diagrams are arranged according to the considered tree species combination. Different lowercase letters indicate significant differences among stand types within regions (i.e., at least 975 of 1000 pairwise comparisons showed higher number of exclusive species of one stand type compared to the others), S-C: Schorfheide-Chorin, NW: Northwestern Germany, SO: Solling, ALB: Schwäbische Alb.

3.3. Diversity Patterns of Bryophytes and Lichens

Stand types of Schorfheide-Chorin showed no significant difference in alpha or gamma diversity for bryophytes and lichens (Table 3). The share of beech and pine on the tree layer within mixtures also had no effect on alpha diversity of both taxonomic groups (Figure A5). Similar to vascular plants, maximum mean gamma diversity per resampling was detected for a combination of pure beech and pure pine stands within a landscape but with no significant difference to the minimum (=100% mixed stands for bryophytes and 100% pure pine stands for lichens; Figure 5a,c).

In Schwäbische Alb, we found highest alpha diversity of bryophytes in conifer stands and highest alpha diversity of lichens in beech stands. The share of beech and spruce within mixtures had no significant effect on alpha diversity of both taxonomic groups (Figure A5). At the landscape scale, an equal share of pure beech and pure spruce stands supported maximum gamma diversity of bryophytes (Figure 5b). Hypothetical landscapes composed of 100% mixed stands or 100% pure spruce stands were significantly less diverse (Table 3). Lichens showed a clear minimum in a pure conifer landscape and were mainly associated with beech stands (Figure 5d). A mean maximum diversity was detected for a combination of 90% pure beech and 10% pure spruce stands, but with no difference to a landscape composed of 100% pure beech stands.

Table 3. Alpha and gamma diversity (species richness) of bryophytes and lichens in pure beech, pure conifer and mixed stands (=corners of the triangle in Figure 1), and the composition of pure and mixed stands that supports maximum and minimum gamma diversity of bryophytes and lichens in Schorfheide-Chorin and Schwäbische Alb. Given are mean values of 1000 resamplings based on the number of plots (*n*) per resampling. Minimum and maximum resampling values are given in parenthesis.[a]

	Schorfheide-Chorin (Beech(Be)/Pine(Pi))		Schwäbische Alb (Beech(Be)/Spruce(Spr))	
	Bryophytes	**Lichens**	**Bryophytes**	**Lichens**
n	36	22	26	10
	Alpha Diversity			
Beech	9.8b	5.0a	13.4b	19.5a
	(8.8–10.9)	(3.8–6.3)	(11.3–15.7)	(15.4–24.0)
Conifer	11.2a	5.4a	16.1a	9.0c
	(10.8–11.7)	(5.0–5.8)	(14.1–17.8)	(6.8–11.4)
Mixed	10.7ab	5.5a	14.1b	15.1b
	(9.9–11.7)	(4.6–6.4)	(13.0–14.9)	(12.7–17.4)
	Gamma Diversity			
Beech	38.0a	20.9a	60.7a	55.7a
	(29–48)	(12–32)	(49–69)	(44–64)
Conifer	38.1a	17.5a	57.0a *	30.6b *
	(35–39)	(15–18)	(50–63)	(23–37)
Mixed	36.1a	19.0a	58.8a *	49.7a
	(31–38)	(13–22)	(53–60)	(41–54)
	Stand Type Composition			
Maximum gamma diversity	50%Be-50%Pi-0%Mix	60%Be-40%Pi-0%Mix	50%Be-50%Spr-0%Mix	90%Be-10%Spr-0%Mix
	41.3	21.6	68.7	55.7
	(34–50)	(15–31)	(57–79)	(44–67)
Minimum gamma diversity	0%Be-0%Pi-100%Mix	0%Be-100%Pi-0%Mix	0%Be-100%Spr-0%Mix	0%Be-100%Spr-0%Mix
	36.1	17.5	57.0	30.6
	(31–38)	(15–18)	(50–63)	(23–37)

[a] Significant differences among stand types and minimum and maximum diversity stand type combinations were inferred by pairwise comparison of resamplings (i.e., for two-sided $p < 0.05$ at least 975 of 1000 comparisons showed larger values for one stand type compared to the others). For lichens in Schwäbische Alb only 10 out of 12 plots were resampled up to 66 times. Different lowercase letters mark significant differences in alpha and gamma diversity among the three stand types. * marks a significant difference from the stand type composition supporting maximum gamma diversity.

Mixed stands showed the lowest number of exclusive bryophyte species in Schorfheide-Chorin (with no significant difference to exclusive species numbers in pure beech stands) and Schwäbische Alb (Figure 6). For Schorfheide-Chorin, the total pool of exclusive bryophyte species within pure stands has not been reached yet, indicating a stabilization of the detected stand type composition pattern with increasing sampling effort (Figure A6). For lichens, beech stands showed the highest number of exclusive species, but with no significant difference to pure pine stands in Schorfheide-Chorin and to mixed stands in Schwäbische Alb.

Threatened and protected species of bryophytes and lichens rarely occurred in surveyed plots of Schorfheide-Chorin (in less than 20% of surveyed plots per stand type) not allowing for a robust resampling (Tables A3 and A4). In Schwäbische Alb, threatened and protected lichen species responded similarly as overall lichen diversity to stand type composition (Figure A4c,f). Threatened bryophyte species were promoted by a high share of pure beech stands (Figure A4b) or by pure beech and pure spruce stands when single occurrences were not considered (Shannon diversity, Figure A4e).

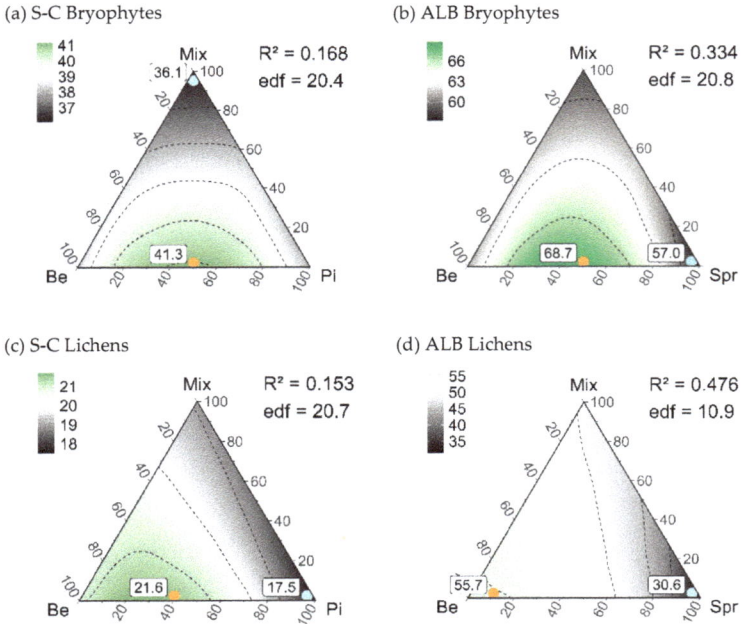

Figure 5. Gamma diversity (species richness) of bryophytes (**a**,**b**) and lichens (**c**,**d**) along compositional gradients of pure beech (Be), pure conifer (Pi = Pine, Spr = Spruce) and mixed stands (Mix) in Schorfheide-Chorin (S-C; (**a**,**c**)) and Schwäbische Alb (ALB; (**b**,**d**)). The stand type composition was varied in steps of 10% using 1000 resamplings of 36 (bryophytes S-C), 26 (bryophytes ALB), 22 (lichens S-C) plots per step (66 unique stand type compositions). For lichens of ALB 10 plots out of 12 were resampled for 66 times. The diversity response to stand type composition is characterized by R^2 and estimated degrees of freedom (edf). Labelled dots mark the maximum (orange) and minimum (light blue) gamma diversity.

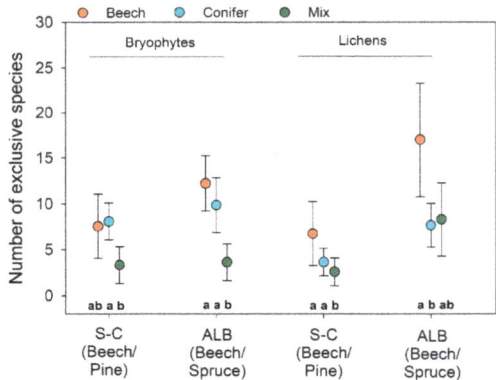

Figure 6. Mean number of exclusive bryophyte and lichen species in pure beech, pure conifer and mixed stands of Schorfheide-Chorin (S-C) and Schwäbische Alb (ALB) based on pairwise comparisons of 1000 resamplings of plots per stand type and region (number of resampled plots see Table 3). Error bars indicate the 95% quantile of resamplings. Different lowercase letters indicate significant differences among stand types within regions (i.e., at least 975 of 1000 pairwise comparisons showed higher number of exclusive species of one stand type compared to the others).

3.4. Characteristic Species within Stand Types

While differences in exclusive species numbers among stand types can explain the contribution of each stand type to gamma diversity, this measure gives no information on the identity of species or their specifity to or frequency in a stand type. Using indicator species analyses, we therefore identified those species that were significantly more strongly associated with one or two stand types compared to the third (see Tables 4–6). These characteristic species were presumably often exclusive per resampling in the specific stand type.

Among vascular plants, we found no characteristic species for mixed stands compared to both pure stands except for *Pteridium aquilinum* in NW-Germany when compared to pure beech and pine stands. The same species, though, characterized pure pine stands in Schorfheide-Chorin. In contrast, 25 species were significantly more strongly associated with pure conifer stands compared to pure beech and mixed stands across regions and tree species combinations (Table 4). In addition, 19 species were characteristic for mixed and pure conifer stands at the same time compared to pure beech stands in at least one region (=C/M; Table 4).

Despite a high number of exclusive species in beech stands of Schorfheide-Chorin, only few species occurred with a sufficient average frequency of more than 20% per resampling to be categorized as being characteristic for beech stands in Schorfheide-Chorin. However, six recorded species (including *Anemone nemorosa*) were significantly more strongly associated with beech stands compared to pure pine stands and beech/pine mixtures. Compared to pure spruce stands and beech/spruce mixtures, no species was significantly more strongly associated with beech stands. With *A. nemorosa* and *A. ranunculoides*, two species were associated with pure beech stands and with beech/spruce mixtures (B/M; Table 4).

Species characteristic of pure conifer stands are more light demanding (higher mean light indicator value) compared to species of beech stands or compared to species also associated with mixtures. A lower share of species closely associated to forests also characterizes them (=forest affinity categories). Species assigned to the categories 1.1 (closed forests species) and 1.2 (species of edges and clearings) within the list of forest species [63] made up 24% among conifer associated species, 80% among beech associated species and 50% among those species also associated with mixtures in at least one region (Table 4).

Table 4. Vascular plant species associated with pure conifer (C), pure beech (B) or mixed stands (M) in the four investigated regions based on indicator species analyses. [a] S-C = Schorfheide-Chorin, NW = Northwestern Germany, SO = Solling, ALB = Schwäbische Alb. Species order follows a decreasing Ellenberg indicator value (EIV) for light (L); x = indifferent species; Ø = mean EIV values.

Characteristic Species of	Beech/Pine		Beech/ Douglas Fir	Beech/Spruce		EIV				FA
	S-C	NW	NW	SO	ALB	L	M	R	N	
pure conifer stands										
Calluna vulgaris		C				8	x	1	1	2.1
Epilobium angustifolium		C		C		8	5	5	8	1.2
Rumex acetosella	C					8	3	2	2	2.2
Rubus fruticosus agg. [b]		C	C			7.7	5.0	4.8	4.6	2.1
Agrostis capillaris				C		7	x	4	4	2.1
Betula pendula	C	C				7	x	x	x	2.1
Cirsium palustre					C	7	8	4	3	2.1
Digitalis purpurea				C		7	5	3	6	1.2
Galeopsis tetrahit				C		7	5	x	6	2.1
Galium mollugo agg. [c]					C	7	5	7	5	2.2
Galium saxatile		C		C		7	5	2	3	2.1
Molinia caerulea		C				7	7	x	2	2.1
Pinus sylvestris		C				7	x	x	x	2.1
Quercus robur		C				7	x	x	x	2.1

Table 4. *Cont.*

Characteristic Species of	Beech/Pine		Beech/Douglas Fir	Beech/Spruce		EIV				FA
	S-C	NW	NW	SO	ALB	L	M	R	N	
Taraxacum sect. *Ruderalia*				C		7	5	x	8	2.1
Cerastium holosteoides		C				6	5	x	5	2.2
Frangula alnus	C					6	8	4	x	2.1
Myosotis sylvatica					C	6	5	x	7	1.2
Stellaria media				C		6	x	7	8	2.2
Veronica officinalis					C	6	4	3	4	2.1
Ceratocapnos claviculata		C				5	5	3	6	1.2
Dryopteris carthusiana		C				5	x	4	3	2.1
Epilobium montanum					C	4	5	6	6	2.1
Athyrium filix-femina			C			3	7	x	6	1.1
Galium rotundifolium					C	2	5	5	4	1.1
Ø						6.3	5.4	4.1	4.8	
pure stands										
Moehringia trinervia		C/B				4	5	6	7	1.1
pure beech stands										
Carex pilulifera		B				5	5	3	3	2.1
Viola riviniana	B					5	4	4	x	1.1
Impatiens parviflora	B					4	5	x	6	1.1
Milium effusum	B					4	5	5	5	1.1
Carex remota		B				3	8	x	x	1.1
Ø						4.2	5.4	4.0	4.7	
pure and mixed stands in at least one region										
Urtica dioica				C/M		x	6	7	9	2.1
Euphorbia cyparissias					C/M	8	3	x	3	2.1
Hypericum perforatum					C/M	7	4	6	4	2.1
Rubus idaeus		C	C/M		C/M	7	x	x	6	1.2
Sambucus nigra					C/M	7	5	x	9	2.1
Sambucus racemosa					C/M	6	5	5	8	2.1
Deschampsia flexuosa	C	C	C/M	C		6	x	2	3	2.1
Pteridium aquilinum	C	M				6	5	3	3	1.1
Sorbus aucuparia	C	C	C/M	C	C	6	x	4	x	2.1
Geranium robertianum					C/M	5	x	x	7	2.1
Picea abies			C/M	C/M		5	x	x	x	2.1
Trientalis europaea		C	C/M	C		5	x	3	2	1.1
Vaccinium myrtillus	C/M	C	C	C		5	x	2	3	2.1
Dryopteris dilatata		C	C/M	C/M		4	6	x	7	1.1
Impatiens noli-tangere					C/M	4	7	7	6	1.1
Mycelis muralis			C/M	C	C/M	4	5	x	6	2.1
Abies alba			C/M			3	x	x	x	1.1
Pseudotsuga menziesii [d]			C/M			3	6	3	3	1.1
Oxalis acetosella			C/M			1	5	4	6	1.1
Anemone nemorosa	B				B/M	x	5	x	x	1.1
Anemone ranunculoides					B/M	3	6	8	8	1.1
Fagus sylvatica		B/M	B/M			3	5	x	x	1.1
Ø						5.4	5.2	4.7	5.6	

[a] Species selection was restricted to species with a mean specifity of S > 40% and a mean frequency of F > 20% per resampling in at least one stand type. A significant association is based on pairwise comparisons of indicator values (=S × F) among stand types across resamplings (e.g., for a species characteristic of pure conifer stands (C), indicator values were higher in 975 of 1000 resamplings for conifer stands compared to beech and mixed stands; B = species characteristic for beech stands; M = characteristic in mixed stands). Species with higher indicator values in one pure and the mixed stand were categorized as conifer/mixed (C/M) or beech/mixed characteristic species (B/M); Ellenberg indicator values (EIV) for light (L), moisture (M), reaction/acidity (R) and nutrients (N). Forest affinity (FA; 1.1 = species of closed forests, 1.2 = species of edges and clearings, 2.1 = species of forests and open sites, 2.2 = species predominantly in open sites). [b] indicator values represent the mean of all *Rubus* species assigned to this taxonomic aggregate. [c] indicator values for *Galium album* as the most common species of this aggregate used. [d] Indicator values according to Landolt et al. [66] adjusted to the Ellenberg-scale.

Twelve bryophyte species were characteristic of pure conifer stands compared to the other stand types in at least one region, while four species were characteristic of pure beech stands. Ten species were also associated with mixed stands in combination with one of the pure stand types (C/M and B/M: 5 species each). Most of the identified bryophyte species are able to grow on different substrates. Among species that were significantly more strongly associated with pure beech stands or beech stands and mixtures, four species do not use soil as substrate (Table 5).

Characteristic lichen species of pure and mixed stands were mainly identified for Schwäbische Alb (Table 6). For this region, three species were significantly more strongly associated with conifer stands and four species with beech stands. Five additional species were characteristic in pure beech stands and beech/spruce mixtures compared to pure spruce stands. In Schorfheide-Chorin, three lichen species were associated with mixed stands in combination with one pure stand type. All identified lichen species grow either epiphytic or on deadwood. However, 80% of the identified species associated with beech either in pure stands or in mixtures are obligatory epiphytes. For the five species that were more strongly associated with conifers, only two are obligatory epiphytes.

Differences in light indicator values among stand types for characteristic cryptogam species are not as pronounced as for vascular plants (see Tables 5 and 6). However, conifer associated cryptogam species supported a lower share of closely associated forest species (categories 1.1 and 1.2; 26.7%) compared to characteristic species of pure beech (37.5%) or of pure and mixed stands (38.8%).

Table 5. Bryophyte species associated with pure conifer, pure beech or mixed stands in Schorfheide-Chorin (S-C) and Schwäbische Alb (ALB). For details see Table 4. SUB gives the preferred substrate species grow on (S = soil, B = bark, D = deadwood) as listed by Schmidt et al. [63]. Rocks as substrate was not included. * marks threatened and protected species (see Table A3).

Characteristic Species of	Beech/Pine	Beech/Spruce	EIV			FA	SUB
	S-C	ALB	L	M	R		
pure conifer stands							
Hypnum jutlandicum	C		7	2	2	2.1	S,B,D
Lophocolea bidentata		C	7	6	5	2.1	S,B,D
Rhytidiadelphus squarrosus		C	7	6	5	2.2	S,D
Hylocomium splendens *		C	6	4	5	2.1	S,D
Pleurozium schreberi	C	C	6	4	2	2.1	S,D
Scleropodium purum	C	C	6	4	5	2.1	S,D
Dicranella heteromalla		C	5	4	2	2.1	S,B,D
Eurhynchium angustirete		C	5	4	7	1.1	S,B,D
Orthodontium lineare	C		4	5	2	2.1	S,B,D
Plagiochila asplenioides *		C	4	6	6	1.1	S,D
Plagiomnium undulatum		C	4	6	6	2.1	S,B,D
Thuidium tamariscinum		C	4	6	4	2.1	S,B,D
		Ø	5.4	4.8	4.3		
pure beech stands							
Amblystegium subtile *		B	7	5	6	1.1	B
Atrichum undulatum	B		6	6	4	2.1	S,D
Dicranum montanum	B		6	5	2	2.1	S,B,D
Isothecium alopecuroides *		B	5	5	6	1.1	S,B,D
		Ø	6.0	5.3	4.5		
pure and mixed stands in at least one region							
Rhytidiadelphus triquetrus *		C/M	7	4	5	2.1	S,D
Dicranum polysetum	C	C/M	6	4	5	2.1	S
Dicranum scoparium	C/M		5	4	4	2.1	S,B,D
Herzogiella seligeri		C/M	5	5	4	1.1	S,B,D
Rhytidiadelphus loreus		C/M	4	6	3	1.1	S,D
Frullania dilatata *		B/M	8	4	5	2.1	B
Radula complanata *		B/M	7	5	7	2.1	S,B,D
Pterigynandrum filiforme		B/M	6	5	4	1.1	B
Metzgeria furcata *		B/M	5	4	6	2.1	S,B,D
Ulota bruchii *		B/M	4	5	4	2.1	B,D
		Ø	5.7	4.6	4.7		

Table 6. Lichen species associated with pure conifer, pure beech and mixed stands in Schorfheide-Chorin (S-C) and Schwäbische Alb (ALB). For details see Table 4. nd = no data available. SUB gives the preferred substrate species grow on (B = bark, D = deadwood) as listed by Schmidt et al. [63]. Rocks as substrate was not included. * marks threatened and protected species (see Table A4).

Characteristic Species of	Beech/Pine	Beech/Spruce	EIV				FA	SUB
	S-C	ALB	L	M	R	N		
pure conifer stands								
Platismatia glauca		C	7	5	2	2	1.1	B,D
Scoliciosporum chlorococcum		C	6	3	3	6	2.1	B
Micarea prasina		C	3	4	4	4	1.1	B,D
		Ø	5.3	4.0	3.0	4.0		
pure beech stands								
Xanthoria polycarpa		B	7	3	7	8	2.1	B,D
Arthonia radiata *		B	3	4	5	4	1.1	B
Lecanora subcarpinea *		B	nd	nd	nd	nd	2.2	B
Lecanora subrugosa *		B	nd	nd	nd	nd	2.1	B
		Ø	5.0	3.5	6.0	6.0		
pure and mixed stands in at least one region								
Lecanora conizaeoides	C/M		7	3	2	5	2.2	B
Hypocenomyce scalaris	C/M		6	3	2	2	2.1	B,D
Lecanora chlarotera		B/M	6	3	6	5	2.2	B
Phlyctis argena		B/M	5	3	4	5	2.1	B
Pertusaria leioplaca *		B/M	4	4	5	2	1.1	B
Graphis scripta *		B/M	3	4	5	3	1.1	B
Porina aenea	B/M		3	4	5	4	1.1	B
Arthonia spadicea		B/M	2	4	4	3	1.1	B,D
		Ø	4.5	3.5	4.1	3.6		

3.5. Composition of Stand Types for Highest Regional Diversity

Combining gamma diversity of the three investigated taxonomic groups into a regional diversity measure for Schorfheide-Chorin and Schwäbische Alb resulted in an optimized stand type composition for biodiversity of 50% pure beech and 50% pure pine stands or 60% pure spruce and 40% pure beech stands, respectively (Figure 7).

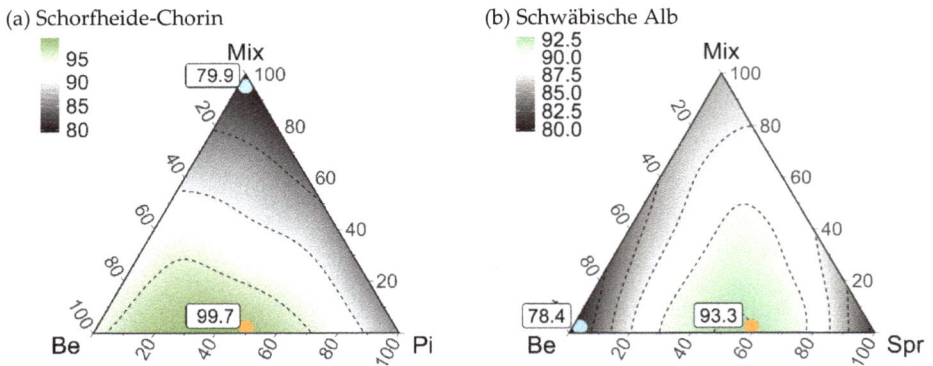

Figure 7. Regional diversity (species richness) in % for (**a**) Schorfheide-Chorin and (**b**) Schwäbische Alb combining vascular plants, bryophytes and lichens along compositional gradients of pure beech (Be), pure conifer (Pi = Pine, Spr = Spruce) and mixed stands (Mix). The stand type composition was varied in steps of 10% using 1000 resamplings of 22 plots per step for Schorfheide-Chroin and using 66 resamplings of 10 plots per step for Schwäbische Alb. The regional diversity quantifies the mean relative gamma diversity of taxonomic groups accounting for the absolute diversity within groups (log weighting of diversity). Labelled dots mark the maximum (orange) and minimum (light blue) regional diversity.

With 99.7% of regional diversity, a combination of pure beech and pure pine stands supported almost the total diversity of the three groups that was sampled in Schorfheide-Chorin, whereas a hypothetical forest landscape with 100% mixed stands reduced the maximum diversity by 19.8%. 100% pine stands supported a regional diversity of 86.1% (−13.6% of maximum gamma diversity), a hypothetical pure beech landscape of 93.8% (−5.9%). For Schwäbische Alb, a maximum regional diversity of 93.3% was reached. 100% mixed stands reduced the maximum diversity by 7.3%. This reduction was lower compared to hypothetical forest landscapes of 100% pure beech (−14.9%) or 100% pure spruce stands (−12.7%). Results for both regions were consistent for Shannon diversity (=down-weighting of infrequent species) with a slight shift of maximum diversity towards a higher share of pure beech stands in Schorfheide-Chorin (optimized composition of 40% pure pine and 60% pure beech stands; Figure A7).

4. Discussion

In contrast to our expectations we did not find higher gamma diversity in mixed compared to pure stands showing that neither additive nor interactive effects between the investigated tree species within a stand promoted the diversity of vascular plants, bryophytes or lichens better than pure stands. For no investigated taxonomic group, study region or tree species combination, a hypothetical forest landscape composed of only mixed stands was significantly more diverse than a landscape composed only of pure stands of a single or of different tree species. This was also the case for threatened and protected species. In addition, we found no stand type combination supporting maximum gamma diversity that included a share of mixed stands.

In the following, we discuss our findings for the different taxonomic groups separately and end with findings on regional diversity combining the investigated groups.

4.1. An Optimized Stand Type Composition for Vascular Plant Diversity

Pure beech and pure pine stands showed a certain degree of complementarity between their species assemblages (compare [65]). Thus, mixing of beech and pine stands at the landscape scale is effective for the gamma diversity of vascular plants (landscape scale diversity), mixing the two tree species at the stand level (=stand scale mixture) led to a reduced gamma diversity. Species that were more associated with beech stands (e.g., *Anemone nemorosa*, *Carex remota*, *Milium effusum* and the neophyte *Impatiens parviflora*) are species of closed forests, whereas species characteristic of pine stands can occur both in forests and at open sites (e.g., *Rubus fruticosus* agg.) or at edges and clearings (e.g., *Epilobium angustifolium*). Thus, differences in canopy cover and consequently light availability are most likely responsible for the detected complementarity between beech and pine stands in the study regions Schorfheide-Chorin and NW-Germany. In addition, species such as *Calluna vulgaris* or *Vaccinium myrtillus*, characteristic for pine stands, are adapted to nutrient deficient habitats and may be relict species of former agricultural and heathland management of sites now dominated by pine [43]. Some characteristic species of beech stands, on the other hand, have been identified as ancient woodland indicators for northern Germany (e.g., *Carex remota*, *Milium effusum*, *Viola riviniana*, [67]). While pure pine stands supported a higher gamma diversity of vascular plants in NW-Germany than pure beech stands, the latter were more diverse than pine stands in Schorfheide-Chorin. In the latter region, beech stands often grow on slightly more favorable soil conditions compared to pine and benefit from a heterogeneous geomorphology of the young Pleistocene landscape [68] compared to the homogeneous sandy sedimentations across northwestern Germany.

For both regions, though, beech/pine mixtures resulted in minimum gamma diversity. Vegetation composition in mixtures of beech and pine has already been described as being similar to beech due to the competitive strength of beech with no differences in the overall species pool [13,69]. Our results, however, show that beech/pine mixtures can neither accumulate beech and pine associated species nor can they fulfill the habitat function of pure beech stands. In a mixture with pine, beech trees may benefit from the lower intraspecific competition [70] and respond with an increased horizontal crown expansion leading to higher structural heterogeneity [71] as well as light absorption [72]. While an optimized use of canopy space can increase forest productivity or maintain a moist microclimate supporting decomposition processes or mitigating effects of global warming (summarized in [73]), the reduced light availability is detrimental for the understory diversity in the studied regions. In addition, belowground complementarity of the two tree species may negatively affect understory diversity. Brassard et al. [27] showed that mixtures of complementary tree species (e.g., early and late-successional, coniferous and deciduous species) enhance fine-root productivity by a more complete filling of the environment, including a higher horizontal volume filling [74,75]. This may increase root competition for the understory. The species that are partly associated with beech/pine mixtures in the two regions, *Vaccinium myrtillus*, *Pteridium aquilinum*, can grow in nutrient deficient habitats. In addition, *Pteridium aquilinum* indicates disturbances. Thus, the unidirectional loss of plant species in mixtures that are associated with both, pure beech and pure pine stands, indicates negative interactive effects among the tree species on a neighborhood scale that can decrease vascular plant understory diversity up to 37%.

No complementarity effects on species assemblages were found for tree species combinations with beech and spruce. Hypothetical forest landscapes of only pure spruce stands supported maximum gamma diversity in both study regions independent of soil type. This shows that pure spruce stands can support almost the total pool of vascular plant species associated with the investigated tree species within the regions. In contrast to Máliš et al. [76], our results give no evidence for a negative effect on typical beech forest species in the understory by pure spruce stands neither on acidic nor on calcareous soils. In fact, we found the lowest number of exclusive species within beech stands when contrasted to spruce stands and beech/spruce mixtures. *Anemone nemorosa* and *A. ranunculoides* were the only species mainly associated with beech both in pure stands and in mixtures (B/M in Table 4). Both spring geophytes are adapted to the high light availability before leaf unfolding and to a protective litter layer [22]. The minimum gamma diversity in beech forest stands can be explained by the low light availability (highest tree layer cover in beech forests), a thick impeding leaf litter layer [77] and a dense root system [76,78]. In addition, across the management cycle of beech, the canopy is intentionally kept relatively dense to ensure stem quality. In contrast, thinning in spruce stands starts early with effects on light availability and soil disturbance by management operations [79]. An admixture of spruce to beech stands presumably increased light availability as shown by Lücke and Schmidt [20] with a positive effect on gamma diversity compared to pure beech stands. However, mixtures have not reached the maximum diversity values of the hypothetical pure spruce landscapes. The resource availability in terms of light seems to be not sufficient in these mixtures. They rather represent intermediate conditions as described by Cavard et al. [25]. This is confirmed by the Ellenberg light indicator values of species associated with pure conifer (mean value 6.3), pure beech (4.2) or pure and mixed stands (5.4; see Table 4).

Besides a higher light availability in pure spruce stands, their maximum gamma diversity both on acidic and on calcareous soils shows that acidification (the pH was significantly lower under spruce compared to beech) has not led to biodiversity loss as stated by different authors [37,76]. While amelioration by liming on acidic soils may have counteracted negative acidification effects in the past [42], the high buffer capacity of calcareous soils has reduced the acidifying impact of spruce [80]. It rather seems that pure spruce stands can support wider acidity gradients compared to mixed stands. Species characteristic of spruce stands showed acidity indicator values ranging from very acidic to weakly basic sites (EIV R 2–7), whereas species also associated with mixed stands indicate moderate acidic to basic conditions (EIV R = 5–8). Thus, resource quantities in terms of light and resource heterogeneity in terms of soil conditions seem to be more effectively provided in pure spruce stands than in mixtures [41].

The tree layer structure of beech/Douglas fir mixtures resembles beech/pine mixtures with a two-layered closed canopy. The gamma diversity pattern, however, is similar to beech/spruce mixtures. In contrast to the spare crown of pine trees, Douglas fir trees may cast more shade on the slower growing beech leading to gaps in the beech cover [43]. This can have a positive effect on other (tree) species in the understory. In addition, Douglas fir has a less acidifying litter with lower C/N ratios compared to pine which can increase nutrient availability [21]. This higher heterogeneity in light and soil conditions may allow for a larger coexistence of plant species compared to beech/pine mixtures. Thus, mixtures of beech and Douglas fir seem to be able to support similar plant diversity than the pure conifer variant. This view is underlined by a number of species characteristic in pure Douglas fir and in mixed stands. In fact, the number of exclusive species did not significantly differ between pure Douglas fir and mixed stands. Concerning the establishment of the non-native Douglas fir, nature conservation calls for mixtures with beech instead of pure Douglas fir stands to counter potential negative effects on native flora and fauna [81]. Our results show that vascular plant diversity will not necessarily differ between pure and mixed stands of Douglas fir.

4.2. An Optimized Composition of Forest Stands for Bryophyte and Lichen Diversity

Bryophytes showed complementarity in species composition between pure conifer and pure beech stands at a hypothetical landscape level, whereas mixed stands were significantly less diverse. Lichens were promoted by a high share of pure beech stands.

Schorfheide-Chorin in eastern Germany is generally less diverse in bryophytes and lichens than the region Schwäbische Alb due to differences in elevation, precipitation (higher in Schwäbische Alb, see Table 1) and former air pollution with its acidic depositions (higher in Schorfheide-Chorin; [45,46]). Thus, in Schorfheide-Chorin we could only find non-significant tendencies indicating that bryophytes and lichens are not as sensitive as vascular plants to the composition of stand types in this region or have not been sufficiently covered by the available number of plots.

For bryophytes in Schwäbische Alb, we found a significant complementarity between pure beech and pure spruce stands. In particular, bryophytes growing on mineral soil and on deadwood can benefit from a coniferous canopy [46]. They are promoted by a thin leaf litter layer [82] and by the higher light availability linking ground dwelling bryophyte and vascular plant species diversity [46,83]. Bryophyte species significantly associated with conifer stands in the investigated regions are predominantly growing on soil (e.g., *Pleurozium schreberi*, *Scleropodium purum*, *Rhytidiadelphus loreus*) or are able to grow on needle litter (e.g., *Orthodontium lineare*, [84]). These species are often characterized by high abundances in coniferous forests [69] with important implications for carbon and nutrient cycling [85,86] and providing important resources for animals [87]. On the other hand, epiphytic bryophyte diversity was shown to be negatively affected by conifer stands due to lower forest continuity [88,89]. Stand age was generally lower in pure conifer stands compared to pure beech and mixed stands. In addition, management in conifer stands generally starts earlier compared to beech stands with conifer trees having shorter rotation periods [79]. Bryophyte species associated with beech stands were often found as epiphytes on beech presumably benefiting from the higher stand age but also from moist conditions in the lower stem region as a consequence of stem flow. Some characteristic species of beech stands were also associated with mixed stands. Thus, it seems that mixtures of spruce and beech can mainly promote species associated with beech due to the presence of the specific host tree species, while conifer specialists require a higher resource quantity either in substrate (=open soil or soil or deadwood not covered by leaf litter) or in light availability. In addition, similar to vascular plants, spruce stands of Schwäbische Alb seem to offer a higher heterogeneity in soil conditions promoting a co-existence of species preferring limestone (e.g., *Eurhynchium angustirete*, a typical species for spruce forests growing on soil over limestone) as well as acidophytic species (e.g., *Dicranella heteromalla*, *Pleurozium schreberi*) that are promoted by the acidifying needle litter. Mixed stands, on the other hand, were similar to beech with regard to their soil pH.

Epiphytic lichens constitute the main part of the lichen diversity in both study regions [45]. Their gamma diversity was negatively affected by pure conifer stands that were mainly younger than investigated beech stands or beech/conifer mixtures leading to lower habitat continuity compared to pure beech stands [88]. Differences between pure beech and mixed stands were, however, not significant, as some species within mixed stands are either able to colonize both broadleaved and coniferous trees (e.g., *Arthonia spadicea*, *Phlyctis argena*) or just require beech (or other broadleaved tree species) as a host tree within mixtures (e.g., *Pertusaria leioplaca*). Characteristic species within the investigated stand types generally cover large gradients in terms of light and moisture conditions. They are partly early colonizers of trees (e.g., *Graphis scripta*, *Porina aenea*) or can indicate disturbance or eutrophication (*Scoliciosporum chlorococcum*, *Xanthoria polycarpa*). Thus, they are rather generalists able to tolerate the heterogeneous habitat conditions created by forest management either in pure or mixed stands. It has also to be noted here that restricting cryptogam sampling to the lower stem region presumably has caused an underestimation of lichen alpha diversity both in beech and in conifer stands [47]. Assuming that beech and conifer crowns harbor different species assemblages due to differences in stand continuity but also in microclimate, detected gamma diversity patters for lichens might strongly change with a more complete sampling.

4.3. An Optimized Composition of Pure and Mixed Stands for Regional Diversity

Combining the three taxonomic groups studied here resulted in a general diversity pattern. It seems that combining pure beech and pure conifer stands at the landscape scale maximizes the regional diversity of vascular plants, bryophytes and lichens in contrast to tree species mixtures at the stand scale. The latter either were most species poor (beech/pine) or took an intermediate position. This shows that interactions among tree species at the stand scale can be too small to affect diversity positively or can even be negative when complementary tree species such as beech and pine are mixed. Results also show that the quantity of resources in terms of light or suitable substrate, particularly important for conifer specialists, is only adequately provided within pure conifer stands, while mixtures are often more similar to beech forests. This finding is in accordance with the area heterogeneity tradeoff, indicating that a decreasing quantity of resources may reduce the size of local populations and may increase the likelihood of local extinctions [90]. With regard to the ongoing fragmentation and loss of non-woodland habitats within intensively used landscapes, forests may represent important refuges also for species that are not solely associated with forests [91]. Our results show that pure beech stands, complemented by pure conifer stands can maintain these species better than stand scale tree species mixtures. In lowland landscapes with beech and pine, almost the complete biodiversity of vascular plants, bryophytes and lichens associated with these tree species can be supported by such a mosaic of different pure stands. In montane regions with higher diversity of epiphytic bryophytes and lichens, including many threatened and protected species, a regional diversity maximum of only ca. 93% was reached by combining the different pure stand types. Epiphytic species seem to benefit from a higher stand age that is provided by beech stands in contrast to conifer stands. This indicates, that the maintenance of large and old trees within managed (conifer) forest types [92] or a certain share of unmanaged beech forests [45,93] may be able to complement the existing managed forests in particular for epiphytic species as they allow for habitat continuity. Pure spruce stands will mainly promote vascular plants and soil-dwelling bryophytes. Both groups are important biomass components within forests and important resources for other trophic levels [86].

5. Conclusions

Our results show that assuming an increase in tree species diversity and, consequently, an increase of structural heterogeneity at the stand scale will, per se, result in an increased biodiversity appears to be too general. Actually, mixtures with beech can reduce biodiversity of vascular plants, bryophytes and lichens by up to 20%. Instead, our results suggest that higher tree species diversity matters at larger spatial scales when beech is involved.

The result that regional diversity benefits, if beech and conifers are available at the landscape scale is in line with recent findings that a single tree species or forest type cannot fulfill all forest functions in temperate managed forests of Europe [4]. For supporting the diversity of the studied taxonomic groups, tree species should be arranged in different pure stands combined at the landscape scale and not in stand scale mixtures that are often dominated by the highly competitive beech. Therefore, future research should focus more on the spatial configuration of different pure stands in order to support overall biodiversity. However, spatial configurations of pure stands may be associated with unintended effects on other ecosystem functions. Productivity and stability, for example, have been shown to be positively influenced by within stand tree species mixtures [7,32,94]. Thus, where pure stands are for example susceptible to extreme events [35,94], trade-offs between high landscape biodiversity and high productivity or stability may occur. Future studies may also explore which key structural attributes (including tree crowns as habitats for epiphytes) are related to biodiversity at different spatial scales and how dispersal abilities of certain species are influenced by different configurations [95]. This should also include the evaluation of the effect of maintaining old trees and unmanaged stands for epiphytic species, many of them being protected. In addition, the investigation of a broader portfolio of mixture types of conifers with broadleaved species other than European beech or of different conifers will give further insights on the effect of tree species diversity and tree species identity on associated biodiversity as well as on other important ecosystem functions, and it may provide alternative future management options within forest landscapes in the light of global change.

Author Contributions: Conceptualization, S.H., P.S., M.M., C.A.; Methodology, P.S. and S.H; Formal analysis, S.H. and P.S.; Data sampling and investigation, S.B. (Steffen Boch), J.M., S.B. (Sabine Budde), M.W., I.S., M.F., E.-D.S., W.S.; Writing—original draft preparation, S.H.; Writing—review and editing, P.S., C.A., M.M., W.S., E.-D.S., S.B (Steffen Boch), J.M., I.S., S.B. (Sabine Budde), M.W.; Project administration, C.A., P.S., M.F., W.S.; Funding acquisition, C.A., M.F., W.S.

Funding: The work has been partly funded by the German Science Foundation (DFG) Priority Program 1374 "Infrastructure-Biodiversity-Exploratories" (Fi-1246/6-1; Am 149/17-1; Am 149/16-2) and by other projects funded by the DFG (Schm-319/13) and the German Federal Ministry of Education and Research (BMBF; FKZ 0339474C/3). Fieldwork permits were issued by the responsible state environmental offices of Baden-Württemberg, Brandenburg and Lower Saxony. We also acknowledge the support by the Open Access Publication Funds of the Göttingen University.

Acknowledgments: We thank A. Bauer, L. Beenken, A. Hemp, A. Parth, U. Pommer, H. Rubbert, S. Socher and B. Witt for support during data collection in all regions; A. Hemp, K. Wells and S. Pfeiffer and their teams for maintaining the plot and project infrastructure during data sampling in Schorfheide-Chorin and Schwäbische Alb; J. Nieschulze, M. Owonibi and A. Ostrowski for managing the central database of the Biodiversity Exploratories (BE), and K.E. Linsenmair, D. Hessenmöller, D. Prati, F. Buscot, W.W. Weisser and the late E. Kalko for their role in setting up the BE project.

Conflicts of Interest: The authors declare no conflict of interest.

Appendix A

Table A1. Number of 400 m^2 plots (n) for the analyses of vascular plants for the study regions and stand types. For mixed stands, also the plot distribution across dominance classes of beech or conifer dominance or an equal share of both species on accumulated tree layer cover is given. For the regions Schorfheide-Chorin and Schwäbische Alb bryophytes (Bry) and lichens (Li) were sampled in a smaller subset of vegetation survey plots. Also given is the share of plots younger or older than 90 years for Northwestern Germany and Solling [42,43] and the share of plots representing the immature (dbh > 15 ≤ 30 cm) and mature (dbh > 30 cm) developmental phase in Schorfheide-Chorin and Schwäbische Alb [50].

	Northwestern Germany			Schorfheide-Chorin			Solling			Schwäbische Alb		
		Stand Age [% Share]			Developmental Phase [% Share]			Stand Age [% Share]			Developmental Phase [% Share]	
	n	<90 yrs	>90 yrs	n	Immature	Mature	n	<90 yrs	>90 yrs	n	Immature	Mature
Pure stand types												
Beech	20	20.0	80.0	141 Bry: 108 Li: 76	5.1	94.9	43	37.2	62.8	95 Bry: 64 Li: 34	31.8	68.2
Pine	20	100.0	0.0	60 Bry: 39 Li: 25	50.0	50.0	-	-	-	-	-	-

Table A1. *Cont.*

| | Northwestern Germany | | | Schorfheide-Chorin | | | Solling | | | Schwäbische Alb | | |
| | Stand Age [% Share] | | | Developmental Phase [% Share] | | | Stand Age [% Share] | | | Developmental Phase [% Share] | | |
	n	<90 yrs	>90 yrs	n	Immature	Mature	n	<90 yrs	>90 yrs	n	Immature	Mature
Spruce	-	-	-	-	-	-	40	62.5	37.5	61 Bry: 40 Li: 20	31.6	68.4
Douglas fir	20	100.0	0.0	-	-	-	-	-	-	-	-	-
Mixed stand types												
Beech/Pine	20	10.0	90.0	77 Bry: 48 Li: 34	2.6	97.4	-	-	-	-	-	-
Beech dominance (>60% share)	20	10.0	90.0	61 Bry: 39 Li: 29	1.6	98.4	-	-	-	-	-	-
Pine dominance (>60% share)	-	-	-	6 Bry: 5 Li: 3	16.7	83.3	-	-	-	-	-	-
Equal share	-	-	-	10 Bry: 4 Li: 2	0.0	100.0	-	-	-	-	-	-
Beech/Douglas fir	20	20.0	80.0	-	-	-	-	-	-	-	-	-
Beech dominance (>60% share)	13	15.4	84.6	-	-	-	-	-	-	-	-	-
Douglas fir dominance (>60% share)	0	-	-	-	-	-	-	-	-	-	-	-
Equal share	7	28.6	71.4	-	-	-	-	-	-	-	-	-
Beech/Spruce	-	-	-	-	-	-	84	54.8	45.2	40 Bry: 29 Li: 12	33.3	66.7
Beech dominance (>60% share)	-	-	-	-	-	-	46	52.2	47.8	17 Bry: 16 Li: 5	43.8	56.3
Spruce dominance (>60% share)	-	-	-	-	-	-	8	87.5	12.5	5 Bry: 4 Li: 3	20	80
Equal share	-	-	-	-	-	-	30	50.0	50.0	18 Bry: 9 Li: 4	26.7	73.3

Table A2. Number of occurrences of threatened and protected vascular plant species in the survey plots (n = total number of plots) of pure and mixed stands in the Schwäbische Alb (ALB), Schorfheide-Chorin (S-C), Solling (SO) and NW-Germany. Be = pure beech stands, Spr = pure spruce stands, Pi = pure pine stands, Be/Spr = beech/spruce mixture, Be/Pi = beech/pine mixture, Be/Dgl = beech/Douglas fir mixture. Species with single occurrences are listed at the end of the table. * marks species not listed in the red list of vascular plants of Germany but protected by law.

| | ALB | | | S-C | | | SO | | | NW-Germany | | | | |
	Be	Be/Spr	Spr	Be	Be/Pi	Pi	Be	Be/Spr	Spr	Be	Be/Dgl	Be/Pi	Dgl	Pi
n	95	40	61	141	77	60	43	84	40	20	20	20	20	20
Abies alba		1									5	2	8	
Carex canescens						1	8	8						
Cephalanthera damasonium*	4	3	6											
Daphne mezereum*	30	11	10											
Epipactis helleborine*	2		3					1		1				
Epipactis purpurata			2											
Erica tetralix														2
Helleborus foetidus*	12	9	5											
Ilex aquifolium*										2	6		4	1
Juniperus communis			1			8								
Lilium martagon*	3													
Listera ovata*	1		2											
Neottia nidus-avis*	2	1	1											
Platanthera bifolia	1	1	2											
Primula elatior	19	11	12											
Ulmus minor				2										
Taxus baccata	3		1											
Asplenium scolopendrium*		1												
Cephalanthera rubra		1												

Table A2. *Cont.*

n	ALB			S-C			SO			NW-Germany			
	Be	Be/Spr	Spr	Be	Be/Pi	Pi	Be	Be/Spr	Spr	Be	Be/Dgl	Be/Pi	Dgl	Pi
	95	40	61	141	77	60	43	84	40	20	20	20	20	20
Corallorhiza trifida	1													
Hypericum montanum			1											
Liparis loeselii			1											
Lunaria rediviva		1												
Luzula campestris									1					
Lycopodium annotinum								1						
Monotropa hypophegea	1													
Orthilia secunda						1								
Platanthera chlorantha	1	1	1											
Total species number	13	11	14	1	0	2	1	3	2	2	2	1	2	2

Table A3. Number of occurrences of threatened and protected bryophyte species in the survey plots (*n* = total number of plots) of pure and mixed stands in Schwäbische Alb and Schorfheide-Chorin. Species with single occurrences are listed at the end of the table. All species are listed in the red list of bryophytes of Germany.

Bryophytes	Schwäbische Alb			Schorfheide-Chorin		
	Pure Beech	Mixed	Pure Spruce	Pure Beech	Mixed	Pure Pine
n	64	29	40	108	48	39
Amblystegium subtile	20	2				
Anomodon viticulosus	2					
Brachythecium oedipodium				5	3	3
Ctenidium molluscum	2	1	2			
Fissidens exilis	2					
Frullania dilatata	22	6	1		1	
Homalia trichomanoides	19	3				
Hylocomium splendens	3	10	33	2		3
Isothecium alopecuroides	55	18	4			
Leucobryum glaucum				2	7	1
Leucodon sciuroides	5					
Metzgeria furcata	31	9	2			
Mnium marginatum	4					
Nowellia curvifolia		1	3			
Orthotrichum affine	38	11	8	14	3	
Orthotrichum lyellii	9					
Orthotrichum pallens	5					
Orthotrichum patens	2	1				
Orthotrichum pumilum	18	3	3	3		
Orthotrichum speciosum	13	3	1			
Orthotrichum stramineum	8	4			1	
Orthotrichum striatum	9	2				
Plagiochila asplenioides	4	3	20			
Plagiothecium undulatum			6			1
Platygyrium repens	8	2	1	13	4	
Porella platyphylla	4					
Pseudoleskeella nervosa	13	2	1			
Pterigynandrum filiforme	28	9				
Ptilium crista-castrensis			2	1		
Pylaisia polyantha	7	5				
Radula complanata	46	21	4		1	
Rhytidiadelphus loreus	3	8	20	1		
Rhytidiadelphus triquetrus	4	10	24			
Tortella tortuosa	3					
Ulota bruchii	57	24	8	7	2	
Ulota crispa	7					

Table A3. *Cont.*

Bryophytes	Schwäbische Alb			Schorfheide-Chorin		
	Pure Beech	Mixed	Pure Spruce	Pure Beech	Mixed	Pure Pine
n	64	29	40	108	48	39
Anomodon attenuatus	1					
Campylium calcareum		1				
Dicranum flagellare						1
Homomallium incurvatum	1		1			
Marchantia polymorpha						1
Metzgeria temperata			1			
Orthotrichum obtusifolium	1	1				
Pleuridium acuminatum					1	
Pleuridium subulatum	1					
Polytrichum commune				1		
Ptilidium ciliare						1
Rhodobryum roseum			1			
Sanionia uncinata	1		1			
Total species number	36	25	22	10	9	7

Table A4. Number of occurrences of threatened and protected lichen species in the survey plots (*n* = total number of plots) of pure and mixed stands in the Schwäbische Alb and Schorfheide-Chorin. Species with single occurrences are listed at the end of the table. * marks species not listed in the red list of lichens of Germany but protected by law.

Lichens	Schwäbische Alb			Schorfheide-Chorin		
	Pure Beech	Mixed	Pure Spruce	Pure Beech	Mixed	Pure Pine
n	34	12	20	76	34	25
Arthonia punctiformis	2					
Arthonia radiata	30	5				
*Bacidina arnoldiana**	12	2	4	15	2	
*Bacidina delicata**	3	1				
Chaenotheca trichialis				2		
Graphis scripta	28	8	1	6	1	
*Lecania cyrtella**	4			1		
Lecanora intumescens	9	1				
Lecanora subcarpinea	9					
Lecanora subrugosa	5					
Mycobilimbia epixanthoides	2					
Opegrapha rufescens	2					
*Parmelia saxatilis**	2					
*Parmelia sulcata**	23	6	9			
Peltigera praetextata	4					
Pertusaria leioplaca	29	5				
Pertusaria pertusa	2					
Physcia aipolia	3					
Pyrenula nitida	7	2				
Pyrrhospora quernea	10	1				
*Ramalina farinacea**	5	1	2			
Strigula stigmatella	2					
Bacidia arceutina	1					
Bacidia rubella	1					
*Bacidina chloroticula**				1		
Calicium salicinum		1				
Cetrelia olivetorum	1					
Chaenotheca chrysocephala			1	1	1	
Chaenotheca stemonea				1		
Chrysothrix candelaris	1					

Table A4. *Cont.*

Lichens	Schwäbische Alb			Schorfheide-Chorin		
	Pure Beech	Mixed	Pure Spruce	Pure Beech	Mixed	Pure Pine
n	34	12	20	76	34	25
Cladonia ramulosa			1			
Cliostomum griffithii	1					
Fellhanera bouteillei			1			
Lecanora albella	1					
Lecanora argentata	1	1				
Leptogium plicatile	1					
Melanelia exasperata	1	1				
Parmelia ernstiae*	1					
Pertusaria hymenea	1					
Placynthiella uliginosa						1
Sarea resinae			1			
Usnea filipendula		1	1			
Total species number	32	14	9	7	3	1

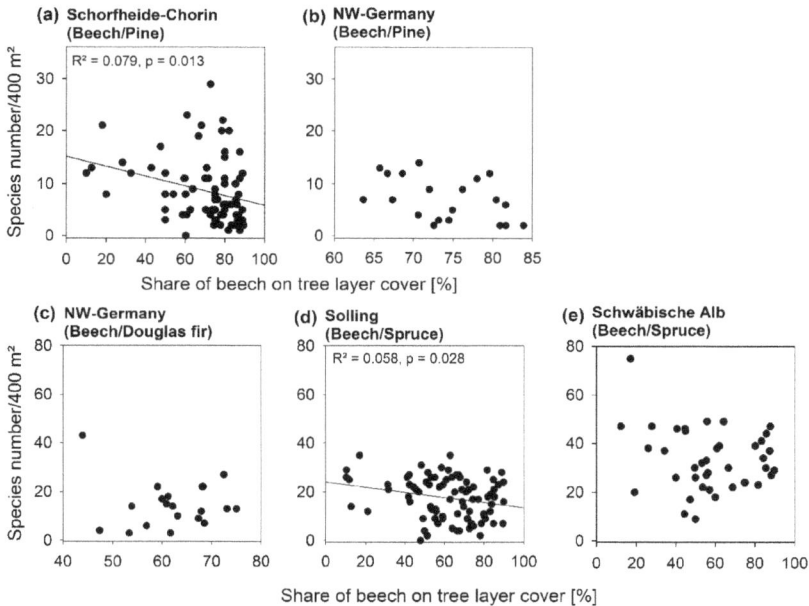

Figure A1. Vascular plant species numbers per 400 m² plots in relation to the percentage share of beech on tree layer cover within mixtures (**a**) of beech and pine in Schorfheide-Chorin and (**b**) NW-Germany, (**c**) of beech and Douglas fir in NW-Germany, (**d**) beech and spruce in Solling and (**e**) Schwäbische Alb. Plots were characterized as mixtures when both target species (beech and conifer) had a minimum of 10% or maximum of 90% share on accumulated tree layer cover. Regression lines and results are printed when relationships were significant. Note that the scale of the y-axis differs for Beech/Pine and Beech/Spruce(Douglas fir) combinations.

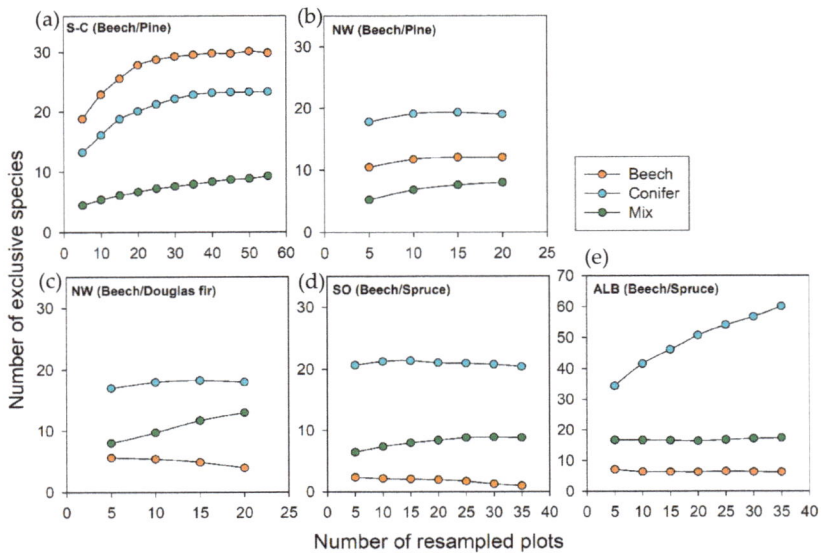

Figure A2. Number of exclusive vascular plant species in relation to the number of resampled plots in pure beech and conifer stands and respective mixtures in (**a**) Schorfheide-Chorin (S-C) and (**b,c**) northwestern Germany (NW), (**d**) Solling (SO) and (**e**) Schwäbische Alb (ALB). Note that y-axis of ALB is double the size of the other regions.

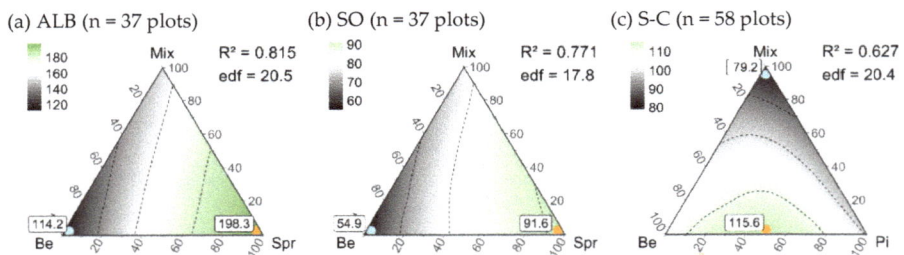

	ALB	SO	S-C
100% Beech	114.2 *	54.9 *	106.6
100% Conifer	198.3	91.6	99.1 *
100% Mixed	153.7 *	77.3 *	79.2 *
Max	198.3	91.6	115.6

Figure A3. Gamma diversity (species richness) of vascular plants along compositional gradients of pure beech (Be), pure conifer (Pi = Pine, Spr = Spruce) and mixed stands (Mix) in (**a**) Schwäbische Alb (ALB), (**b**) Solling (SO) and (**c**) Schorfheide-Chorin (S-C). Stand type composition varied in steps of 10% using 1000 resamplings of 37 (ALB, SO) and 58 plots (S-C) per step (66 unique compositions), the maximum number of plots allowing for more than 1000 resamplings at each corner of the triangle. The diversity response to composition is characterized by R^2 and estimated degrees of freedom (edf). Labelled dots mark the maximum (orange) and minimum (light blue). Numbers below the triangles show gamma diversity values for 100% beech stands, 100% conifer stands (either spruce (Spr) or pine (Pi)), 100% mixed stands and the composition of stand types supporting maximum gamma diversity (Max). * mark significant differences to the maximum gamma diversity.

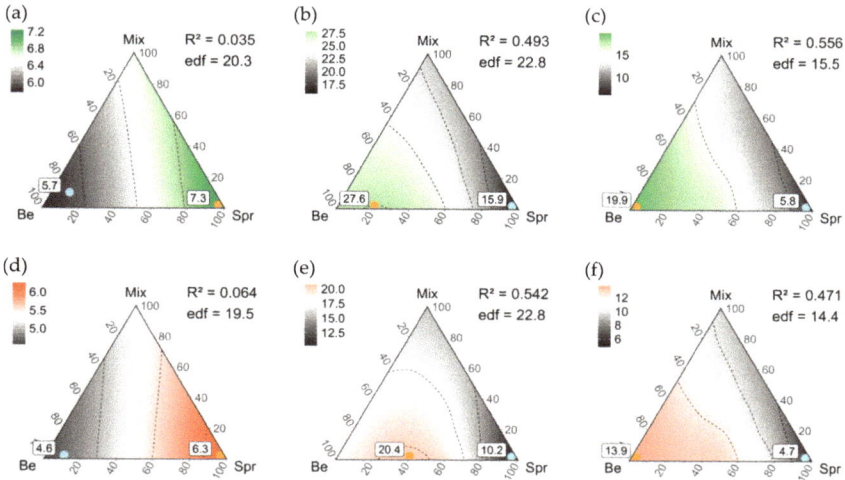

Figure A4. Gamma diversity (**a–c**: species richness; **d–f**: Shannon diversity) for threatened and protected (**a,d**) vascular plant species, (**b,e**) bryophyte species and (**c,f**) lichen species in Schwäbische Alb along compositional gradients of pure beech (Be), pure spruce (Spr) and mixed stands (Mix). The composition of stand types was varied in steps of 10% using 1000 resamplings of 17 (vascular plants) and 26 (bryophytes) plots per step or 66 resamplings of 10 plots per step (lichens). Labelled dots mark the maximum (orange) and minimum (light blue) gamma diversity.

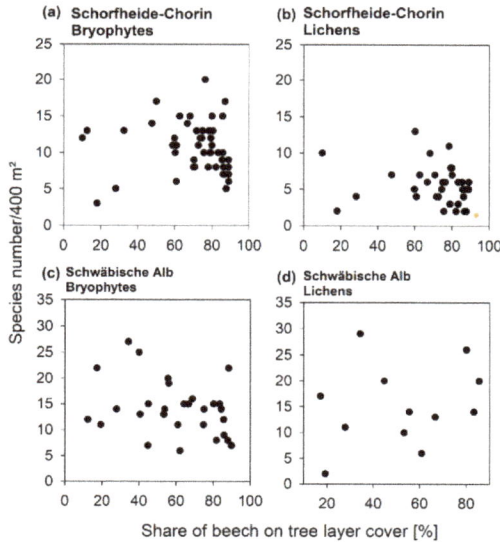

Figure A5. Bryophyte and lichen species numbers per 400 m^2 plots in relation to the percentage share of beech on tree layer cover within mixtures of (**a,b**) beech and pine in Schorfheide-Chorin and (**c,d**) beech and spruce in Schwäbische Alb. Plots were characterized as mixtures when both target species (beech and conifer) had a minimum of 10% or maximum of 90% share on accumulated tree layer cover. Note that the scale of the y-axis differs for Beech/Pine in Schorfheide-Chorin and the Beech/Spruce combination in Schwäbische Alb.

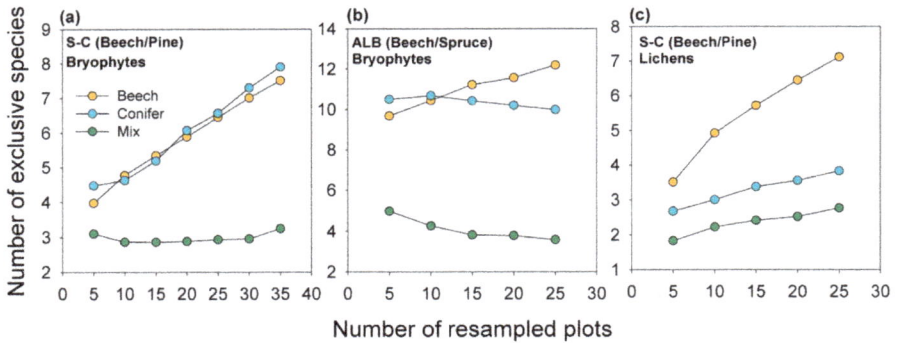

Figure A6. Number of exclusive bryophyte (**a,b**) and lichen (**c**) species in relation to the number of resampled plots in pure beech and conifer stands and respective mixtures in (**a,c**) Schorfheide-Chorin (S-C) and (**b**) Schwäbische Alb (ALB).

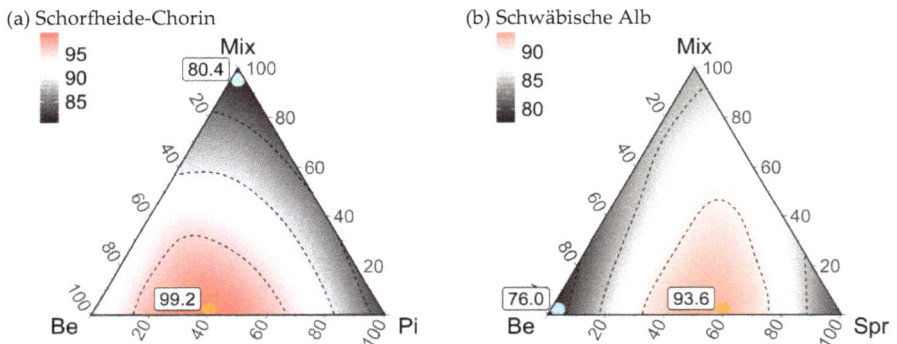

Figure A7. Regional diversity (Shannon diversity) in % for (**a**) Schorfheide-Chorin and (**b**) Schwäbische Alb combining vascular plants, bryophytes and lichens along compositional gradients of pure beech (Be), pure conifer (Pi = Pine, Spr = Spruce) and mixed stands (Mix). The composition of stand types was varied in steps of 10% using 1000 resamplings of 22 plots per step for Schorfheide-Chorin and using 66 resamplings of 10 plots per step for Schwäbische Alb. The regional diversity quantifies the mean relative gamma diversity of taxonomic groups accounting for the absolute diversity within groups (log weighting of diversity). Labelled dots mark the maximum (orange) and minimum (light blue) regional diversity.

References

1. Gamfeldt, L.; Snall, T.; Bagchi, R.; Jonsso, M.; Gustafsson, L.; Kjellander, P.; Ruiz-Jaen, M.C.; Froberg, M.; Stendahl, J.; Philipson, C.D.; et al. Higher levels of multiple ecosystem services are found in forests with more tree species. *Nat. Commun.* **2013**, *4*, 1340. [CrossRef] [PubMed]
2. Schuler, L.J.; Bugmann, H.; Snell, R.S. From monocultures to mixed-species forests: Is tree diversity key for providing ecosystem services at the landscape scale? *Landsc. Ecol.* **2017**, *32*, 1499–1516. [CrossRef]
3. Ratcliffe, S.; Wirth, C.; Jucker, T.; van der Plas, F.; Scherer-Lorenzen, M.; Verheyen, K.; Allan, E.; Benavides, R.; Bruelheide, H.; Ohse, B.; et al. Biodiversity and ecosystem functioning relations in European forests depend on environmental context. *Ecol. Lett.* **2017**, *20*, 1414–1426. [CrossRef]
4. Felipe-Lucia, M.R.; Soliveres, S.; Penone, C.; Manning, P.; van der Plas, F.; Boch, S.; Prati, D.; Ammer, C.; Schall, P.; Gossner, M.M.; et al. Multiple forest attributes underpin the provision of multiple ecosystem services. *Nat. Commun.* **2018**, *9*, 4839. [CrossRef] [PubMed]

5. Chamagne, J.; Tanadini, M.; Frank, D.; Matula, R.; Paine, C.E.T.; Philipson, C.D.; Svátek, M.; Turnbull, L.A.; Volařík, D.; Hector, A. Forest diversity promotes individual tree growth in central European forest stands. *J. Appl. Ecol.* **2017**, *54*, 71–79. [CrossRef]

6. Jucker, T.; Bouriaud, O.; Avacaritei, D.; Coomes, D.A. Stabilizing effects of diversity on aboveground wood production in forest ecosystems. Linking patterns and processes. *Ecol. Lett.* **2014**, *17*, 1560–1569. [CrossRef] [PubMed]

7. Ammer, C. Diversity and forest productivity in a changing climate. *New Phytol.* **2019**, *221*, 50–66. [CrossRef] [PubMed]

8. Mina, M.; Huber, M.O.; Forrester, D.I.; Thürig, E.; Rohner, B. Multiple factors modulate tree growth complementarity in central European mixed forests. *J. Ecol.* **2018**, *106*, 1106–1119. [CrossRef]

9. Pretzsch, H.; Dieler, J.; Seifert, T.; Rötzer, T. Climate effects on productivity and resource-use efficiency of Norway spruce (*Picea abies* [L.] Karst.) and European beech (*Fagus sylvatica* [L.]) in stands with different spatial mixing patterns. *Trees* **2012**, *26*, 1343–1360. [CrossRef]

10. Fichtner, A.; Härdtle, W.; Bruelheide, H.; Kunz, M.; Li, Y.; von Oheimb, G. Neighbourhood interactions drive overyielding in mixed-species tree communities. *Nat. Commun.* **2018**, *9*, 1144. [CrossRef]

11. Pretzsch, H.; Schütze, G.; Uhl, E. Resistance of European tree species to drought stress in mixed versus pure forests. Evidence of stress release by inter-specific facilitation. *Plant Biol.* **2013**, *15*, 483–495. [CrossRef]

12. Metz, J.; Annighöfer, P.; Schall, P.; Zimmermann, J.; Kahl, T.; Schulze, E.-D.; Ammer, C. Site-adapted admixed tree species reduce drought susceptibility of mature European beech. *Glob. Chang. Biol.* **2016**, *22*, 903–920. [CrossRef]

13. Simmons, E.A.; Buckley, G.P. Ground vegetation under planted mixtures of trees. In *The Ecology of Mixed-Species Stands of Trees*; Cannell, M.G.R., Malcolm, D.C., Robertson, P.A., Eds.; Blackwell Scientific Publications: Oxford, UK, 1992; pp. 211–231.

14. Felton, A.; Lindbladh, M.; Brunet, J.; Fritz, Ö. Replacing coniferous monocultures with mixed-species production stands: An assessment of the potential benefits for forest biodiversity in northern Europe. *For. Ecol. Manag.* **2010**, *260*, 939–947. [CrossRef]

15. Scherer-Lorenzen, M. The functional role of biodiversity in forests. In *Integrative Approaches as an Opportunity for the Conservation of Forest Biodiversity*; Kraus, D., Krumm, F., Eds.; European Forest Institute: Freiburg, Germany, 2013; pp. 216–223.

16. Gilliam, F.S. The ecological significance of the herbaceous layer in temperate forest ecosystems. *BioScience* **2007**, *57*, 845–858. [CrossRef]

17. Scherber, C.; Eisenhauer, N.; Weisser, W.W.; Schmid, B.; Voigt, W.; Fischer, M.; Schulze, E.-D.; Roscher, C.; Weigelt, A.; Allan, E.; et al. Bottom-up effects of plant diversity on multitrophic interactions in a biodiversity experiment. *Nature* **2010**, *468*, 553–556. [CrossRef] [PubMed]

18. Handa, I.T.; Aerts, R.; Berendse, F.; Berg, M.P.; Bruder, A.; Butenschoen, O.; Chauvet, E.; Gessner, M.O.; Jabiol, J.; Makkonen, M.; et al. Consequences of biodiversity loss for litter decomposition across biomes. *Nature* **2014**, *509*, 218–221. [CrossRef]

19. Ódor, P.; Király, I.; Tinya, F.; Bortignon, F.; Nascimbene, J. Patterns and drivers of species composition of epiphytic bryophytes and lichens in managed temperate forests. *For. Ecol. Manag.* **2013**, *306*, 256–265. [CrossRef]

20. Lücke, K.; Schmidt, W. Vegetation und Standortsverhältnisse in Buchen-Fichten Mischbeständen des Sollings. *Forstarchiv* **1997**, *68*, 135–143.

21. Augusto, L.; Dupouey, J.; Ranger, J. Effects of tree species on understory vegetation and environmental conditions in temperate forests. *Ann. For. Sci.* **2003**, *60*, 823–831. [CrossRef]

22. Sydes, C.; Grime, J.P. Effects of tree leaf litter on herbaceous vegetation in deciduous woodland. I. Field investigations. *J. Ecol.* **1981**, *69*, 237–248. [CrossRef]

23. Wilson, S.D. Heterogeneity, diversity and scale in plant communities. In *The Ecological Consequences of Environmental Heterogeneity*; Hutchings, M.J., John, E.A., Stewart, A.J.A., Eds.; Blackwell Science: Oxford, UK, 2000; pp. 53–69.

24. Stein, A.; Gerstner, K.; Kreft, H. Environmental heterogeneity as a universal driver of species richness across taxa, biomes and spatial scales. *Ecol. Lett.* **2014**, *17*, 866–880. [CrossRef]

25. Cavard, X.; Macdonald, S.E.; Bergeron, Y.; Chen, H.Y.H. Importance of mixedwoods for biodiversity conservation: Evidence for understory plants, songbirds, soil fauna, and ectomycorrhizae in northern forests. *Environ. Rev.* **2011**, *19*, 142–161. [CrossRef]

26. Gosselin, M.; Fourcin, D.; Dumas, Y.; Gosselin, F.; Korboulewsky, N.; Toïgo, M.; Vallet, P. Influence of forest tree species composition on bryophytic diversity in mixed and pure pine (*Pinus sylvestris* L.) and oak (*Quercus petraea* (Matt.) Liebl.) stands. *For. Ecol. Manag.* **2017**, *406*, 318–329. [CrossRef]

27. Brassard, B.W.; Chen, H.Y.H.; Cavard, X.; Langanière, J.; Reich, P.B.; Bergeron, Y.; Paré, D.; Yuan, Z. Tree species diversity increases fine root productivity through increased soil volume filling. *J. Ecol.* **2013**, *101*, 210–219. [CrossRef]

28. Barbier, S.; Gosselin, F.; Balandier, P. Influence of tree species on understory vegetation diversity and mechanisms involved—A critical review for temperate and boreal forests. *For. Ecol. Manag.* **2008**, *254*, 1–15. [CrossRef]

29. MacDonald, S.E.; Fenniak, T.E. Understory plant communities of boreal mixedwood forests in western Canada: Natural patterns and response to variable-retention harvesting. *For. Ecol. Manag.* **2007**, *242*, 34–48. [CrossRef]

30. Spiecker, H. Silvicultural management in maintaining biodiversity and resistance of forests in Europe—temperate zone. *J. Environ. Manag.* **2003**, *67*, 55–65. [CrossRef]

31. Spathelf, P.; Ammer, C. Forest management of Scots pine (*Pinus sylvestris* L.) in northern Germany—A brief review of the history and current trends. *Forstarchiv* **2015**, *86*, 59–66.

32. Knoke, T.; Ammer, C.; Stimm, B.; Mosandl, R. Admixing broadleaved to coniferous tree species: A review on yield, ecological stability and economics. *Eur. J. For. Res.* **2008**, *127*, 89–101. [CrossRef]

33. Brang, P.; Spathelf, P.; Larsen, J.B.; Bauhus, J.; Bončϊna, A.; Chauvin, C.; Drössler, L.; García-Güemes, C.; Heiri, C.; Kerr, G.; et al. Suitability of close-to-nature silviculture for adapting temperate European forests to climate change. *Forestry* **2014**, *87*, 492–503. [CrossRef]

34. Von Lüpke, B.; Ammer, C.; Braciamacchie, M.; Brunner, A.; Ceitel, J.; Collet, C.; Deuleuze, C.; Di Placido, J.; Huss, J.; Jankovic, J.; et al. Silvicultural strategies for conversion. In *Norway Spruce Conversion—Options and Consequences*; European Forest Institute Research Report 18; Spiecker, H., Hansen, J., Klimo, E., Skovsgaard, J.P., Sterba, H., von Teuffel, K., Eds.; Brill: Leiden, The Netherlands, 2004; pp. 121–164.

35. Ammer, C.; Bickel, E.; Kölling, C. Converting Norway spruce stands with beech—A review on arguments and techniques. *Austrian J. For. Sci.* **2008**, *125*, 3–26.

36. Leuschner, C.; Ellenberg, H. *Ecology of Central European Forests*; Springer International Publishing: Cham, Switzerland, 2017; ISBN 978-3-319-43040-9.

37. Emmer, I.M.; Fanta, J.; Kobus, A.T.; Kooijman, A.; Sevink, J. Reversing borealization as a means to restore biodiversity in Central-European mountain forests—An example from the Krkonoše Mountains, Czech Republic. *Biodivers. Conserv.* **1998**, *7*, 229–247. [CrossRef]

38. MLUR-Ministerium für Landwirtschaft, Umweltschutz und Raumordnung des Landes Brandenburg. *Waldbaurichtlinie 2004—"Grüner Ordner" der Landesforstverwaltung Brandenburg*; Ministerium für Landwirtschaft, Umweltschutz und Raumordnung: Berlin, Germany, 2004; Available online: http://forst. brandenburg.de/media_fast/4055/wbr2004.pdf (accessed on 18. January 2019).

39. Borrass, L.; Kleinschmit, D.; Winkel, G. The "German model" of integrative multifunctional forest management—Analysing the emergence and political evolution of a forest management concept. *For. Policy Econ.* **2017**, *77*, 16–23. [CrossRef]

40. Vitali, V.; Büntgen, U.; Bauhus, J. Silver fir and Douglas fir are more tolerant to extreme droughts than Norway spruce in south-western Germany. *Glob. Chang. Biol.* **2017**, *23*, 5108–5119. [CrossRef] [PubMed]

41. Bartels, S.F.; Chen, H.Y.H. Is understory plant species diversity driven by resource quantity or resource heterogeneity? *Ecology* **2010**, *91*, 1931–1938. [CrossRef] [PubMed]

42. Weckesser, M. *Die Bodenvegetation von Buchen-Fichten-Mischbeständen im Solling—Struktur, Diversität und Stoffhaushalt*; Cuvillier-Verlag: Göttingen, Germany, 2003; ISBN 3898737373.

43. Budde, S. *Auswirkungen des Douglasienanbaus auf die Bodenvegetation im Nordwestdeutschen Tiefland*; Cuvillier-Verlag: Göttingen, Germany, 2006; ISBN 978-3-86727-079-3.

44. Boch, S.; Prati, D.; Müller, J.; Socher, S.; Baumbach, H.; Buscot, F.; Gockel, S.; Hemp, A.; Hessenmöller, D.; Kalko, E.K.V.; et al. High plant species richness indicates management-related disturbances rather than the conservation status of forests. *Basic Appl. Ecol.* **2013**, *14*, 496–505. [CrossRef]

45. Boch, S.; Prati, D.; Hessenmöller, D.; Schulze, E.-D.; Fischer, M. Richness of lichen species, especially of threatened ones, is promoted by management methods furthering stand continuity. *PLoS ONE* **2013**, *8*, e55461. [CrossRef] [PubMed]

46. Müller, J.; Boch, S.; Prati, D.; Socher, S.A.; Pommer, U.; Hessenmöllder, D.; Schall, P.; Schulze, E.-D.; Fischer, M. Effects of forest management on bryophyte species richness in Central European forests. *For. Ecol. Manag.* **2019**, *432*, 850–859. [CrossRef]

47. Boch, S.; Müller, J.; Prati, D.; Blaser, S.; Fischer, M. Up in the tree—The over-looked richness of bryophytes and lichens in tree crowns. *PLoS ONE* **2013**, *8*, e84913. [CrossRef]

48. Kiebacher, T.; Keller, C.; Scheidegger, C.; Bergamini, A. Hidden crown jewels: The role of tree crowns for bryophyte and lichen species richness in sycamore maple wooded pastures. *Biodivers. Conserv.* **2016**, *25*, 1605–1624. [CrossRef]

49. Gauer, J.; Aldinger, E. Waldökologische Naturräume Deutschlands—Forstliche Wuchsgebiete und Wuchsbezirke. *Mitt. Ver. Forstl. Standortskde. Forstpflanzenz.* **2005**, *43*, 1–324.

50. Hessenmöller, D.; Nieschulze, J.; von Lüpke, N.; Schulze, E.-D. Identification of forest management types from ground-based and remotely sensed variables and the effects of forest management on forest structure and composition. *Forstarchiv* **2011**, *82*, 171–183.

51. Schall, P.; Schulze, E.-D.; Fischer, M.; Ayasse, M.; Ammer, C. Relations between forest management, stand structure and productivity across different types of Central European forests. *Basic Appl. Ecol.* **2018**, *32*, 39–52. [CrossRef]

52. Ganz, M. Entwicklung von Baumartenzusammensetzung und Struktur der Wälder vom Schwarzwald bis auf die Schwäbische Alb—mit besonderer Berücksichtigung der Buche. Dissertation Thesis, University of Freiburg, Freiburg im Breisgau, Germany, 2004. Available online: https://freidok.uni-freiburg.de/data/1616 (accessed on 18. January 2019).

53. R Core Team. *R: A Language and Environment for Statistical Computing*; R Foundation for Statistical Computing: Vienna, Austria, 2018; Available online: https://www.R-project.org/ (accessed on 18. January 2019).

54. Wisskirchen, R.; Haeupler, H. *Standardliste der Farn-und Blütenpflanzen Deutschlands*; Ulmer: Stuttgart, Germany, 1998; ISBN 978-3800133604.

55. Koperski, M.; Sauer, M.; Braun, W.; Gradstein, S.R. *Referenzliste der Moose Deutschlands*; Bundesamt für Naturschutz: Bonn, Germany, 2000; ISBN 3-7843-3504-7.

56. Wirth, V. *Flechtenflora*, 2nd ed.; Ulmer: Stuttgart, Germany, 1995; ISBN 3-82521062-6.

57. Crowley, P.H. Resampling methods for computation-intensive data analysis in ecology and evolution. *Annu. Rev. Ecol. Syst.* **1992**, *23*, 405–447. [CrossRef]

58. Chao, A.; Gotelli, N.J.; Hsieh, T.C.; Sander, E.L.; Ma, K.H.; Colwell, R.K.; Ellison, A.M. Rarefaction and extrapolation with Hill numbers: A framework for sampling and estimation in species diversity studies. *Ecol. Monogr.* **2014**, *84*, 45–67. [CrossRef]

59. Wood, S.N. Fast stable restricted maximum likelihood and marginal likelihood estimation of semiparametric generalized linear models. *J. R. Stat. Soc. B* **2011**, *73*, 3–36. [CrossRef]

60. Hamilton, N. Ggtern: An Extension to 'ggplot2', for the Creation of Ternary Diagrams. R package version 2.2.1. 2017. Available online: https://CRAN.R-project.org/package=ggtern (accessed on 18. January 2019).

61. Dufrêne, M.; Legendre, P. Species assemblages and indicator species: The need for a flexible asymetrical approach. *Ecol. Monogr.* **1997**, *67*, 345–366.

62. Ellenberg, H.; Weber, H.E.; Düll, R.; Wirth, V.; Werner, W. Zeigerwerte von Pflanzen in Mitteleuropa. *Scr. Geobot.* **2001**, *18*, 1–262.

63. Schmidt, M.; Kriebitzsch, W.-U.; Ewald, J. *Waldartenlisten der Farn-und Blütenpflanzen, Moose und Flechten Deutschlands*; Bundesamt für Naturschutz: Bonn, Germany, 2011; ISBN 978-3-89624-034-7.

64. Allan, E.; Bossdorf, O.; Dormann, C.F.; Prati, D.; Gossner, M.M.; Tscharntke, T.; Blüthgen, N.; Bellach, M.; Birkhofer, K.; Boch, S.; et al. Interannual variation in land-use intensity enhances grassland multidiversity. *Proc. Natl. Acad. Sci. USA* **2014**, *111*, 308–313. [CrossRef]

65. Colwell, R.K.; Coddington, J.A. Estimating terrestrial biodiversity through extrapolation. *Philos. Trans. R. Soc. B* **1994**, *345*, 101–118.

66. Landolt, E.; Bäumler, B.; Erhardt, A.; Hegg, O.; Klötzli, F.; Lämmler, W.; Nobis, M.; Rudmann-Maurer, K.; Schweingruber, F.H.; Theurillat, J.-P.; et al. *Flora Indicative—Ecological Indicator values and Biological Attributes of the Flora of Switzerland and the Alps*; Haupt Verlag: Bern, Switzerland, 2010; ISBN 978-3-258-07461-0.

67. Schmidt, M.; Mölder, A.; Schönfelder, E.; Engel, F.; Schmiedel, I.; Culmsee, H. Determining ancient woodland indicator plants for practical use: A new approach developed in northwest Germany. *For. Ecol. Manag.* **2014**, *330*, 228–239. [CrossRef]

68. Hofmann, G.; Pommer, U. *Potentielle Natürliche Vegetation von Brandenburg und Berlin*; Eberswalder Forstliche Schriftenreihe 24: Berlin, Germany, 2005; ISBN 3-933352-62-2.

69. Budde, S.; Schmidt, W.; Weckesser, M. Impact of the admixture of European beech (*Fagus sylvatica* L.) on plant species diversity and naturalness of conifer stands in Lower Saxony. *Wald. Landsch. Nat.* **2011**, *11*, 49–61.

70. Metz, J.; Seidel, D.; Schall, P.; Scheffer, D.; Schulze, E.-D.; Ammer, C. Crown modeling by terrestrial laser scanning as an approach to assess the effect of aboveground intra- and interspecific competition on tree growth. *For. Ecol. Manag.* **2013**, *310*, 275–288. [CrossRef]

71. Pretzsch, H.; del Río, M.; Schütze, G.; Ammer, C.; Annighöfer, P.; Avdagic, A.; Barbeito, I.; Bielak, K.; Brazaitis, G.; Coll, L.; et al. Mixing of Scots pine (*Pinus sylvestris* L.) and European beech (*Fagus sylvatica* L.) enhances structural heterogeneity, and the effect increases with humidity. *For. Ecol. Manag.* **2016**, *373*, 149–166. [CrossRef]

72. Forrester, D.; Ammer, C.; Annighöfer, P.; Barbeito, I.; Bielak, K.; Bravo-Oviedo, A.; Coll, L.; del Río, M.; Drössler, L.; Heym, M.; et al. Effects of crown architecture and stand structure on light absorption in mixed and monospecific *Fagus sylvatica* and *Pinus sylvestris* forests along a productivity and climate gradient through Europe. *J. Ecol.* **2018**, *106*, 746–760. [CrossRef]

73. Jucker, T.; Bouriaud, O.; Coomes, D.A. Crown plasticity enables trees to optimize canopy packing in mixed-species forests. *Funct. Ecol.* **2015**, *29*, 1078–1086. [CrossRef]

74. Bolte, A.; Villanueva, I. Interspecific competition impacts on the morphology and distribution of fine roots in European beech (*Fagus sylvatica* L.) and Norway Spruce (*Picea abies* (L.) Karst. *Eur. J. For. Res.* **2006**, *125*, 15–26. [CrossRef]

75. Ma, Z.; Chen, H.Y.H. Effects of species diversity on fine root productivity increase with stand development and associated mechanisms in a boreal forest. *J. Ecol.* **2017**, *105*, 237–245. [CrossRef]

76. Máliš, F.; Ujházy, K.; Vodálová, A.; Barka, I.; Čaboun, V.; Sitková, Z. The impact of Norway spruce planting on herb vegetation in the mountain beech forests on two bedrock types. *Eur. J. For. Res.* **2012**, *131*, 1551–1569. [CrossRef]

77. Mölder, A.; Bernhardt-Römermann, M.; Schmidt, W. Herb-layer diversity in deciduous forests: Raised by tree richness or beaten by beech? *For. Ecol. Manag.* **2008**, *256*, 272–281. [CrossRef]

78. Schmid, I.; Leuschner, C. Warum fehlt den Gipsbuchenwäldern des Kyffhäusers (Thüringen) eine Krautschicht? *Forstwiss. Centralbl.* **1998**, *117*, 277–288. [CrossRef]

79. Röhrig, E.; Bartsch, N.; von Lüpke, B. *Waldbau auf Ökologischer Grundlage*; Ulmer: Stuttgart, Germany, 2006; ISBN 3825283100.

80. Ulrich, B. Stabilität von Waldökosystemen unter dem Einfluss des Sauren Regens. *Allg. Forstz.* **1983**, *26/27*, 670–677.

81. Höltermann, A.; Klingenstein, F.; Ssymank, A. Naturschutzfachliche Bewertung der Douglasie aus Sicht des Bundesamtes für Naturschutz (BfN). *LWF Wissen* **2008**, *59*, 74–81.

82. Márialigeti, S.; Németh, B.; Tinya, F.; Ódor, P. The effects of stand structure on ground-floor bryophyte assemblages in temperate mixed forests. *Biodivers. Conserv.* **2009**, *18*, 2223–2241. [CrossRef]

83. Tinya, F.; Márialigeti, S.; Király, I.; Németh, B.; Ódor, P. The effect of light conditions on herbs, bryophytes and seedlings of temperate mixed forests in Őrség, Western Hungary. *Plant Ecol.* **2009**, *204*, 69–81. [CrossRef]

84. Nebel, M.; Philippi, G. *Die Moose Baden-Württembergs, Band 2*; Verlag Eugen Ulmer: Stuttgart, Germany, 2001; ISBN 3800135302.

85. Turetsky, M.R. The role of bryophytes in carbon and nitrogen cycling. *Bryologist* **2003**, *106*, 395–409. [CrossRef]

86. Woziwoda, B.; Parzych, A.; Kopeć, D. Species diversity, biomass accumulation and carbon sequestration in the understorey of post-agricultural Scots pine forests. *Silva Fenn.* **2014**, *48*, 1119. [CrossRef]

87. Boch, S.; Berlinger, M.; Fischer, M.; Knop, E.; Nentwig, W.; Türke, M.; Prati, D. Fern and bryophyte endozoochory by slugs. *Oecologia* **2013**, *172*, 817–822. [CrossRef] [PubMed]

88. Fritz, Ö.; Gustafsson, L.; Larsson, K. Does forest continuity matter in conservation?—A study of epiphytic lichens and bryophytes in beech forests of southern Sweden. *Biol. Conserv.* **2008**, *141*, 655–668. [CrossRef]

89. Brunialti, G.; Frati, L.; Aleffi, M.; Marignani, M.; Rosati, L.; Burrascano, S.; Ravera, S. Lichens and bryophytes as indicators of old-growth features in Mediterranean forests. *Plant Biosyst.* **2010**, *144*, 221–233. [CrossRef]

90. Allouche, O.; Kalyuzhny, M.; Moreno-Rueda, G.; Pizarro, M.; Kadmon, R. Area–heterogeneity tradeoff and the diversity of ecological communities. *Prod. Natl. Acad. Sci. USA* **2012**, *109*, 17495–17500. [CrossRef] [PubMed]

91. Kriebitzsch, W.-U.; Bültmann, H.; von Oheimb, G.; Schmidt, M.; Thiel, H.; Ewald, J. Forest-specific diversity of vascular plants, bryophytes, and lichens. In *Integrative Approaches as an Opportunity for the Conservation of Forest Biodiversity*; Kraus, D., Krumm, F., Eds.; European Forest Institute: Freiburg, Germany, 2013; pp. 158–169.

92. Hofmeister, J.; Hošek, J.; Malíček, J.; Palice, Z.; Syrovátková, L.; Steinová, J.; Černajová, I. Large beech (*Fagus sylvatica*) trees as 'lifeboats' for lichen diversity in central European forests. *Biodivers. Conserv.* **2016**, *25*, 1073–1090. [CrossRef]

93. Schall, P.; Gossner, M.M.; Heinrichs, S.; Fischer, M.; Boch, S.; Prati, D.; Jung, K.; Baumgartner, V.; Blaser, S.; Böhm, S.; et al. The impact of even-aged and uneven-aged forest management on regional biodiversity of multiple taxa in European beech forests. *J. Appl. Ecol.* **2018**, *55*, 267–278. [CrossRef]

94. Neuner, S.; Albrecht, A.; Cullmann, D.; Engels, F.; Griess, V.C.; Hahn, W.A.; Hanewinkel, M.; Härtl, F.; Kölling, C.; Staupendahl, K.; Knoke, T. Survival of Norway spruce remains higher in mixed stands under a dryer and warmer climate. *Glob. Chang. Biol.* **2015**, *21*, 935–946. [CrossRef] [PubMed]

95. Ammer, C.; Fichtner, A.; Fischer, A.; Gossner, M.M.; Meyer, P.; Seidl, R.; Thomas, F.M.; Annighöfer, P.; Kreyling, J.; Ohse, B.; Berger, U.; et al. Key ecological research questions for Central European forests. *Basic Appl. Ecol.* **2018**, *32*, 3–25. [CrossRef]

© 2019 by the authors. Licensee MDPI, Basel, Switzerland. This article is an open access article distributed under the terms and conditions of the Creative Commons Attribution (CC BY) license (http://creativecommons.org/licenses/by/4.0/).

forests

MDPI

Article

Changes in Soil Arthropod Abundance and Community Structure across a Poplar Plantation Chronosequence in Reclaimed Coastal Saline Soil

Yuanyuan Li [1,2], Han Y. H. Chen [3], Qianyun Song [1], Jiahui Liao [1], Ziqian Xu [1], Shide Huang [1] and Honghua Ruan [1,*]

[1] College of Biology and the Environment, Joint Center for Sustainable Forestry in Southern China, Nanjing Forestry University, Nanjing 210037, China; lyy_njfu@163.com (Y.L.); 18705190508@163.com (Q.S.); liaojiahui@njfu.edu.cn (J.L.); zqxu@njfu.edu.cn (Z.X.); hsd9876@126.com (S.H.)

[2] School of Food Science, Nanjing Xiaozhuang University, Nanjing 211171, China

[3] Faculty of Natural Resources Management, Lakehead University, 955 Oliver Road, Thunder Bay, ON P7B 5E1, Canada; hchen1@lakeheadu.ca

* Correspondence: hhruan@njfu.edu.cn; Tel.: +86-25-854-27-312

Received: 9 September 2018; Accepted: 13 October 2018; Published: 15 October 2018

Abstract: Poplar plantations have the capacity to improve the properties of soils in muddy coastal areas; however, our understanding of the impacts of plantation development on soil arthropods remains limited. For this study, we determined the community dynamics of soil dwelling arthropods across poplar plantations of different ages (5-, 10-, and 21-years) over the course of one year in Eastern Coastal China. The total abundance of soil arthropods differed with stand development. Further, there were some interactions that involved the sampling date. On average, total abundance was highest in the 10-year-old stands and lowest in the 5-year-old stands. Total abundance exhibited strong age-dependent trends in June and September, but not in March or December. The abundance of Prostigmata and Oribatida increased in the 5- to 21-year-old stands, with the highest levels being in the 10-year-old stands. The abundance of Collembola increased with stand development; however, the stand age had no significant impact on the abundance of epedapic, hemiedaphic, and euedaphic Collembola. Order richness (Hill number $q = 0$) curve confidence intervals overlapped among three stand ages. Shannon and Simpson diversity (Hill numbers $q = 1$ and $q = 2$) differed between 10- and 21-year-old stand age. They showed almost similar trends, and the highest and lowest values were recorded in the 21- and 10-year-old stand ages, respectively. Permutational multivariate analysis of variance demonstrated that composition also varied significantly with the sampling date and stand age, and the 10-year-old stands that were sampled in June stood well-separated from the others. Indicator analysis revealed that Scolopendromorpha and Prostigmata were indicators in June for the 10-year-old stands, while Collembola were indicators for the 21-year-old stands sampled in September. Our results highlight that both stand development and climate seasonality can significantly impact soil arthropod community dynamics in the reclaimed coastal saline soils of managed poplar plantations.

Keywords: microarthropod; diversity; seasonal variations; stand development

1. Introduction

Soil arthropods are critical to many forest ecosystem processes, and may be employed as bioindicators of ecosystem soil conditions and changes, such as soil fertility, levels of pollutants, and stand development phases [1–3]. They play an essential role in multiple soil functions, including organic matter decomposition, nutrient mineralization and redistribution, and the stimulation of the growth of mycorrhizal and other fungi [4–7]. Plant diversity and productivity may exert positive effects on soil arthropod abundance and diversity [8–10]; in turn, soil arthropods facilitate plant community succession while enhancing local plant diversity [11].

In 2014, it was estimated that the saline soil area in China was about 34.5 Mha [12], primarily distributed on the eastern coast between the Yellow and Yangzi Rivers. These soils are often restored for agricultural use through the construction of dikes to ensure a steady and reliable food supply. However, these newly reclaimed coastal alkali soils are not suitable for immediate agricultural use [13]. Stand development following afforestation can reduce the soil pH through the accumulation of biomass [14,15]. There are wide variations in condition during different stand development phases, such as canopy closure, humidity, and other abiotic and biotic factors [16–18], which directly or indirectly impact soil arthropod activity [19–22].

Soil arthropod communities are highly dynamic in forests [23], and following their establishment, they are modified by stand development. In European subalpine spruce forests, the abundance of Collembola was higher in young fertile stands, while that of Oribatida was more abundant in mature stands with increased litter input [24,25]. The abundance of Oribatida in spruce forests was observed to increase from young (25-year-old) to mature (170-year-old) stands [24], while the highest abundance was observed in 25-year-old stands along a 5- to 95-year chronosequence [26]. These studies suggest that the effects of forest stand development on the abundance of arthropods are regulated through changes in soil nutrients [9,27,28]. Most previous studies have focused on one taxon, such as collembola, mites, beetles, centipedes, or millipedes [17,24,25,29,30], each of which contributes in various ways to soil functionality [31–33]. Our understanding of overall soil arthropod communities in response to forest stand development, however, remains elusive.

Soil arthropod communities undergo seasonal changes [34]. In tropical forests, soil arthropods are more abundant during the rainy season than the dry season [35]. In high latitude forests, drought reduces the abundance of soil arthropods [36,37]. Changing seasonal precipitation and temperature influences the composition of soil arthropod communities not only directly, but also indirectly, through the influence of understory shrub communities [36]. Studies that have examined the effects of plant communities and seasonality found that changes in resident arthropod populations are often more influenced by seasonality than changes in plant communities [19,21,37], while others have reported more pronounced influences from shifts in tree species composition over seasonality [1]. Nevertheless, knowledge of the seasonal dynamics of soil arthropods in plantations with diverse stand ages remains limited.

Poplar plantations are an important contributor to bioenergy resources and carbon sequestration [38]. In China, poplar trees are widely cultivated not only due to their rapid growth and wood supplies, but also for improving soil properties in muddy coastal areas [39]. For this study, we sampled soil arthropods across three poplar plantation ages (5-, 10-, and 21-years) on four sampling dates (in March, June, September, and December) in Eastern Coastal China. We sought to determine the structures of soil arthropod assemblages as relating to stand development in terms of total abundance, major group abundance, diversity, and composition. We hypothesized that (1) the abundance, diversity, and composition of soil arthropods would change with stand development, as aboveground and belowground conditions in forests are altered with stand aging processes [28,39]; and (2) if changes occurred, the patterns would not be consistent across seasons in these subtropical plantations as a result of variable seasonal rainfall and temperatures.

2. Materials and Methods

2.1. Site Description

This study was conducted at the Yellow Sea State Forest Park in Eastern Coastal China (32°33′–32°57′ N, 102°07′–102°53′ E), which is located in a warm temperate subtropical transition zone influenced by a monsoon climate. Seasonal changes in precipitation and temperature were recorded at the local Dongtai meteorological station (Figure S1). The average annual temperature and rainfall in this area over the past ten years (2005–2015) were 15.4 °C and 1494.0 mm, respectively, with an annual mean relative humidity of 76.0%. The terrain includes the middle and lower reaches of the alluvial Yangtze River plain, with desalted sandy loam meadow soil. The Forest Park consists of an area of approximately 3000 ha of pure poplar (*Populus deltoides* Marsh) plantations, with stand ages that range from three to 23 years.

2.2. Experimental and Sampling Design

For this study, three stands (5-, 10-, and 21-years) of pure poplar plantations under similar site conditions were sampled in triplicate, with a spatial interspersion of ~500–600 m for stands of the same age. All stands contained the same basalt parent material, similar elevations (less than 5 m difference), and original field management with the same initial plant community composition prior to the establishment of the plantations (based on the management records provided by the State Forest Park). We established experimental plots (20 m × 30 m) in each stand, for a total of nine plots.

In September, within each plot, tree trunks were measured at the height of 1.3 m above ground and diameters at breast height (DBH) of >5 cm were recorded. Understory vegetation richness surveys were conducted by counting all species found in the 1 m² of each plot. Coverage of individual layers of shrub and herb species was visually estimated as the percentage cover of the plot area [40].

During September 2014, ten 2.5-cm diameter soil cores were extracted from random locations at a 0–20 cm depth (10 cm intervals) in each plot. Five random soil cores from the ten locations were combined to represent a specific soil depth (0–10 cm or 10–20 cm) for the determination of soil moisture, while the other five cores were employed to quantify its chemical properties. Each of the samples was sieved through a mesh (2 mm) to remove coarse fragments prior to analysis. Soil moisture was calculated as:

$$\text{Soil moisture} = [(\text{wet} - \text{dry mass}) \times 100]/\text{dry mass} \qquad (1)$$

The dry mass of the soil was determined following oven drying at 105 °C for 24 h. The soil samples for chemical analyses were air dried and the pH was determined using a glass electrode in a 1:2.5 soil/water solution (w/v). Soil carbon and nitrogen were measured using an element analyzer (Elementar, Vario ELIII, Elementar Analysen Systeme GmbH, Hanau, Germany).

2.3. Soil Arthropod Sampling and Extraction

Soil arthropod sampling was conducted seasonally, i.e., in December 2013 and in March, June, and September 2014. On each sampling date, three random replicate soil samples, excluding the litter layer, were extracted from each plot. Three random replicate soil samples were extracted from a 0–20 cm soil depth with soil cores (4 cm in diameter) and averaged to sample soil arthropods using Tullgren extractors (Tullgren Funnel Unit, Burkard, UK) over 24 h [17]. All core samples were immediately placed in plastic bags, sealed, and transferred to the laboratory for further processing. On each sampling date, we also dug one pit that measured 25 cm × 25 cm × 20 cm to sample macroarthropods (Polydesmida, Scolopendromorpha, Coleoptera, Diptera (larvae), Araneida, Lepidoptera, Siphonaptera, Thysanoptera, and Hymenoptera (Formicidae)) in each plot [41]. All extracted arthropods were preserved in 75% ethanol and subsequently sorted under a dissecting microscope (Eclipse E200, Nikon, Tokyo, Japan).

The soil arthropods were identified to an ordinal level [42], which is often used for the rapid assessment of arthropod diversity [43,44]. In particular, it can be a useful method in the early stages of investigation for assessments of biodiversity [45]. The habitat preference structure of Collembola was organized into epedaphic, hemiedaphic, and euedaphic types [46,47]. The biodiversity of the soil arthropod communities was estimated through the abundance of soil arthropods (ind. m^{-2}), and diversity was computed with Hill numbers (q = 0, 1, 2) using the 'iNEXT' package [48–50].

2.4. Statistical Analysis

We tested the effects of stand age (*A*, years) and sampling date (*D*) on soil arthropod abundance, using the following model:

$$Y_{ijkl} = A_i + D_{j(k)} + A_i \times D_{j(k)} + \pi_k + \varepsilon_{l(ijk)} \tag{2}$$

where Y_{ijkl} is the total abundance of soil arthropods, the abundance of each major group (i.e., Prostigmata, Oribatida, Hymenoptera, Collembola, or Diptera); A_i ($i = 1, 2, 3$) is the stand age; $D_{j(k)}$ ($j = 1, 2, 3, 4$) is the sampling date (March, June, September, and December); π_k is the random plot effect ($k = 1, 2, \ldots 9$) to account for temporal autocorrelation among sampling dates within each plot; and $\varepsilon_{l(ijk)}$ ($l = 1, 2, 3$) is the random sampling error. We conducted mixed effect analysis using maximum likelihood with the *lme4* package in R 3.4.1 [51]. Shapiro-Wilk's tests of model residuals indicated that the assumption of normality was not met at $\alpha = 0.05$ for most models. Hence, we bootstrapped parameter estimates by 1000 iterations using the 'boot' [52] and 'ggplot2' packages [53].

To examine the effects of stand age and sampling date on arthropod composition, we used permutational multivariate analysis of variance (PerMANOVA) [54]. When analyzing counts of abundances (which are often overdispersed), we calculated lg($x + 1$)-transformed data prior to the perMANOVA [55]. The perMANOVA was implemented using the *Adonis* function of the 'vegan' package in R with the Bray–Curtis dissimilarity measure and 999 permutations for the compositional data [56]. Subsequently, we visualized the compositional data using non-metric multidimensional scaling of the Bray–Curtis dissimilarity measure. Moreover, we used indicator analyses to identify the arthropod orders that were associated with particular stand ages and sampling date combinations [57], computing the specificity (i.e., the positive predictive value of species as an indicator of a group) and sensitivity (i.e., the probability of identifying an arthropod belonging to the group) associated with each indicator value [57]. Spearman correlation analysis was used to examine the relationships between community or major group-level variables for soil arthropods, and environmental variables for the samples extracted in September.

3. Results

Stand age had a significant impact on soil arthropod total abundance, which was significantly higher in the 10-year-old stand (Table 1). Furthermore, stand age had an interaction with sampling date (Table S1). In June, the total abundance was highest in the 10-year-old stands and lowest in the 5-year-old stands. In September, the total abundance increased in the 5- to 21-year-old stands; however, the stand-age-dependent trends were less apparent in March and December (Figure 1a).

Dominant soil arthropod orders included Prostigmata mites (40%) and Oribatida mites (26.4%), followed by Hymenoptera (11.8%), Collembola (9.6%), and Diptera (7.2%). The others were Mesostigmata, Coleoptera, Araneae, Scolopendromorpha, Polydesmida, Thysanoptera, Siphonaptera, and Lepidoptera. The abundance of Prostigmata, Oribatida, and Diptera increased from the 5- to 21-year-old stands, with the highest level being in the 10-year-old stands, while the abundance of Collembola increased with stand development (Table 2). In addition, there were some interactions with the sampling dates. The abundance of Prostigmata revealed similar stand age and sampling date trends to those for total abundance. The abundance of Oribatida was higher in the 10-year-old stands than in the other stands for the March, June, and September sampling dates; however, this was not the case in December. The abundance of Hymenoptera (Formicidae) was highest in the 10-year-old stands in June; however, in September, it peaked in the 21-year-old stands. The abundance of Collembola increased with stand age in September and December; however, the effect of stand age was less apparent in March or June. The stand age and sampling date interaction had weak effects on the abundance of epedaphic Collembola (Table S1). The stand age or sampling date had no impact on the abundance of euedaphic and hemiedaphic Collembola. The abundance of Diptera (larvae) did not differ significantly with stand age.

Table 1. The abundance (ind m^{-2}) of soil arthropod groups in the three stand ages.

Parameters		5 Year Old	10 Year Old	21 Year Old
Total abundance		7433.3 ± 1823.4 a	27,271.8 ± 5837.8 b	18,113.7 ± 2935.7 ab
Prostigmata		2454.9 ± 762.0 a	13,402.3 ± 3569.8 b	5241.5 ± 856.9 a
Oribatida		3118.4 ± 992.0 a	7099.3 ± 810.9 b	3781.8 ± 955.8 a
Mesostigmata		66.3 ± 66.3	530.8 ± 315.0	66.3 ± 66.3
Collembola		597.1 ± 199.0 a	1791.4 ± 450.4 ab	2653.9 ± 816.0 b
	Epedaphic	265.4 ± 113.2	663.5 ± 165.0	1194.3 ± 443.7
	Hemiedaphic	199.0 ± 142.9	597.1 ± 221.9	729.8 ± 206.9
	Euedaphic	133.0 ± 89.5	530.8 ± 149.7	729.8 ± 267.6
Diptera		796.5 ± 240.0	1659.4 ± 494.8	1327.0 ± 436.1
Coleoptera		133.4 ± 89.4	199.0 ± 142.9	531.1 ± 246.8
Hymenoptera		0	2056.8 ± 1071.1	4179.9 ± 2165.2
Araneae		0	0	66.4 ± 66.3
Scolopendromorpha		66.7 ± 66.3	256.7 ± 149.6	0
Polydesmida		0.67 ± 0.45	133.4 ± 89.9	0.33 ± 0.33
Thysanoptera		66.3 ± 66.3	0.33 ± 0.33	66.3 ± 66.3
Siphonaptera		132.7 ± 89.4	0.33 ± 0.33	199.0 ± 199.0
Lepidoptera		0.33 ± 0.33	133.0 ± 89.4	0

Different lowercases letters indicate significant differences between stand ages at α = 0.05.

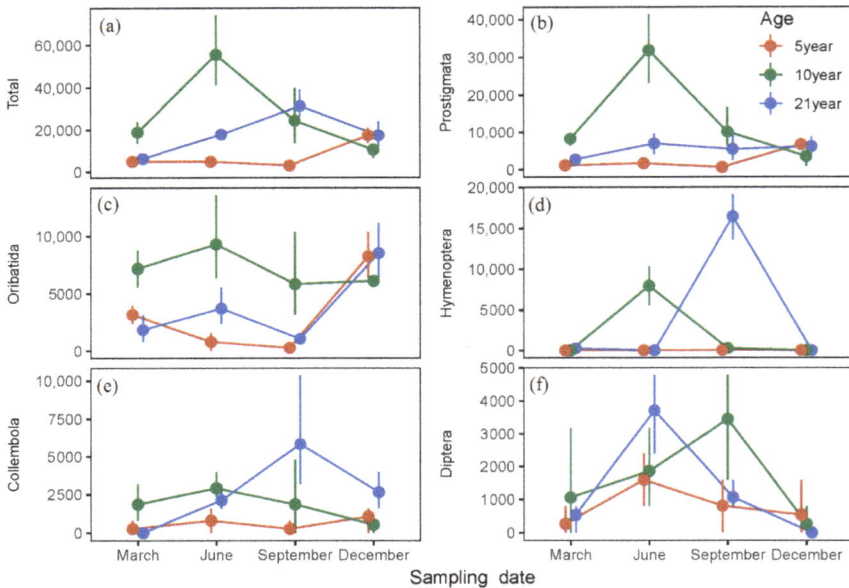

Figure 1. The effects of stand age and sampling date on the abundance (ind. m^{-2}) of soil arthropods. Filled circles are bootstrapped means with 95% bootstrapped confidence intervals (error bars). Difference is statistically significant at α = 0.05 when 95% bootstrapped confidence intervals do not overlap other means. (**a–f**) stands for the abundance of Total, Prostigmata, Oribatida, Hymeoptera, Collembola and Diptera, respectively.

In the 5-year-old stands (Figure 2), Oribatida, Prostigmata, and Diptera comprised the most abundant groups, accounting for 85.7% of the total abundance, followed by Collembola, Coleoptera, and Siphonaptera, which together accounted for 11.6% of the total abundance. Less abundant groups included (in a decreasing order of abundance), Scolopendromorpha, Mesostigmata, Thysanoptera, Polydesmida, Lepidoptera, Hymeoptera, and Araneae. In contrast to the 5-year-old stands, the abundance levels of Prostigmata increased more than fivefold and twofold in the 10- and 21-year-old stands, respectively, and ranked highest among arthropod levels in the 10- and 21-year-old stands. Accordingly, the ranks of Oribatida transitioned to second and third from the top in the 10- and 21-year-old stands, respectively. The rank of Diptera was altered from third position in the 5-year-old stands, to fifth position in the 10- and 21-year-old stands. Hymenoptera emerged and ranked in third and second positions in the 10- and 21-year-old stands, respectively. The rank order of Collembola did not change in the three stand ages.

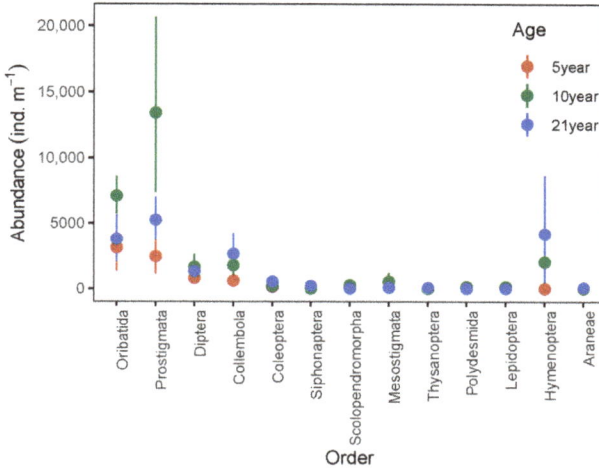

Figure 2. Rank-abundance of different arthropod orders. Filled circles are bootstrapped means, with 95% bootstrapped confidence intervals (error bars).

The diversity differences among the three stand ages can be seen in rarefaction and extrapolation curves with 95% bootstrapped confidence intervals (Figure 3). The similarities of richness ($q = 0$) among the three ages can be seen in the fact that the 95% confidence intervals all overlapped. The Shannon diversity ($q = 1$) and Simpson diversity ($q = 2$) differed between the 10- and 21-year-old stand ages. They showed an almost similar trend, and the highest and lowest values were recorded in the 21- and 10- year-old stands, respectively.

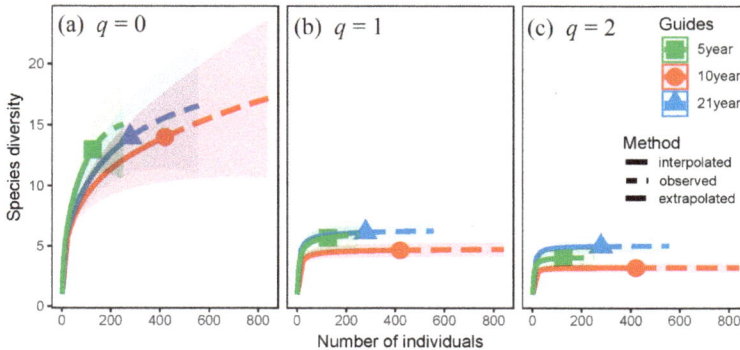

Figure 3. Sample size based rarefaction and extrapolation of soil arthropod diversity for Hill number (**a**) richness ($q = 0$), (**b**) Shannon diversity ($q = 1$), and (**c**) Simpson diversity ($q = 2$) with 95% bootstrapped confidence intervals (shaded areas).

PerMANOVA analysis revealed that the composition of arthropods differed significantly among stand ages, sampling dates, and their interactions, accounting for 56% of the observed variation (Table 2). The nonmetric multidimensional scaling ordination (NMDS) with a stress of 0.105 indicated that the arthropod composition of the 10-year-old stands in June was well-separated from that of the other stands (Figure 4). Among the sampling dates, the arthropod communities found in June and September were distinct from those found in March and December. Indicator analysis revealed that Scolopendromorpha and Prostigmata were indicators in June for the 10-year-old stands, while Collembola were indicators for the 21-year-old stands sampled in September (Table 3).

Table 2. Results of permutation multivariate analysis of variance testing (perMANOVA) for the effects of stand age (A), sampling date (D), and their interactions on soil arthropod Order composition.

Source	df	Sum Squares	F	R^2	p
A	2	0.26	2.05	0.08	0.04
D	3	1.01	5.27	0.29	0.001
A × D	6	0.69	1.79	0.19	0.015
Residuals	24	1.53		0.44	

Table 3. Indicators for stand ages and overstory types.

Stand Age (Years)	Date	Indicator	Indicator Value	Specificity	Sensitivity	p
10	June	Scolopendromorpha	0.800	1.00	0.894	0.005
10	June	Prostigmata	0.377	1.00	0.614	0.001
21	September	Collembola	0.290	1.00	0.538	0.035

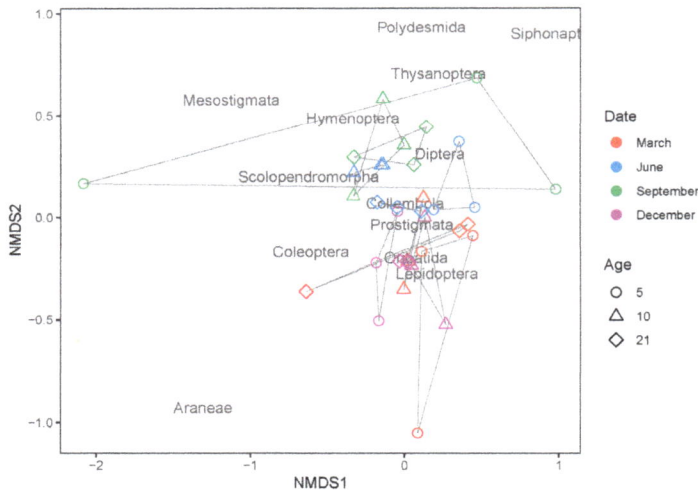

Figure 4. Nonmetric multidimensional scaling ordination (NMDS) showing a two-dimensional representation of the soil arthropod composition. The best NMDS solution was attained at a stress of 0.105. The circles nearest each other in the ordination space have similar assemblages, whereas those located farther apart are less similar.

Environmental variables differed with stand development (Table 4). Significant differences were observed in the top 0–10 cm of soil. Aboveground plant species richness and plant coverage were higher in the 10-year-old stand (Table 4). The Spearman correlation (Table 5) showed that the total abundance of Oribatida was correlated with soil moisture and plant coverage ($p < 0.05$). Prostigmata abundance was significantly correlated with soil moisture and nitrogen, while Collembola was correlated with nitrogen. Multiple correlation results revealed that the abundance of the soil arthropod community was primarily related to soil variables.

Table 4. Stand characteristics (n = 3) along an age chronosequence of poplar plantations in a coastal region of Eastern China.

Dimension	Characteristics	5 Year Old	10 Year Old	21 Year Old
	DBH (cm)	16.50 ± 1.63 c	26.04 ± 0.46 b	32.77 ± 1.48 a
Aboveground	Plant species richness	19.0 ± 0.58 ab	23.0 ± 1.52 a	17.3 ± 3.79 b
	Plant coverage (%)	75.37 ± 3.87 a	91.1 ± 2.74 b	77.5 ± 4.22 a
	Moisture (%)	19.74 ± 0.79 b	24.88 ± 0.36 a	23.61 ± 1.43 a
0–10 cm soil	pH	8.25 ± 0.12	8.09 ± 0.11	8.03 ± 0.09
	N (g kg^{-1})	1.69 ± 0.04 b	2.10 ± 0.06 a	2.24 ± 0.02 a
	C (g kg^{-1})	17.05 ± 0.23	17.25 ± 0.30	18.26 ± 0.35
	Moisture (%)	19.25 ± 0.2 b	24.6 ± 0.27 a	23.59 ± 1.11 a
10–20 cm soil	pH	8.35 ± 0.06	8.31 ± 0.13	8.29 ± 0.13
	N (g kg^{-1})	1.66 ± 0.02 b	1.93 ± 0.05 a	1.97 ± 0.07 a
	C (g kg^{-1})	16.07 ± 0.23	16.27 ± 0.30	16.09 ± 0.35

Different lowercase letters indicate significant differences between stand ages at α = 0.05. DBH: Diameters at breast height.

Table 5. Correlation between abundance of soil arthropods and soil characteristics. Values are Spearman correlation coefficients.

Parameters	Plant Species Richness	Plant Coverage	Moisture	pH	C	N
Total abundance	0.11	0.10	0.55 *	−0.37	0.51 *	0.75 ***
Prostigmata	0.24	0.31	0.59 **	−0.18	0.08	0.47 *
Oribatida	0.39	0.54 *	0.52 *	−0.02	0.04	0.33
Hymenoptera	−0.38	−0.15	0.22	−0.38	0.71 **	0.63 **
Collembola	−0.39	−0.29	0.28	−0.40	0.41	0.56 *
Diptera	0.29	0.33	0.34	0.23	−0.53 *	−0.1

Statistical significance: * $p < 0.05$, ** $p < 0.01$, *** $p < 0.001$.

4. Discussion

We found that the abundance and composition of soil arthropod communities were strongly affected by stand age and sampling date in the poplar plantations. On average, the total abundance increased from the 5-year-old to the 10-year-old stands, while remaining similar, or declining, in the 21-year-old stands. The soil arthropod assemblage composition in the 10-year-old stands was well-separated from that of the other two stand ages. The sampling dates explained a similar or greater variation in the abundance and composition of the soil arthropods with their interactions, which accounted for more than the sum of the primary effects and revealed that stand age-associated trends were strongly dependent on the sampling date. These findings supported our expectations that soil arthropod abundance and composition were altered with stand development.

Our findings partially supported previous results on the abundance and diversity of soil arthropods that were associated with forest stand development [20,25–27]. For our study, only Shannon and Simpson diversity of soil arthropods changed across the plantation chronosequence. The distribution of mites, particularly Oribatida, was also in contrast with that recorded in a spruce plantation in northern Germany [58], where no differences in Oribatida abundance between development phases were observed, while [26] found that the Oribatida density increased from that in 5- to that in 25-year-old stands in Tharandter Wald, Germany. However, a direct comparison of stand-age-associated trends, between our study and those published earlier, was rather difficult due to the available stand ages for sampling and the role of vegetation types in soil resources. The physical conditions and quality of soil and organic matter are known to impact soil faunal communities [33]. Soil fauna populations appeared to be lower when the uptake of water and nutrients were greatest during the stem exclusion stage, leaving fewer resources available for soil fauna [28,59].

In our study system, the available nitrogen tended to be higher in the 10-year-old stands than in the other stands (Table 1), and published data have demonstrated that microbial biomass carbon exhibited the same trend [16]. There was also a strong relationship between the total abundance and soil properties [17], but we lacked the litter data required to analyze the correlation between the litter quantity and soil arthropod activities. Their abundance may have been more strongly impacted by soil properties than by the plant community [22,60]. For this study, plant species richness and plant coverage were higher in the 10-year-old stands (Table 4), which not only prevented UV damage to soil arthropods, but also supplied them with improved living conditions. Thus, the soil properties and habitat conditions may have contributed to the higher total abundance of soil arthropods observed in the 10-year-old stands than in the 5- and 21-year-old stands. Formicidae were only captured in the 10- and 21-year-old stands in June and September, respectively. Ants are not only herbivores and predominant in older fragments [20,61], but are also predators [62]. Our finding of peak total abundance in the 10-year-old stands suggested that arthropod abundance could peak at an intermediate stage of stand development in these poplar plantations. The peak of abundance in an intermediate stand age may be the result of compositional shifts, from pioneer to mature species, or in conjunction with stand development [63]. Indeed, our compositional analysis supported the premise that the composition of soil arthropods shifted along the chronosequence.

Supporting our second hypothesis, the responses of soil arthropod assemblages to stand development varied widely by sampling date. The effects of stand age were more pronounced during the warm and wet seasons than in the cold and dry seasons. These findings are consistent with the results from desertified steppes [20] and dry tropical forests [64]. As poplar forests are deciduous, there was little difference in the canopy cover among the three age groups of stands in winter; thus, the microclimatic conditions were likely to be similar during this period. However, favorable climatic conditions in summer, combined with differences in soil moisture and nutrient availabilities [22,65,66], contributed to the higher abundance and distinct community composition in the 10-year-old stands.

The compositions of arthropod communities in June and September were distinct from those surveyed in March and December. Previous studies reported that soil arthropod abundance was directly or indirectly influenced by changing climate factors, such as solar radiation [24], temperature [67,68], precipitation [69], and moisture [70,71], among which soil moisture was found to be the most important factor in the control of soil arthropod communities [64]. In our study area, the precipitation and temperature both increased from December to June, and attained their highest levels in September (Figure S1). This observation suggests that June or September provides the most suitable climate conditions for soil arthropods in the study area. However, we could not attribute the observed seasonal dynamics of soil arthropods to strong climate seasonality due to a lack of sufficient field data for testing in our study.

The increase in total abundance with stand development largely resulted from the higher number of Collembola in September. This result is similar to previous findings in beech and spruce forests [72]. Collembola feed predominantly on plant detritus and fungi, with a preference for fungi [73,74], whose biomass increases with stand development [24]. In this study, the increased abundance of euedaphic Collembola with stand age in September may be attributable to the N content, as high N availability might increase the availability of potential food sources (e.g., fungal biomass) for euedaphic Collembola [46]. Further, the population of centipedes increased with the higher accumulation of litter and prey [75]. Since centipedes prey on Collembola [76,77], they were an indicator for the 10-year-old stands sampled in June for this study. It appeared that predation by centipedes was responsible for the fewer Collembola in the 10-year-old stands compared with in the 21-year-old stands by the next sampling date (September).

5. Conclusions

In summary, the total abundance was higher, on average, in the 10-year-old stands during the summer months, with potent interactive effects between the stand age and sampling date. Further, the soil arthropod community composition differed strongly with changing sampling date and stand age, with the 10-year-old stands sampled in June being well-separated from the others. The responses of soil arthropod assemblages to stand development were most pronounced during the warm and wet summer seasons. Indicator analysis revealed that Scolopendromorpha and Prostigmata were indicators for the 10-year-old stands sampled in June, while Collembola were indicators for the 21-year-old stands sampled in September. To elucidate the mechanisms that initiated these changes, more detailed work should be done in the next phases of study, particularly on individual taxonomic soil arthropod groups or functional groups; however, these findings will provide valuable guidelines for plantation managers, considering the conservation of biodiversity in reclaimed coastal saline soil hosting managed poplar plantations.

Supplementary Materials: The following are available online at http://www.mdpi.com/1999-4907/9/10/644/s1, Figure S1: Total precipitation (mm) and mean temperature (°C) three months prior to sampling dates in the study area, during 2013–2014. The data are from the local meteorological station, Table S1: The effects of stand age (A) and sampling date (D) on the abundance of total groups and major groups.

Author Contributions: H.R. and Y.L. conceived and designed the experiments; Y.L., Q.S., J.L., Z.X., and S.H performed the experiments; H.Y.H.C., Y.L, and J.L. analyzed the data; Y.L. and H.Y.H.C. wrote the manuscript; and all co-authors edited numerous drafts.

Funding: This study was supported by the National Key Research and Development Program of China (No. 2016YFD0600204) and Priority Academic Program Development of Jiangsu Higher Education Institutions (PAPD).

Acknowledgments: We gratefully acknowledge and thank David C. Coleman for his review and revision of this paper.

Conflicts of Interest: The authors declare no conflict of interest.

References

1. Grgič, T.; Kos, I. Influence of forest development phase on centipede diversity in managed beech forests in Slovenia. *Biodivers. Conserv.* **2005**, *14*, 1841–1862. [CrossRef]
2. Cakir, M.; Makineci, E. Humus characteristics and seasonal changes of soil arthropod communities in a natural sessile oak (*Quercus petraea* L.) stand and adjacent Austrian pine (*Pinus nigra* Arnold) plantation. *Environ. Monit. Assess.* **2013**, *185*, 8943–8955. [CrossRef] [PubMed]
3. Cortet, J.; Vauflery, A.G.; Poinsotbalaguer, N.; Gomot, L.; Texier, C.; Cluzeau, D. The use of invertebrate soil fauna in monitoring pollutant effects. *Eur. J. Soil Biol.* **1999**, *35*, 115–134. [CrossRef]
4. Wardle, D.A.; Bardgett, R.D.; Klironomos, J.N.; Setala, H.; van der Putten, W.H.; Wall, D.H. Ecological linkages between aboveground and belowground biota. *Science* **2004**, *304*, 1629–1633. [CrossRef] [PubMed]
5. Mace, G.M.; Norris, K.; Fitter, A.H. Biodiversity and ecosystem services: A multilayered relationship. *Trends Ecol. Evol.* **2012**, *27*, 19–26. [CrossRef] [PubMed]
6. De Deyn, G.B.; Van der Putten, W.H. Linking aboveground and belowground diversity. *Trends Ecol. Evol.* **2005**, *20*, 625–633. [CrossRef] [PubMed]
7. Fu, S.; Coleman, D.C.; Hendrix, P.F.; Crossley, D.A. Responses of trophic groups of soil nematodes to residue application under conventional tillage and no-till regimes. *Soil Biol. Biochem.* **2000**, *32*, 1731–1741. [CrossRef]
8. Siemann, E. Experimental tests of effects of plant productivity and diversity on grassland arthropod diversity. *Ecology* **1998**, *79*, 2057–2070. [CrossRef]
9. Hansen, R.A. Effect of habitat complexity and composition on a diverse litter microarthropod assemblage. *Ecology* **2000**, *81*, 1120–1132. [CrossRef]
10. Callaham, M.A.; Richter, D.D.; Coleman, D.C.; Hofmockel, M. Long-term land-use effects on soil invertebrate communities in Southern Piedmont soils, USA. *Eur. J. Soil Biol.* **2006**, *42*, S150–S156. [CrossRef]
11. De Deyn, G.B.; Raaijmakers, C.E.; Zoomer, H.R.; Berg, M.P.; de Ruiter, P.C.; Verhoef, H.A.; Bezemer, T.M.; van der Putten, W.H. Soil invertebrate fauna enhances grassland succession and diversity. *Nature* **2003**, *422*, 711–713. [CrossRef] [PubMed]

12. Li, J.; Pu, L.; Zhu, M.; Zhang, J.; Li, P.; Dai, X.; Xu, Y.; Liu, L. Evolution of soil properties following reclamation in coastal areas: A review. *Geoderma* **2014**, *226–227*, 130–139. [CrossRef]

13. Sun, Y.; Li, X.; Mander, Ü.; He, Y.; Jia, Y.; Ma, Z.; Guo, W.; Xin, Z. Effect of reclamation time and land use on soil properties in Changjiang River Estuary, China. *Chin. Geogr. Sci.* **2011**, *21*, 403–416. [CrossRef]

14. Brais, S.; Camiré, C.; Bergeron, Y.; Parécd, D. Changes in nutrient availability and forest floor characteristics in relation to stand age and forest composition in the southern part of the boreal forest of northwestern Quebec. *For. Ecol. Manag.* **1995**, *76*, 181–189. [CrossRef]

15. Binkley, D.; Valentine, D.W.; Wells, C.G.; Valentine, U. An empirical analysis of the factors contributing to 20-year decrease in soil pH in an old-field plantation of loblolly pine. *Biogeochemistry* **1989**, *8*, 39–54. [CrossRef]

16. Ge, Z.; Fang, S.; Chen, H.; Zhu, R.; Peng, S.; Ruan, H. Soil aggregation and organic carbon dynamics in poplar plantations. *Forests* **2018**, *9*, 508. [CrossRef]

17. Wang, S.; Chen, H.Y.H.; Tan, Y.; Fan, H.; Ruan, H. Fertilizer regime impacts on abundance and diversity of soil fauna across a poplar plantation chronosequence in coastal Eastern China. *Sci. Rep.* **2016**, *6*, 20816. [CrossRef] [PubMed]

18. Chen, H.Y.H.; Biswas, S.R.; Sobey, T.M.; Brassard, B.W.; Bartels, S.F.; Mori, A. Reclamation strategies for mined forest soils and overstorey drive understorey vegetation. *J. Appl. Ecol.* **2018**, *55*, 926–936. [CrossRef]

19. Jucevica, E.; Melecis, V. Global warming affect Collembola community: A long-term study. *Pedobiologia* **2006**, *50*, 177–184. [CrossRef]

20. Liu, R.; Zhu, F.; Song, N.; Yang, X.; Chai, Y. Seasonal distribution and diversity of ground arthropods in microhabitats following a shrub plantation age sequence in desertified steppe. *PLoS ONE* **2013**, *8*, e77962. [CrossRef]

21. Basset, Y.; Cizek, L.; Cuenoud, P.; Didham, R.K.; Novotny, V.; Odegaard, F.; Roslin, T.; Tishechkin, A.K.; Schmidl, J.; Winchester, N.N.; et al. Arthropod distribution in a tropical rainforest: Tackling a four dimensional puzzle. *PLoS ONE* **2015**, *10*, e0144110. [CrossRef] [PubMed]

22. Wu, P.; Liu, X.; Liu, S.; Wang, J.; Wang, Y. Composition and spatio-temporal variation of soil microarthropods in the biodiversity hotspot of northern Hengduan Mountains, China. *Eur. J. Soil Biol.* **2014**, *62*, 30–38. [CrossRef]

23. Eisenbeis, G.; Wichard, W. *Atlas on the Biology of Soil Arthropods*; Springer Science & Business Media: New York, NY, USA, 2012.

24. Salmon, S.; Artuso, N.; Frizzera, L.; Zampedri, R. Relationships between soil fauna communities and humus forms: Response to forest dynamics and solar radiation. *Soil Biol. Biochem.* **2008**, *40*, 1707–1715. [CrossRef]

25. Salmon, S.; Mantel, J.; Frizzera, L.; Zanella, A. Changes in humus forms and soil animal communities in two developmental phases of Norway spruce on an acidic substrate. *For. Ecol. Manag.* **2006**, *237*, 47–56. [CrossRef]

26. Zaitsev, A. Oribatid mite diversity and community dynamics in a spruce chronosequence. *Soil Biol. Biochem.* **2002**, *34*, 1919–1927. [CrossRef]

27. Bokhorst, S.; Wardle, D.A.; Nilsson, M.-C.; Gundale, M.J. Impact of understory mosses and dwarf shrubs on soil micro-arthropods in a boreal forest chronosequence. *Plant Soil* **2014**, *379*, 121–133. [CrossRef]

28. Miller, H.G. Forest fertilization: Some guiding concepts. *Forestry* **1981**, *54*, 157–167. [CrossRef]

29. Scheu, S.; Albers, D.; Alphei, J.; Buryn, R.; Klages, U.; Migge, S.; Platner, C.; Salamon, J.-A. The soil fauna community in pure and mixed stands of beech and spruce of different age: Trophic structure and structuring forces. *Oikos* **2003**, *101*, 225–238. [CrossRef]

30. Magura, T.; Bogyó, D.; Mizser, S.; Nagy, D.D.; Tóthmérész, B. Recovery of ground-dwelling assemblages during reforestation with native oak depends on the mobility and feeding habits of the species. *For. Ecol. Manag.* **2015**, *339*, 117–126. [CrossRef]

31. Scheu, S. Plants and generalist predators as links between the below-ground and above-ground system. *Basic Appl. Ecol.* **2001**, *2*, 3–13. [CrossRef]

32. Seastedt, T.R. The role of microarthropods in decomposition and mineralization processes. *Ann. Rev. Entomol.* **1984**, *29*, 25–46. [CrossRef]

33. Vasconcellos, R.L.F.; Segat, J.C.; Bonfima, J.A.; Baretta, D.; Cardoso, E.J.B.N. Soil macrofauna as an indicator of soil quality in an undisturbed riparian forest and recovering sites of different ages. *Eur. J. Soil Biol.* **2013**, *58*, 105–112. [CrossRef]

34. Kardol, P.; Reynolds, W.N.; Norby, R.J.; Classen, A.T. Climate change effects on soil microarthropod abundance and community structure. *Appl. Soil Ecol.* **2011**, *47*, 37–44. [CrossRef]

35. Wiwatwitaya, D.; Takeda, H. Seasonal changes in soil arthropod abundance in the dry evergreen forest of north-east Thailand, with special reference to collembolan communities. *Ecol. Res.* **2004**, *20*, 59–70. [CrossRef]

36. Lindberg, N.; Bengtsson, J. Population responses of oribatid mites and collembolans after drought. *Appl. Soil Ecol.* **2005**, *28*, 163–174. [CrossRef]

37. Pflug, A.; Wolters, V. Influence of drought and litter age on Collembola communities. *Eur. J. Soil Biol.* **2001**, *37*, 305–308. [CrossRef]

38. Sartori, F.; Lal, R.; Ebinger, M.H.; Eaton, J.A. Changes in soil carbon and nutrient pools along a chronosequence of poplar plantations in the Columbia Plateau, Oregon, USA. *Agric. Ecosyst. Environ.* **2007**, *122*, 325–339. [CrossRef]

39. Wang, H.; Xu, Y.; Zhang, J.; Zhang, Y.; Yang, X. Study on soil improvement effect of the *Populus tomentos* pure forest of saline land on the muddy sea-coast in northern Jiangsu Province. *Sci. Soil Water Conserv.* **2013**, *11*, 65–68. (In Chinese)

40. Mueller-Dumbois, D.; Ellenberg, H. *Aims and Methods of Vegetation Ecology*; John Wiley & Sons: Toronto, ON, Canada, 1974.

41. Decaëns, T.; Dutoitb, T.; Alardb, D.; Lavelle, P. Factors influencing soil macrofaunal communities in post-pastoral successions of westem France. *Appl. Soil Ecol.* **1998**, *9*, 361–367. [CrossRef]

42. Yin, W.Y. *Pictorial Keys to Soil Animals of China*; Science Press: Beijing, China, 1998. (In Chinese)

43. Gkisakis, V.D.; Kollaros, D.; Bàrberi, P.; Livieratos, I.C.; Kabourakis, E.M. Soil arthropod diversity in organic, integrated, and conventional olive orchards and different agroecological zones in Crete, Greece. *Agroecol. Sustain. Food Syst.* **2015**, *39*, 276–294. [CrossRef]

44. Cotes, B.; Campos, M.; Pascual, F.; García, P.A.; Ruano, F. Comparing taxonomic levels of epigeal insects under different farming systems in Andalusian olive agroecosystems. *Appl. Soil Ecol.* **2010**, *44*, 228–236. [CrossRef]

45. Biaggini, M.; Consorti, R.; Dapporto, L.; Dellacasa, M.; Paggetti, E.; Corti, C. The taxonomic level order as a possible tool for rapid assessment of arthropod diversity in agricultural landscapes. *Agric. Ecosyst. Environ.* **2007**, *122*, 183–191. [CrossRef]

46. Salamon, J.-A.; Wissuwa, J.; Moder, K.; Frank, T. Effects of *Medicago sativa*, *Taraxacum officinale* and *Bromus sterilis* on the density and diversity of Collembola in grassy arable fallows of different ages. *Pedobiologia* **2011**, *54*, 63–70. [CrossRef]

47. Chauvat, M.; Zaitsev, A.S.; Wolters, V. Successional changes of Collembola and soil microbiota during forest rotation. *Oecologia* **2003**, *137*, 269–276. [CrossRef] [PubMed]

48. Chao, A.; Gotelli, N.J.; Hsieh, T.C.; Sander, E.L.; Ma, K.H.; Colwell, R.K.; Ellison, A.M. Rarefaction and extrapolation with Hill numbers: A framework for sampling and estimation in species diversity studies. *Ecol. Monogr.* **2014**, *84*, 45–67. [CrossRef]

49. Colwell, R.K.; Chao, A.; Gotelli, N.J.; Lin, S.Y.; Mao, C.X.; Chazdon, R.L.; Longino, J.T. Models and estimators linking individual-based and sample-based rarefaction, extrapolation and comparison of assemblages. *J. Plant Ecol.* **2012**, *5*, 3–21. [CrossRef]

50. Hsieh, T.C.; Ma, K.H.; Chao, A.; McInerny, G. iNEXT: An R package for rarefaction and extrapolation of species diversity (Hill numbers). *Methods Ecol. Evol.* **2016**, *7*, 1451–1456. [CrossRef]

51. Bates, D.; Bolker, B.; Walker, S.; Christensen, R.H.B.; Singmann, H.; Dai, B.; Grothendieck, G. lme4: Linear Mixed-Effects Models Using Eigen and S4. R Package; Version 1.1-13. Available online: https://cran.r-project.org/web/packages/lme4/index.html (accessed on 1 September 2016).

52. Canty, A.; Ripley, B. *Package 'boot'*, Version 1.3-19; Available online: https://cran.r-project.org/web/packages/boot/index.html (accessed on 11 February 2017).

53. Wickham, H.; Chang, W. *ggplot2: Create Elegant Data Visualisations Using the Grammar of Graphics*; Version 2.2.1; Springer: New York, NY, USA, 2016.

54. Anderson, M.J. A new method for non-parametric multivariate analysis of variance. *Austral Ecol.* **2001**, *26*, 32–46.

55. Anderson, M.J.; Crist, T.O.; Chase, J.M.; Vellend, M.; Inouye, B.D.; Freestone, A.L.; Sanders, N.J.; Cornell, H.V.; Comita, L.S.; Davies, K.F.; et al. Navigating the multiple meanings of beta diversity: A roadmap for the practicing ecologist. *Ecol. Lett.* **2011**, *14*, 19–28. [CrossRef] [PubMed]

56. Oksanen, J.; Blanchet, F.G.; Friendly, M.; Kindt, R.; Legendre, P.; McGlinn, D.; Minchin, P.R.; O'Hara, R.B.; Simpson, G.L.; Solymos, P.; et al. vegan: Community Ecology Package. Available online: https://cran.r-project.org/web/packages/vegan/index.html (accessed on 7 April 2017).

57. De Caceres, M.; Jansen, F. Package 'Indicspecies'. Available online: https://cran.r-project.org/web/packages/indicspecies/indicspecies.pdf (accessed on 30 August 2016).

58. Miggea, S.; Maraun, M.; Scheu, S.; Schaefer, M. The oribatid mite community (Acarina) of pure and mixed stands of beech (*Fagus sylvatica*) and spruce (*Picea abies*) of different age. *Appl. Soil Ecol.* **1998**, *9*, 115–121. [CrossRef]

59. Berg, M.P.; Hemerik, L. Secondary succession of terrestrial isopod, centipede, and millipede communities in grasslands under restoration. *Biol. Fertil. Soils* **2004**, *40*, 163–170. [CrossRef]

60. Wu, P.; Liu, S.; Liu, X. Composition and spatio-temporal changes of soil macroinvertebrates in the biodiversity hotspot of northern Hengduanshan Mountains, China. *Plant Soil* **2012**, *357*, 321–338. [CrossRef]

61. Sanders, N.J.; Gotelli, N.J.; Wittman, S.E.; Ratchford, J.S.; Ellison, A.M.; Jules, E.S. Assembly rules of ground-foraging ant assemblages are contingent on disturbance, habitat and spatial scale. *J. Biogeogr.* **2007**, *34*, 1632–1641. [CrossRef]

62. Letourneau, D.K.; Dyer, L.A.; Burslem, D.; Pinard, M.; Hartley, S. Multi-trophic interactions and biodiversity: Beetles, ants, caterpillars, and plants. In *Biotic Interactions in the Tropics: Their Role in the Maintenance of Species Diversity*; Cambridge University Press: Cambridge, UK, 2005; pp. 366–385.

63. Connell, J.H. Diversity in tropical rain forests and coral reefs—High diversity of trees and corals is maintained only in a non-equilibrium state. *Science* **1978**, *199*, 1302–1310. [CrossRef] [PubMed]

64. Peña-Peña, K.; Irmler, U. Moisture seasonality, soil fauna, litter quality and land use as drivers of decomposition in Cerrado soils in SE-Mato Grosso, Brazil. *Appl. Soil Ecol.* **2016**, *107*, 124–133. [CrossRef]

65. González, G.; Seastedt, T.R. Comparison of the abundance and composition of litter fauna in tropical and subalpine forests. *Pedobiologia* **2000**, *44*, 545–555. [CrossRef]

66. Wang, S.; Tan, Y.; Fan, H.; Ruan, H.; Zheng, A. Responses of soil microarthropods to inorganic and organic fertilizers in a poplar plantation in a coastal area of eastern China. *Appl. Soil Ecol.* **2015**, *89*, 69–75. [CrossRef]

67. Berthe, S.C.F.; Derocles, S.A.P.; Lunt, D.H.; Kimball, B.A.; Evans, D.M. Simulated climate-warming increases Coleoptera activity-densities and reduces community diversity in a cereal crop. *Agric. Ecosyst. Environ.* **2015**, *210*, 11–14. [CrossRef]

68. Bokhorst, S.; Huiskes, A.H.; Convey, P.; Van Bodegom, P.M.; Aerts, R. Climate change effects on soil arthropod communities from the Falkland Islands and the Maritime Antarctic. *Soil Biol. Biochem.* **2008**, *40*, 1547–1556. [CrossRef]

69. Frampton, G.K.; Brink, P.J.V.D.; Gould, P.J.L. Effects of spring drought and irrigation on farmland and arthropods in southern Britain. *J. Appl. Ecol.* **2000**, *37*, 865–883. [CrossRef]

70. Kaneda, S.; Kaneko, N. Influence of Collembola on nitrogen mineralization varies with soil moisture content. *Soil Sci. Plant Nutr.* **2011**, *57*, 40–49. [CrossRef]

71. Chikoski, J.M.; Ferguson, S.H.; Meyer, L. Effects of water addition on soil arthropods and soil characteristics in a precipitation-limited environment. *Acta Oecol.* **2006**, *30*, 203–211. [CrossRef]

72. Salamon, J.-A.; Scheu, S.; Schaefer, M. The Collembola community of pure and mixed stands of beech (*Fagus sylvatica*) and spruce (*Picea abies*) of different age. *Pedobiologia* **2008**, *51*, 385–396. [CrossRef]

73. Rotheray, T.D.; Boddy, L.; Jones, T.H. Collembola foraging responses to interacting fungi. *Ecol. Entomol.* **2009**, *34*, 125–132. [CrossRef]

74. Jonas, J.L.; Wilson, G.W.T.; White, P.M.; Joern, A. Consumption of mycorrhizal and saprophytic fungi by Collembola in grassland soils. *Soil Biol. Biochem.* **2007**, *39*, 2594–2602. [CrossRef]

75. Chen, B.R.; Wise, D.H. Bottom-up limitation of predaceous arthropods in a detritus-based terrestrial food web. *Ecology* **1999**, *80*, 761–772. [CrossRef]

76. Lewis, J.G.E. *The Biology of Centipedes*; Cambridge University Press: Cambridge, UK, 1981.

77. Gao, M.; Taylor, M.K.; Callaham, M.A. Trophic dynamics in a simple experimental ecosystem: Interactions among centipedes, Collembola and introduced earthworms. *Soil Biol. Biochem.* **2017**, *115*, 66–72. [CrossRef]

© 2018 by the authors. Licensee MDPI, Basel, Switzerland. This article is an open access article distributed under the terms and conditions of the Creative Commons Attribution (CC BY) license (http://creativecommons.org/licenses/by/4.0/).

forests

MDPI

Article

Herbaceous Vegetation Responses to Gap Size within Natural Disturbance-Based Silvicultural Systems in Northeastern Minnesota, USA

Nicholas W. Bolton [1] and Anthony W. D'Amato [2,*]

[1]   Department of Forest Resources, University of Minnesota, St., Paul, MN 55108, USA; nwbolton@mtu.edu
[2]   Rubenstein School of Environment and Natural Resources, University of Vermont, Burlington,
       VT 05405, USA
*   Correspondence: awdamato@uvm.edu; Tel.: +1-802-656-8030

Received: 31 December 2018; Accepted: 24 January 2019; Published: 30 January 2019

Abstract: The use of silvicultural systems that emulate aspects of natural disturbance regimes, including natural disturbance severities and scales, has been advocated as a strategy for restoring and conserving forest biodiversity in forests managed for wood products. Nonetheless, key information gaps remain regarding the impacts of these approaches on a wide range of taxa, including understory plant species. We investigated the 6- or 7-year response of herbaceous vegetation to natural disturbance-based silvicultural harvest gaps in a northern hardwood forest in Northeastern Minnesota. These results indicate that harvest gaps are effective in conserving understory plant diversity by promoting conditions necessary for disturbance-dependent understory plant species. However, harvest gaps also contained non-native invasive plant species.

Keywords: understory plant communities; natural disturbance-based silviculture; forest management; species conservation; northern hardwood forests

---

1. Introduction

The forest understory layer contains the majority of species richness within forest ecosystems around the globe [1]. Thus, conservation of understory plant species is an important goal of sustainable forest management [2], particularly in light of growing concerns regarding the loss of native biodiversity from managed systems [3]. Given the role natural disturbances play in the maintenance of biodiversity through effects on resource and propagule availability and microhabitat conditions [4–6], the use of silvicultural systems patterned after the severity and frequency of natural disturbances may serve as a management approach for restoring and maintaining native biodiversity within managed forests [7,8]. Nonetheless, evaluations of the response of understory vegetation community to these management regimes are largely lacking for most forest ecosystems [9–11] hampering our ability to develop systems for maintaining natural patterns of species richness and abundance within the understory layer.

Forest disturbance dynamics within northern hardwood forests in Northeastern North America are dominated by infrequent, low intensity tree fall gaps caused by wind, disease, and insects [12–15]. In many cases, these gaps provide opportunities for understory vegetation development not afforded by dense, close-canopied forest conditions due to the variety of resource conditions occurring across the gap environment [16–18] and the diversity of microhabitat conditions created by disturbance processes [19–21]. The response of understory vegetation to natural and harvest gaps of varying sizes has been well-studied [22–26]; however, the findings from this work are inconsistent. For example, several studies have documented differences in species abundance and composition across varying gaps sizes, whereas other work has found no difference in understory plant communities between gaps and intact forest [27–31]. Reasons for these differences may include the varying gap sizes studied, the intensity of forest disturbance, and the condition of the forests studied (i.e., old-growth versus managed second-growth forests).

Inconsistencies between studies regarding the response of understory vegetation to canopy gaps has also been documented within a given forest type. For example, understory vegetation density differed between gaps and intact forest within northern hardwood forests of Northeastern USA [24]; however, investigations within the same forest type in the upper Great Lakes region have generated mixed results dependent on forest condition. In particular, no compositional differences between small gaps and intact forest conditions were detected in uneven-aged northern hardwood forests in the Upper Peninsula of Michigan [28], whereas species diversity was greater in canopy gaps relative to intact forests in second-growth stands in Northern Wisconsin [26]. Nonetheless, even where differences were not detected, Shields and Webster (2007) concluded that gaps provide opportunities for the regeneration of species absent from intact forest, particularly *Sanguinaria canadensis* L. (bloodroot), *Impatiens capensis* Meerb. (jewelweed), and *Rubus idaeus* L. (red raspberry). As such, the use of harvest gaps patterned after natural canopy gaps may provide an opportunity to enhance and restore understory plant richness in managed northern hardwood forest systems.

In many managed forests, management practices have historically focused on creating relatively uniform, homogeneous conditions with little diversity in the canopy, sapling, shrub, seedling, and herbaceous vegetation layers. Within the context of changing global conditions, these homogeneous systems are viewed as being highly vulnerable to future environmental changes [32]. Correspondingly, enhancing stand resilience in managed forest systems has become an emerging management objective [33]. In particular, management regimes that result in an increase in species and functional richness may improve a given forest's ability to adapt to environmental change [32,34]. In many cases, recommended management regimes for increasing adaptive capacity of managed systems to these changes have built on ecological silvicultural principles associated with creating a diversity of structural and compositional conditions through emulation of historic disturbance patterns [35]. This includes creating a range of canopy openings within the system to introduce spatial heterogeneity that may allow for the development of novel understory vegetation patterns on the landscape [2,31,36].

The present study examined the effects of harvest techniques that emulate natural gap openings on understory vegetation 6- or 7- years post-harvest within second-growth northern hardwood systems in the upper Great Lakes. The objective for this work was to develop an understanding of how understory vegetation responded to harvest gaps compared to the intact forest and the spatial distribution of understory vegetation within harvest gaps (i.e., gap edge and gap center). To achieve this objective, we examined the response of understory vegetation (herbaceous plant species and a singular shrub species (*Rubus idaeus*), to a range of gap sizes patterned after natural disturbances for the region [12]. We hypothesized that (i) large harvest gaps will enhance understory vegetation diversity and abundance relative to smaller canopy gaps and the intact forest and (ii) establishment patterns of understory vegetation will vary spatially across harvest gaps, particularly between gap edges and centers.

2. Materials and Methods

2.1. Study Sites

Study sites were located along the northern shore of Lake Superior in Northeastern Minnesota, USA (Table 1). Elevations within this area range from 381 to 472 m, and soils are loams derived from glacial tills [37]. Mean annual precipitation is 739 mm and annual temperatures range from $-8.5\,°C$ in January to 18.7 °C in July. Forests within the study area are dominated by northern hardwoods (Table 1) and scattered *Thuja occidentalis* L. (northern white cedar). Historically, the dominant overstory tree regeneration technique implemented in these systems were clearfelling. Understory vegetation within the study area includes *Clintonia borealis* (Aiton) Raf. (bluebead lily), *Streptopus lanceolatus* (Aiton) Reveal (twisted rosy-stalk), *Polygonatum pubescens* (Willd.) Pursh (Solomon's seal), *Claytonia caroliniana* Michx. (springbeauty), and *Thelypteris phegopteris* (L.) Sloss. (beech fern).

Table 1. Physiographic and compositional characteristics of the study sites in Northeastern Minnesota, USA.

Site	Lat/Long	Harvest Year	Elevation (m)	Aspect	Slope	Soils	Canopy Composition [†] (% Basal Area)	Number of Gaps	Gap Size Range (ha)
Big Pine (BP)	(47.47, −91.15)	2003	487	162°	8%	Loam	*Acer saccharum* Marshall: 84% *Betula alleghaniensis* Britt.: 3% *Betula papyrifera* Marshall: 3% *Thuja occidentalis* L.: 9%	10	(0.024–0.066)
Birch Cut (BC)	(47.45, −91.19)	2002	455	119°	6%	Loam	*Acer saccharum*: 89% *Betula alleghaniensis*: 3% *Acer rubrum* L.: 3% *Picea glauca* (Moench) Voss: 3% *Thuja occidentalis*: 3%	10	(0.021–0.071)
Power Line (PL)	(47.34, −91.20)	2003	385	142°	10%	Fine loam	*Acer saccharum*: 74% *Fraxinus nigra* Marshall: 24% *Populus grandidentata* Michaux: 2%	10	(0.011–0.069)
Schoolhouse (SH)	(47.46, −91.20)	2002	478	327°	2%	Loam	*Acer saccharum*: 63% *Betula alleghaniensis*: 15% *Picea glauca*: 15% *Thuja occidentalis*: 7%	16	(0.008–0.050)

[†] Canopy composition based upon surrounding intact forest data gathered by basal area swings with a 2.3 m^2/ha factor prism.

2.2. Study Design & Field Procedures

Harvest gaps were created within each site and replicated across four blocks in a completely randomized block design. Harvests took place during the winters of 2002 and 2003. Harvest gaps and intact forest conditions were measured in the summer of 2009 to assess the 6- and 7-year vegetation responses of these communities to harvesting treatments. To ensure adequate representation of gap environments, transects were laid across each gap oriented in subcardinal directions (northeast (NE), northwest (NW), southeast (SE), and southwest (SW)) and extended to the gap edge. Along each transect, 1 m^2 plots were systematically located and used for measuring tree regeneration and understory plant communities. Spacing between plots along each transect was adjusted according to gap size in order to provide an even distribution across each transect. To ensure consistent sampling intensities, each transect contained enough plots to represent 5% of each gap area (e.g., 8.1 m^2 plots for a 0.016 ha gap).

The location of each 1 m^2 plot within the gap was noted in the field to allow for analysis of the impact of spatial location on understory plant community composition. Plots falling within the outer third of a given gap were categorized as gap edge, and plots located in the inner third were categorized as gap center. The density (stems and clumps/m^2) and species of herbaceous understory plants, including herbs, ferns, fern-allies, graminoids, and *R. idaeus*, were recorded within each 1 m^2 plot. The inclusion of the woody shrub, *R. ideaus*, in our characterization of herbaceous understory plant communities was due to the recognized importance of this species in affecting vegetation development in canopy gaps in northern hardwood ecosystems [38,39]. An herbivory index was developed by dividing the number of *A. saccharum* Marshall stems with visible browse damage out of the first ten *A. saccharum* stems along a randomly chosen subcardinal transect within each gap or interior forest plot location. Subcardinal transects were also used for point-line sampling of coarse woody debris (CWD) volume estimates.

A series of control plots were placed in unharvested, intact portions at least 50 m from created gaps to serve as an approximation of pre-harvest vegetation conditions. For all control plots (*n* = 14), all measurements were performed in the same manner as study gaps. With the exception of Power Line (PL), each site contained three intact forest plots that were randomly located within unharvested portions of each stand. Due to a greater degree of variation in canopy composition at the PL site, an additional two plots were included within the intact forest portions of this site, resulting in five control plots at this site.

2.3. Statistical Analyses

Understory plant species densities were averaged for harvest gaps and the intact forest. Analysis of variance (ANOVA) was used to examine the impact of gap size on species richness and densities of understory vegetation [40]. ANOVAs were followed by Tukey-Kramer tests in cases in which significant gap size effects were detected. In addition, multivariate tests were conducted to assess understory vegetation compositional differences among gap size classes and spatial location using blocked multi-response permutation procedures (MRBPs) (PC-ORD version 5; McCune and Mefford 1999). Indicator species analysis was used to identify plant species most likely to be found within gap size classes and different gap locations (gap center, gap edge, or intact forest). Non-metric multidimensional scaling (NMS; [41]) was used to examine patterns in understory community composition within and among gap size classes. Due to differences in site conditions between blocks, NMS was run separately for each study site to allow for an evaluation of the effects of gap conditions on understory communities. Sørensen distances were used for MRBP and NMS to calculate a distance matrix for the harvest gaps and intact forest plots. To reduce noise in the data set, species with fewer than three occurrences were removed from the data matrices and the "slow-and-thorough" autopilot mode of NMS was used in PC-ORD to generate solutions [41]. Ordinations were run until a configuration of lowest stress was found, which for all four study blocks was a three-axis solution. The two axes that explained the most variability in the data are presented in the results section.

The relationship between understory community composition and environmental and forest structural characteristics (gap size, CWD volume, and herbivory index) were explored by using the bi-plot function overlaying a secondary matrix of these variables in PC-ORD [42]. Ordinations were rotated to place the environmental or forest structural variable with the highest correlation to understory community composition on the first axis. Relationships between species density and NMS axis scores were explored using Kendall's tau statistic in SAS [40].

3. Results

3.1. Density and Diversity of Understory Vegetation

Forty-two understory plant species or groups (e.g., *Carex* sp.) occurred across all study blocks. Thirty-three (79%) of those occurred across all harvest gap sizes and the intact forest. Understory vegetation was much denser in harvested gaps compared to intact forest (Figure 1), whereas richness was lower in small gaps compared to intact forest and was slightly higher in larger gaps (Figure 2). Certain understory species only occurred in gaps: *Actaea* sp. (baneberry), *Botrychium virginianum* (L.) Sw. (rattlesnake fern), *Mertensia paniculata* (Aiton) G. Don (northern bluebell), *Rubus idaeus* L. (red raspberry), *Sanguinaria canadensis* (bloodroot), and *Cirsium arvense* (L.) Scop. (Canada thistle). All other species occurred in both the intact forest and at least one harvest gap.

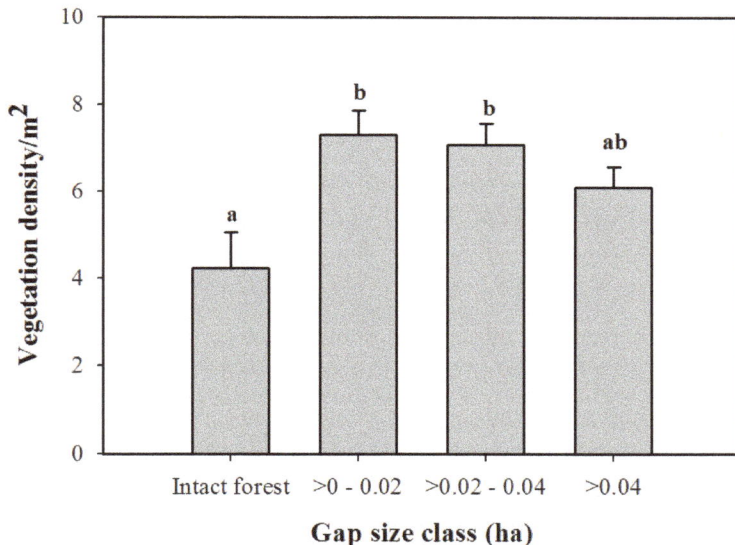

Figure 1. Understory vegetation density (stem count or clump per m^2) by harvest gap size (ha). Understory vegetation densities are averaged for all meter square sampling points throughout harvest gaps and intact forest plots. Error bars represent one standard error, and values with different letters are statistically different ($p \leq 0.05$; Tukey-Kramer test).

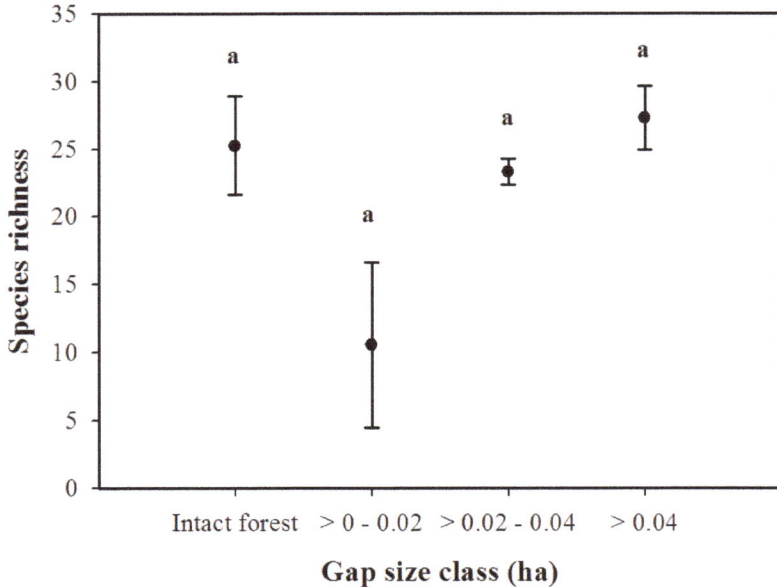

Figure 2. Understory vegetation species richness by harvest gap size (ha). All identified individual species encountered within all meter square sample plots were used to calculate richness. Error bars represent standard error. There were no significant differences between gap size classes ($p > 0.05$).

3.2. Patterns in Understory Plant Composition across Gap Conditions

Understory plant composition differed among gap sizes (MRBP; A-statistic = 0.017, $p = 0.0002$). Species composition differed as gap size increased with the introduction of disturbance-adapted species, weedy plants, and non-native invasive species (Figure 3). In particular, abundances of *Botrychium virginianum*, *Cirsium arvense*, *Mertensia paniculata*, and *Rubus ideaus* increased as gap size increased. In contrast, there was no difference between understory plant communities within gap center and gap edge conditions. A significant indicator species ($p < 0.05$) for the intact forest was *Maianthemum canadense* Desf. (Canada mayflower). Whereas *Polygonatum pubescens* was indicative of gap edge conditions, and *Impatiens capensis* (jewelweed) and *Rubus idaeus* were indicative of gap center conditions, there was no statistical difference detected for understory plant composition between gap edge and gap center conditions. All ordinations had a final stress less than 15.5 and instability <0.0000001.

In addition to gap size being a significant environmental variable that explained the variability of the herbaceous communities within the study gaps, the herbivory index also explained the variability within experimental gaps. We did not specifically investigate the influence of browse on the herbaceous community response to these harvest gaps, but this result does indicate that larger harvest gaps were locations in which herbivores congregated.

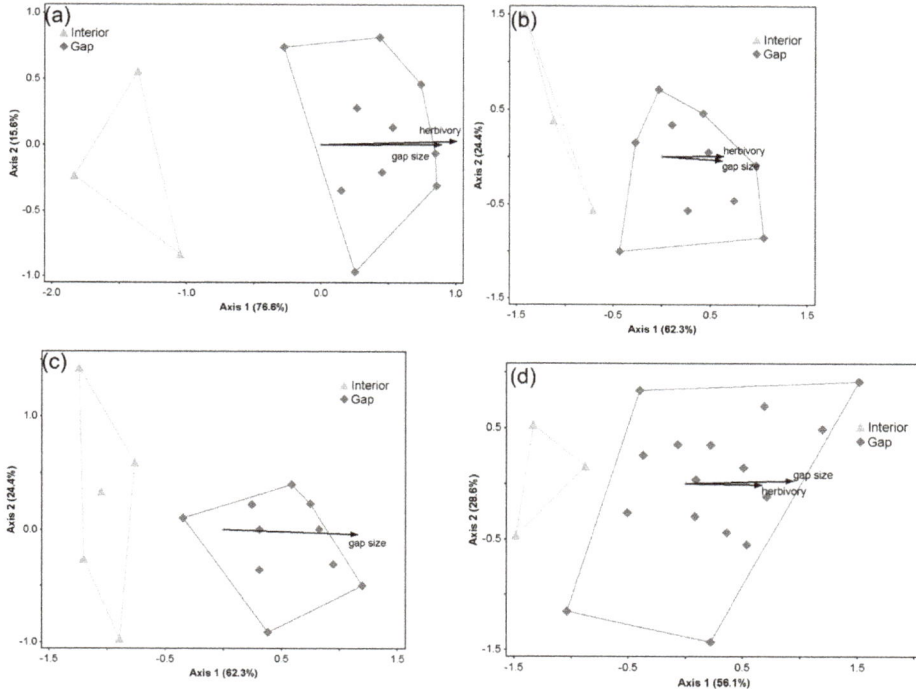

Figure 3. Nonmetric multidimensional scaling (NMS) ordinations of understory plant communities in gap and interior forest locations for the (**a**) Birch Cut, (**b**) Big Pine, (**c**) Power Line, and (**d**) Schoolhouse study areas. Vector length represents explanatory power of environmental variables (only variables with $r^2 > 0.2$ are depicted). Ordination diagrams were rigidly rotated to place the variable "gap size" parallel with the NMS axis 1. See Table 1 for study site information.

4. Discussion

Understory Vegetation Responses

Natural disturbances, including forest canopy gap formation, play an important role in structuring understory plant communities [23,43]. Correspondingly, the use of natural disturbance-based silvicultural (NDBS) regimes has been proposed as a means to maintain and restore native biodiversity to managed landscapes [8]. This study found that harvests patterned after natural canopy gaps increased understory vegetation richness within the northern hardwood systems examined—a result consistent with previous work examining a range of harvest gap sizes in the upper Great Lakes region [26,30]. In larger gaps (>0.02 ha), the occurrence of native species such as *Sanguinaria canadensis* increased. In addition, these gaps served to restore an element of spatial variability in the abundance of understory vegetation across these stands—a stand structural characteristic often lacking from second-growth northern hardwood stands [44].

Compositional changes between the intact forest and harvest gaps were largely due to the presence of disturbance-dependent species, wetland species, and non-native invasive species within harvest gaps. This finding is consistent with other studies following compositional changes post-harvest that have found that harvesting disturbance introduced weedy and non-native species resulting in increased species richness [27,45]. The primary disturbance-dependent species found within the stands examined was *Rubus idaeus*, which was found in greatest abundance in larger canopy gaps. This finding is consistent with other studies examining patterns of understory plant species in northern hardwood forests [28,38,39]; however, work by Donoso and Nyland [38] suggests that the abundance of *R. idaeus* sharply declines in the decade following gap creation as canopy gaps close and advance regeneration overtop this species. Nonetheless, *R. idaeus* can hinder tree regeneration [38] within northern hardwood systems, particularly in areas with elevated herbivore densities, although we did not detect any evidence for this in the analysis of forest regeneration in these stands [46].

Although largely associated with wetter forest conditions, *Impatiens capensis* was found within canopy gaps within these stands. Similar patterns were documented by Shields and Webster [28] within canopy gaps in Northern Michigan, and they postulated that this phenomenon was due to increases in soil moisture stemming from the reduction of evapotranspiration following the removal of canopy trees. The presence of this species may also reflect its ability to establish and occasionally dominate areas following disturbance [47].

An increasing challenge within managed landscapes is the prevalence of non-native invasive species. In many cases, these species are associated with road edges, landings, and other highly disturbed locations [48]; however, many species are now being found within the interior of managed forests. Within this study, the non-native, invasive species *Cirsium arvense* was associated with large canopy gaps (>0.02 ha). The presence of this species in harvested stands has often been linked to the transport of seeds on logging equipment [49], as well as its ability to take advantage of soil compaction created by harvesting practices [50]. Given that the harvests in this study were done during the winter, it is likely that the main factors explaining the abundance of this species in larger gaps are its large dispersal range and aggressive establishment in disturbed areas [51,52].

The density of understory vegetation in harvest gaps was higher than the intact forest, a result similar to findings of other research in northern hardwoods [43,53,54]. Even small gaps had higher densities of understory vegetation per 1 m^2 compared to intact forest conditions, which is the opposite of the findings of Goldblum [24]. These differences are likely due to the difference in gap size classification (i.e., ranges of gap size classes) between the studies as well as time passed since disturbance. Gaps studied in Goldblum (1997) were 1–30 years post-disturbance, which are considerably longer time periods than the age of the gaps investigated in this study. Given these findings, it is possible that the changes in abundance of understory vegetation we documented within smaller gaps may be ephemeral and will decline as gaps close.

Most species encountered within the study gaps were present within the intact forest and increased in abundance as gap size increased. These species most likely took advantage of increased availability of resources within the gaps (e.g., light) [36,55,56]. Consistent with Shields and Webster [28] and Kern et al. [30], certain species that were not found within the intact forest were primarily established in large gaps, which included *Rubus idaeus, Impatiens capensis, Sanguinaria canadensis*, and *Cirsium arvense*.

Consistent with other studies within northern hardwood forests [9], the abundance of a few species did not change from the intact forest or across harvest gap sizes, including *Clintonia borealis* and *Aralia nudicaulis*. These species are rhizomatous [57] and primarily spread across the forest floor at very slow rates [57] as opposed to the species dominating the ground layer within the larger gaps we examined (e.g., *R. idaeus*), which are seed-bankers that often inhabit disturbed areas [58].

The lower diversity of understory plants found within small gaps (<0.02 ha) may be due to a combination of increased sapling growth rates from the harvest [46], damage to vegetation from logging operations [57], and harvest gap closures by border tree encroachment [59,60]. In particular, stimulated sapling growth and gap closure by bordering tree crowns likely minimized the amount of time resource levels were elevated at the forest floor layer following canopy gap formation. Although harvests led to an increase in the abundance of existing understory species (see above), there was limited opportunity for other species to regenerate, a finding consistent with research investigating gap partitioning in northern hardwood and western conifer forests (e.g., [30,61]). In this work, increases in species richness relating to resource availability were detected in larger gaps (e.g., >0.1 ha) within which resource gradients existed.

Research conducted in Northeastern Minnesota by Burton et al. [62] found that understory layers differed between second-growth forests managed by NDBS and old-growth forests. They concluded that second-growth forests similar to the ones examined in this study were higher in species diversity and contained higher levels of understory species abundance when compared to old-growth forests nearby. Given these findings, it is possible that the canopy gaps created in the current study are driving the structure and composition of the understory layers in these second-growth stands further from those found in old-growth forests. In particular, Burton et al. [57] speculated that the presence of a higher diversity of overstory tree species within old-growth stands, including *Betula allegheniensis* Britt., *Picea glauca* (white spruce), and *Thuja occidentalis* (northern white cedar), resulted in a greater diversity of understory environmental conditions and higher levels of heterogeneity in understory plant communities. The gaps examined in this study, although containing increased levels of resource availability, largely released understory species that developed under a fairly mono-specific (i.e., *A. saccharum*) second-growth canopy. Future work aimed at restoring the diversity and structure of these communities may require a greater emphasis on restoring overstory community composition, particularly the currently underrepresented *B. alleghaniensis* and *T. occidentalis*.

5. Conclusions

The natural disturbance-based silvicultural treatment applied to the study area likely increased the richness of understory vegetation within the stand by providing opportunities for native species, disturbance-dependent species, and invasive species to establish. In particular, gap sizes between 0.008 and 0.07 ha allowed for the regeneration or establishment of several native species that were not found in the intact forest. However, the increase in richness also came at the expense of introducing aggressively establishing non-native invasive species and disturbance-dependent species into portions of these stands. The minimal gains in functional diversity need to be weighed in the context of the potential displacement of important native species by more aggressive native and non-native species. Much of the management within northern hardwood forests typically employs single-tree selection methods, which create relatively uniform canopy conditions and understory environments resulting in homogenization of the understory with relatively low species richness and understory structure [44]. This work adds to the growing body of work that suggests that the use of management regimes in which larger canopy gaps are created enhances compositional diversity of understory vegetation in second-growth northern hardwood forests of the upper Great Lakes, but may also introduce non-native invasive species if precautions are not taken [26,34]. Nonetheless, by creating a patchwork of high-density areas of understory herbaceous vegetation within gaps, these treatments restored some of the spatial diversity in understory abundance patterns found in old-growth northern hardwood systems [44].

Author Contributions: Methodology, analysis, writing, figure creation, and editing was conducted by N.W.B. & A.W.D.

Funding: This research was funded by University of Minnesota Department of Forest Resources and The Nature Conservancy. Access to the study locations was provided by the Lake County Minnesota Forestry Department.

Acknowledgments: We thank Meredith Cornett, Chris Dunham, Mark White, and Lee Frelich for their assistance during the experimental design, analysis, and reporting phases of this work. We appreciate Matthew Tyler for his vast knowledge of Minnesota native plants and field assistance.

Conflicts of Interest: The authors declare no conflict of interest.

References

1. Whigham, D.F. Ecology of woodland herbs in temperate deciduous forests. *Annu. Rev. Ecol. Syst.* **2004**, *35*, 583–621. [CrossRef]
2. Roberts, M.R.; Gilliam, F.S. Patterns and mechanisms of plant diversity in forested ecosystems: Implications for forest management. *Ecol. Appl.* **1995**, *5*, 969–977. [CrossRef]
3. Hunter, M.L. *Maintaining Biodiversity in Forest Ecosystems*; Cambridge University Press: Cambridge, UK, 1999.

4. Goldberg, D.E.; Gross, K.L. Disturbance regimes of midsuccessional old fields. *Ecology* **1988**, *69*, 1677–1688. [CrossRef]

5. Sousa, W.P. Intertidal mosaics: Patch size, propagule availability, and spatially variable patterns of succession. *Ecology* **1984**, *65*, 1918–1935. [CrossRef]

6. White, P.S.; Pickett, S.T.A. *Natural Disturbance and Patch Dynamics: An Introduction*; Academic: New York, NY, USA, 1985.

7. Fries, C.; Johansson, O.; Pettersson, B.; Simonsson, P. Silvicultural models to maintain and restore natural stand structures in Swedish boreal forests. *For. Ecol. Manag.* **1997**, *94*, 89–103. [CrossRef]

8. Seymour, R.S.; White, A.S.; de Maynadier, P.G. Natural disturbance regimes in northeastern North America—Evaluating silvicultural systems using natural scales and frequencies. *For. Ecol. Manag.* **2002**, *155*, 357–367. [CrossRef]

9. Schumann, M.E.; White, A.S.; Witham, J.W. The effects of harvest-created gaps on plant species diversity, composition, and abundance in a Maine oak-pine forest. *For. Ecol. Manag.* **2003**, *176*, 543–561. [CrossRef]

10. Smith, K.J.; Keeton, W.S.; Twery, M.J.; Tobi, D.R. Understory plant responses to uneven-aged forestry alternatives in northern hardwoodconifer forests. *Can. J. For. Res.* **2008**, *38*, 1303–1318. [CrossRef]

11. Falk, K.J.; Burke, D.M.; Elliott, K.A.; Holmes, S.B. Effects of single-tree and group selection harvesting on the diversity and abundance of spring forest herbs in deciduous forests in southwestern Ontario. *For. Ecol. Manag.* **2008**, *255*, 2486–2494. [CrossRef]

12. Frelich, L.E.; Lorimer, C.G. Natural disturbance regimes in hemlock-hardwood forests of the upper Great-Lakes region. *Ecol. Monogr.* **1991**, *61*, 145–164. [CrossRef]

13. Zhang, Q.; Pregitzer, K.S.; Reed, D.D. Catastrophic disturbance in the presettlement forests of the Upper Peninsula of Michigan. *Can. J. For. Res.* **1999**, *29*, 106–114. [CrossRef]

14. Webb, S.L.; Scanga, S.E. Windstorm disturbance without patch dynamics: Twelve years of change in a Minnesota forest. *Ecology* **2001**, *82*, 893–897. [CrossRef]

15. Schulte, L.A.; Mladenoff, D.J. Severe wind and fire regimes in northern forests: Historical variability at the regional scale. *Ecology* **2005**, *86*, 431–445. [CrossRef]

16. Bazzaz, F.A. The physiological ecology of plant succession. *Annu. Rev. Ecol. Syst.* **1979**, *10*, 351–371. [CrossRef]

17. Sipe, T.W.; Bazzaz, F.A. Gap Partitioning Among Maples (Acer) in Central New England: Survival and Growth. *Ecology* **1995**, *76*, 1587–1602. [CrossRef]

18. Busing, R.T.; White, P.S. Species diversity and small-scale disturbance in an old-growth temperate forest: A consideration of gap partitioning concepts. *Oikos* **1997**, *78*, 562–568. [CrossRef]

19. Beatty, S.W. Influence of microtopography and canopy species on spatial patterns of forest understory plants. *Ecology* **1984**, *65*, 1406–1419. [CrossRef]

20. Zenner, E.K.; Kabrick, J.M.; Jensen, R.G.; Peck, J.L.E.; Grabner, J.K. Responses of ground flora to a gradient of harvest intensity in the Missouri Ozarks. *For. Ecol. Manag.* **2006**, *222*, 326–334. [CrossRef]

21. Peterson, C.J.; Carson, W.P.; McCarthy, B.C.; Pickett, S.T.A. Microsite variation and soil dynamics within newly created treefall pits and mounds. *Oikos* **1990**, *58*, 39–46. [CrossRef]

22. Naaf, T.; Wulf, M. Effects of gap size, light and herbivory on the herb layer vegetation in European beech forest gaps. *For. Ecol. Manag.* **2007**, *244*, 141–149. [CrossRef]

23. Anderson, K.L.; Leopold, D.J. The role of canopy gaps in maintaining vascular plant diversity at a forested wetland in New York State. *J. Torrey Bot. Soc.* **2002**, *129*, 238–250. [CrossRef]

24. Goldblum, D. The effects of treefall gaps on understory vegetation in New York State. *J. Veg. Sci.* **1997**, *8*, 125–132. [CrossRef]

25. Royo, A.A.; Collins, R.; Adams, M.B.; Kirschbaum, C.; Carson, W.P. Pervasive interactions between ungulate browsers and disturbance regimes promote temperate forest herbaceous diversity. *Ecology* **2010**, *91*, 93–105. [CrossRef] [PubMed]

26. Kern, C.C.; Montgomery, R.A.; Reich, P.B.; Strong, T.F. Harvest-created canopy gaps increase species and functional trait diversity of the forest ground-layer communiiy. *For. Sci.* **2014**, *60*, 335–344. [CrossRef]

27. Metzger, F.; Schultz, J. Understory response to 50 years of management of a northern hardwood forest in upper Michigan. *Am. Midl. Nat.* **1984**, *112*, 209–223. [CrossRef]

28. Shields, J.M.; Webster, C.R. Ground-layer response to group selection with legacy-tree retention in a managed northern hardwood forest. *Can. J. For. Res.* **2007**, *37*, 1797–1807. [CrossRef]

29. Schoonmaker, P.; McKee, A. Species composition and diversity during secondary succession of coniferous forests in the western Cascade Mountains of Oregon. *For. Sci.* **1988**, *34*, 960–979.

30. Kern, C.C.; Montgomery, R.A.; Reich, P.B.; Strong, T.F. Canopy gap size influences niche partitioning of the ground-layer plant community in a northern temperate forest. *J. Plant Ecol.* **2013**, *6*, 101–112. [CrossRef]

31. Kern, C.C.; Burton, J.I.; Raymond, P.; D'Amato, A.W.; Keeton, W.S.; Royo, A.A.; Walters, M.B.; Webster, C.R.; Willis, J.L. Challenges facing gap-based silviculture and possible solutions for mesic northern forests in North America. *For. Int. J. For. Res.* **2017**, *90*, 4–17. [CrossRef]

32. Holling, C.S. Resilience and stability of ecological systems. *Annu. Rev. Ecol. Syst.* **1973**, *4*, 1–23. [CrossRef]

33. Bauhus, J.; Puettmann, K.; Messier, C. Silviculture for old-growth attributes. *For. Ecol. Manag.* **2009**, *258*, 525–537. [CrossRef]

34. Webster, C.R.; Dickinson, Y.L.; Burton, J.I.; Frelich, L.E.; Jenkins, M.A.; Kern, C.C.; Raymond, P.; Saunders, M.R.; Walters, M.B.; Willis, J.L. Promoting and maintaining diversity in contemporary hardwood forests: Confronting contemporary drivers of change and the loss of ecological memory. *For. Ecol. Manag.* **2018**, *421*, 98–108. [CrossRef]

35. Fahey, R.T.; Alveshere, B.C.; Burton, J.I.; D'Amato, A.W.; Dickinson, Y.L.; Keeton, W.S.; Kern, C.C.; Larson, A.J.; Palik, B.J.; Puettmann, K.J. Shifting conceptions of complexity in forest management and silviculture. *For. Ecol. Manag.* **2018**, *421*, 59–71. [CrossRef]

36. Moore, M.R.; Vankat, J.L. Responses of the herb layer to the gap dynamics of a mature beech-maple forest. *Am. Midl. Nat.* **1986**, *115*, 336–347. [CrossRef]

37. Hobbs, H.C.; Goebel, J.E. *Geologic Map of Minnesota, Quarternary Geology (Scale 1:500,000)*; Minnesota Geolgcial Survey, University of Minnesota: Minneapolis, MN, USA, 1982.

38. Donoso, P.J.; Nyland, R.D. Interference to hardwood regeneration in northeastern North America: The effects of raspberries (*Rubus* spp.) following clearcutting and shelterwood methods. *North. J. Appl. For.* **2006**, *23*, 288–296.

39. Kern, C.C.; D'Amato, A.W.; Strong, T.F. Diversifying the composition and structure of managed, late-successional forests with harvest gaps: What is the optimal gap size? *For. Ecol. Manag.* **2013**, *304*, 110–120. [CrossRef]

40. SAS Institute. *SAS Version 9.1*; SAS Institute: Cary, NC, USA, 2008.

41. McCune, B.; Grace, J. *MRPP (Multi-Response Permutation Procedures) and Related Techniques*; MjM Software Design: Gleneden Beach, OR, USA, 2002; pp. 188–197.

42. McCune, B.; Mefford, M. *PC-ORD: Multivariate Analysis of Ecological Data*; Version 4 for Windows; [User's Guide]; MjM Software Design: Gleneden Beach, OR, USA, 1999.

43. Collins, B.S.; Pickett, S.T.A. Demographic responses of herb layer species to experimental canopy gaps in a northern hardwoods forest. *J. Ecol.* **1988**, *76*, 437–450. [CrossRef]

44. Crow, T.R.; Buckley, D.S.; Nauertz, E.A.; Zasada, J.C. Effects of management on the composition and structure of northern hardwood forests in Upper Michigan. *For. Sci.* **2002**, *48*, 129–145.

45. Scheller, R.M.; Mladenoff, D.J. Understory species patterns and diversity in old-growth and managed northern hardwood forests. *Ecol. Appl.* **2002**, *12*, 1329–1343. [CrossRef]

46. Bolton, N.W.; D'Amato, A.W. Regeneration responses to gap size and coarse woody debris within natural disturbance-based silvicultural systems in northeastern Minnesota, USA. *For. Ecol. Manag.* **2011**, *262*, 1215–1222. [CrossRef]

47. USDA National Plant Data Team. *The PLANTS Database*; National Plant Data Team: Greensboro, NC, USA, 2015. Available online: http://plants.usda.gov (accessed on 15 May 2011).

48. Sumners, W.H.; Archibold, O.W. Exotic plant species in the southern boreal forest of Saskatchewan. *For. Ecol. Manag.* **2007**, *251*, 156–163. [CrossRef]

49. Berger, A.L.; Puettmann, K.J.; Host, G.E. Harvesting impacts on soil and understory vegetation: The influence of season of harvest and within-site disturbance patterns on clear-cut aspen stands in Minnesota. *Can. J. For. Res.* **2004**, *34*, 2159–2168. [CrossRef]

50. Buckley, D.S.; Crow, T.R.; Nauertz, E.A.; Schulz, K.E. Influence of skid trails and haul roads on understory plant richness and composition in managed forest landscapes in Upper Michigan, USA. *For. Ecol. Manag.* **2003**, *175*, 509–520. [CrossRef]

51. Heimann, B.; Cussans, G.W. The importance of seeds and sexual reproduction in the population biology of Cirsium arvense-a literature review. *Weed Res.* **1996**, *36*, 493–503. [CrossRef]

52. Wilson, R.G., Jr. Germination and seedling development of Canada thistle (*Cirsium arvense*). *Weed Sci.* **1979**, *27*, 146–151.

53. Mladenoff, D.J. The relationship of the soil seed bank and understory vegetation in old-growth northern hardwood-hemlock treefall gaps. *Can. J. Bot.* **1990**, *68*, 2714–2721. [CrossRef]

54. Ehrenfeld, J.G. Understory response to canopy gaps of varying size in a mature oak forest. *Bull. Torrey Bot. Club* **1980**, *107*, 29–41. [CrossRef]

55. Anderson, M.C. Studies of the Woodland Light Climate: II. Seasonal Variation in the Light Climate. *J. Ecol.* **1964**, *52*, 643–663. [CrossRef]

56. Davison, S.E.; Forman, R.T.T. Herb and shrub dynamics in a mature oak forest: A thirty-year study. *Bull. Torrey Bot. Club* **1982**, *109*, 64–73. [CrossRef]

57. Meier, A.J.; Bratton, S.P.; Duffy, D.C. Possible ecological mechanisms for loss of vernal-herb diversity in logged eastern deciduous forests. *Ecol. Appl.* **1995**, *5*, 935–946. [CrossRef]

58. Whitney, G.G. The productivity and carbohydrate economy of a developing stand of Rubus idaeus. *Can. J. Bot.* **1982**, *60*, 2697–2703. [CrossRef]

59. Hibbs, D.E. Gap dynamics in a hemlock-hardwood forest. *Can. J. For. Res.* **1982**, *12*, 538–544. [CrossRef]

60. Webster, C.R.; Lorimer, C.G. Minimum opening sizes for canopy recruitment of midtolerant tree species: A retrospective approach. *Ecol. Appl.* **2005**, *15*, 1245–1262. [CrossRef]

61. Fahey, R.T.; Puettmann, K.J. Patterns in spatial extent of gap influence on understory plant communities. *For. Ecol. Manag.* **2008**, *255*, 2801–2810. [CrossRef]

62. Burton, J.I.; Zenner, E.K.; Frelich, L.E.; Cornett, M.W. Patterns of plant community structure within and among primary and second-growth northern hardwood forest stands. *For. Ecol. Manag.* **2009**, *258*, 2556–2568. [CrossRef]

© 2019 by the authors. Licensee MDPI, Basel, Switzerland. This article is an open access article distributed under the terms and conditions of the Creative Commons Attribution (CC BY) license (http://creativecommons.org/licenses/by/4.0/).

forests

MDPI

Article

Woody Species Composition, Diversity, and Recovery Six Years after Wind Disturbance and Salvage Logging of a Southern Appalachian Forest

Callie A. Oldfield * and Chris J. Peterson

Department of Plant Biology, University of Georgia, Athens, GA 30602, USA; cjpete@uga.edu
* Correspondence: callieoldfield@uga.edu

Received: 31 December 2018; Accepted: 1 February 2019; Published: 6 February 2019

Abstract: Salvage logging after wind disturbance of a mixed conifer-hardwood forest results in sapling compositional changes but no changes to species diversity six years post-disturbance. Several conceptual frameworks allow for predictions of the effects of forest disturbances on composition, but fewer yield predictions of species diversity. Following compound disturbance, tree species diversity and composition is predicted to shift to early successional species. Because of the greater cumulative severity, diversity should be lower in areas experiencing windthrow + salvage logging than in similar sites experiencing windthrow alone. We examined the effects of wind disturbance and salvage logging on diversity parameters over six years. We hypothesized that the effects of salvage logging on diversity would be short-lived, but that species composition would be altered six years post-disturbance. Sampling plots were established in a mixed-hardwood forest in north Georgia, USA, after a 2011 EF3 tornado and surveyed in 2012 and 2017. Nineteen 20 × 20 m plots were surveyed (10 unsalvaged, 9 salvaged) for parameters including Shannon diversity, species richness, and composition. Ordinations were used to visualize tree and sapling species composition in salvage logged plots. We found that there was no significant difference in Shannon diversity between salvaged and unsalvaged plots before disturbance, <1 post-disturbance, or 6 years post-disturbance. The disturbances altered the tree and sapling species compositions, with salvaged plots having more mid-successional saplings but few true pioneer species. There appears to be an emerging pattern in the wind disturbance + salvaging literature which our study supports– salvaging does not affect tree species diversity but shifts species composition over time.

Keywords: windthrow; tornado; tree species; disturbance severity; tree regeneration; salvaging; salvage logging; succession

---

1. Introduction

While our understanding of individual disturbance effects in forests is well-established, knowledge of and ideas about compounded disturbances (multiple events in a short period of time) is still developing [1,2]. Natural disturbances such as wind, fire or drought may interact in ways that suggest the combined effects can be understood in light of the cumulative severity [3,4]. Forests are a patchwork of previous disturbances with different stages of recovery. There may be many overlapping natural disturbances in a forest—and these can interact with one another to increase severity [5,6]. The amplifying effects of such interactions can lead to ecological tipping points, which when reached, can permanently change aspects of a community, including its function and composition [5,7].

Wind is the most common agent of disturbance in mesic temperate forests and affects thousands of square kilometers annually [1,8,9]. The immediate effects of a windthrow may include tree mortality, changes in the size structure, and reduced species diversity of affected sites [8,10]. Less well-known is how wind may interact with anthropogenic disturbances, such as wind followed by salvage logging of damaged and downed trees [11–13]. We test whether salvage logging compounds the effects of wind disturbance, using a tornado damaged forest that was partially salvage logged.

Several conceptual frameworks can provide expectations of trends in composition, diversity and successional state after wind disturbance and salvage logging. The intermediate disturbance hypothesis (IDH) posits that local species diversity will be highest when disturbance frequency, size, or severity is intermediate due to increased dominance by pioneer (ruderal) species during regeneration as disturbance severity increases [14]. Therefore, under the framework of the intermediate disturbance hypothesis, we expect species composition in wind disturbance + salvaged logged sites to exhibit higher relative abundance of pioneer species compared to wind disturbance alone. Regarding species diversity, the IDH would predict highest diversity following wind damage alone (generally considered a moderate severity event [3]), and lower diversity following wind + salvage, because the combined severity of the two events (*sensu* [4]) is quite high.

Roberts' [15,16] model of disturbance severity provides predictions based on three axes describing the percent damage to the canopy, understory, and forest floor [15]. For instance, a wind disturbance of moderate severity would have moderate canopy removal, limited understory removal, and limited soil disruption, but when combined with salvage logging, would have greater canopy removal, greater understory removal, and greater soil disruption [16]. The Roberts model predicts the more severe the disturbance is on multiple axes, the greater proportion of regeneration will derive from long distance dispersal and regeneration from the seed bank (i.e., pioneer species). The Roberts model, however, does not yield predictions of species diversity.

The cusp catastrophe model [3,17] provides a framework to explain compositional change as a result of disturbance severity and neighborhood effects. In this framework, 'heavy windthrow' alone is considered of moderate severity, implying that windthrow + salvaging would be a high severity combination. Combined with the classification of mixed oak forests as neutral to negative in terms of neighborhood effects, this model predicts a steadily increasing shift to earlier-successional species composition following disturbance combinations of increasing severity but does not make explicit predictions about diversity.

Distinct from expectations derived from the above concepts, numerous empirical studies fuel a vigorous, ongoing controversy over the extent to which post-disturbance salvage logging may be ecologically detrimental (e.g., [18]). Several prominent reviews [19–22] emphasize detrimental effects of salvaging, such as soil compaction, injury to surviving trees, crushing of seedlings and advance regeneration, and altered biogeochemical cycles and trace gas fluxes. For example, Donato et al. [23] reported that post-fire salvaging hindered regeneration in conifer forests of Oregon, while Lindenmayer and Ough [24] found similar effects after fire in eucalypt forests of Australia.

In a potential counterpoint, research into effects of salvaging after wind disturbance often—but not always—finds no detrimental effects. On one hand, Rumbaitis del Rio [25], studying herbaceous layer vegetation, found windthrow + salvage to lower species richness, species diversity, and total cover, as well as tree seedling density, compared to windthrow alone. Most other studies, however, report no difference or higher diversity, basal area, stem density, or size structure [26–29]; the few studies that span more than a decade of recovery show that initial seemingly detrimental salvaging effects are often transitory [11,30,31]. Thorn et al. [12] presents a comprehensive meta-analysis on the effects of salvage logging on different taxonomic groups, concluding that it alters species composition, but not species diversity. Of the studies in the meta-analysis, nearly 75% presented results from five years or fewer after the disturbance. Several studies sampled from situations that did not allow separation of blowdown effects from salvaging effects, but nevertheless found broadly similar patterns [32–34].

Despite the scarcity of detrimental salvaging effects on diversity after wind disturbance, a clear trend in many studies is a shift to an earlier-successional species composition of regeneration when wind is followed by salvaging (e.g., [26–30]), whereas wind disturbance alone often advances the successional state of woody vegetation.

Given the above, we hypothesized that (1) species diversity would decrease immediately after the windthrow and diversity would be even lower in sites that experienced salvage logging. Six years post-disturbance, however, we expect diversity to be indistinguishable between salvage logged and windthrown sites; (2) Species composition will differ between salvage logged and windthrown sites, with salvage logged sites dominated by early successional species.

2. Materials and Methods

2.1. Site Background

In April 2011, an EF3 tornado [35], blew through Northeast Georgia [1]. This tornado affected 5629 hectares (ha) of forest, with 18% of the affected area experiencing a damage severity of over 50% tree basal area felled [1]. To examine the effects of this tornado on tree composition, we established sampling plots within the Chattachoochee-Oconee National Forest (34.698 N, −83.886 W). The elevation at the site ranged from 588–672 m asl. The secondary forests are mixed hardwood, dominated by *Quercus* spp., *Carya* spp., *Tsuga canadensis* (eastern hemlock), *Oxydendron arboreum* (sourwood), *Pinus strobus* (white pine), and *Acer rubrum* (red maple). According to the USDA Web Soil Survey [36], the soils in this area are fine-loamy, mixed, mesic humic Hapludults (Tusquitee, Edneyville, and Porters loams). Temperature ranges seasonally from 3.9–22.8 °C, and annual precipitation is 1580 mm [37].

2.2. Plot Establishment

In July–September 2011, and March–June 2012, 36 20 × 20 m plots were established in tornado damaged areas, and trees surveyed before salvage logging. Plots were established in transects of tornado damage on north and south facing slopes, with half of the plots planned to be salvaged. All trees over 10 cm diameter at breast height (DBH) were surveyed within each plot. For the purposes of this manuscript, each time DBH is mentioned, its units of measurement will be in cm. Trees killed by the tornado were distinguished from previously dead trees so that tornado severity could be calculated. Salvage operations were carried out from mid-2012 through 2013. The spatial extent of the salvage logging was unpredictable, and plots were revisited to determine whether salvaging had occurred. In 2017, 19 plots were relocated and tree DBH and sapling density were recorded. Of those plots, 9 were salvaged, and 10 were unsalvaged. In 2017, saplings, defined as trees >2 m tall and up to 10 cm DBH, were measured in four 2 × 2 m quadrats at each cardinal direction in each plot. Some permanent plots were not included in the 2017 survey due to difficulty of locating plot corner markers, as plots were marked before salvaging.

2.3. Comparison of Plots before Tornado or Salvaging

Using the 2012 data, the pre-disturbance structure, diversity, and species composition could be assessed before the tornado by 'resurrecting' trees killed in the disturbance.

For the purposes of this study, 2012 pre-disturbance (before both tornado and salvage) plots will be labeled as either 'pre-disturbance, unsalvaged' or 'pre-disturbance, salvage' to indicate that they have not yet experienced wind disturbance or salvaging but have been separated based on whether they will be salvaged in the future. All sampling in 2012 was post-tornado, pre-salvage. By the time of the 2017 sampling, 6 years had elapsed since wind disturbance and 4–5 years since salvaging. In the 2017 data, the two types of plots are defined as 'post-tornado, post-salvaging' or 'post-tornado, unsalvaged'. We use the terms <1 and 6 years post-disturbance refer to the time since the wind disturbance, as the salvage logging took place in the years after the wind disturbance.

2.4. Data Analysis Comparing Salvaged Plots with Unsalvaged Plots

To characterize regeneration of woody vegetation, we calculated measures of diversity and size class distribution. We used Welch's *t*-tests and two-way ANOVAs to determine whether there was a difference in the means of diversity and size class parameters between the years and salvage conditions. Mean values are reported ± standard deviation. We also visualized the basal area (m^2ha^{-1}) of tree and sapling species (2017 only) in pre-tornado pre-salvage, pre-tornado unsalvage, and post-tornado post-salvage and post-tornado post unsalvaged areas as a whole.

Non-metric multidimensional scaling (nMDS) was performed on tree species compositions in salvaged ($n = 9$) and unsalvaged ($n = 10$) plots in 2012 and 2017 using R [38] and package vegan [39]. Specifically, pre-tornado pre-salvage, pre-tornado unsalvage, and post-tornado salvage and post-tornado unsalvaged plots were compared, and ordinal hulls were generated.

3. Results

3.1. Tornado Severity

The mean tornado severity, here defined as relative basal area loss, for salvaged and unsalvaged plots was 0.64 ± 0.28. Salvaged and unsalvaged plots did not differ in tornado severity (*t*-test; $p = 0.58$; Figure 1).

Figure 1. A study site in the Chattahoochee-Oconee National Forest shortly after the 2011 windthrow (**a**), and after salvage logging (**b**).

3.2. Tree Diversity, Density, and Basal Area before, <1 Year, and 6 Years after Windthrow and Salvage Logging

Table 1 shows the trajectory of tree diversity pre-disturbance, <1 year post-tornado and pre-salvage logging, and 6 years post-tornado and post-salvage logging (Table 1). Shannon diversity decreased in the one-year to six-year post-disturbance time interval. There were 439 individuals sampled pre-disturbance, compared to 281 individuals surviving the wind disturbance. Six years post-disturbance, 252 surviving individuals were sampled in the same plots.

At the plot scale, pre-tornado, pre-salvaging Shannon diversity for trees was 1.93 ± 0.24 in plots that would later be salvaged, compared to 1.81 ± 0.25 for plots that would remain unsalvaged. Shannon diversity did not differ between plots before experiencing disturbance (*t*-test; $p = 0.31$).

Six years post-tornado and post-salvaging, surviving trees in salvaged plots had a mean Shannon Diversity of 1.29 ± 0.39, and those in unsalvaged plots had a mean diversity of 1.39 ± 0.35. An ANOVA examining the effects of condition (salvage or unsalvage), year, and their interaction on Shannon diversity indicated that year was significant (F = 24.92, $p < 0.001$). Condition (F = 0.017, $p = 0.90$) and the interaction of condition and year were insignificant (F = 1.073, $p = 0.31$).

Species richness decreased immediately after the disturbance but increased to pre-disturbance levels after six years. An ANOVA examining the effects of condition (salvage or unsalvage), year, and their interaction on species richness indicated that year was significant (F = 27.62; $p < 0.001$). Condition (F = 0.28, $p = 0.60$) and the interaction of condition and year was insignificant (F = 1.30, $p = 0.26$).

Stem density and basal area showed similar patterns as Shannon diversity, with year being the only significant factor in an ANOVA examining the effects of condition (salvage or unsalvage), year, and their interaction on the dependent variable. For stem density, year was significant (F = 12.75,

$p = 0.001$), and condition (F = 0.92, $p = 0.34$) and the interaction of condition and year were insignificant (F = 0.97, $p = 0.33$).

Table 1. Shannon diversity, total stems, basal area, and species richness for pre-tornado, <1 year post-tornado, and 6 years post-tornado salvaged and unsalvaged plots (± Standard Deviation) of the remaining stand

Tree Metrics	Pre-Tornado, Pre-Salvage	<1 y Post-Tornado, Post-Salvage	6 y Post-Tornado, Post-Salvage
Shannon			
Salvage	1.93 (0.24)	1.62 (0.19)	1.29 (0.39)
Unsalvage	1.81 (0.25)	1.54 (0.34)	1.39 (0.35)
Total stems (per ha)			
Salvage	650.0 (140.25)	377.75 (130.75)	330.50 (232.0)
Unsalvage	515.0 (151.75)	362.50 (167.25)	332.50 (242.0)
Basal area (m²/ha)			
Salvage	33.75 (9.0)	14.0 (9.0)	12.0 (7.25)
Unsalvage	42.75 (14.25)	27.0 (14.25)	22.5 (27.5)
Species richness			
Salvage	9.22 (1.56)	6.66 (1.58)	9.87 (1.64)
Unsalvage	7.90 (2.02)	6.30 (2.40)	9.20 (2.15)

3.3. Tree Size Class Distributions

The mean pre-disturbance DBH of trees in unsalvaged plots was 28.7 ± 1.3, whereas the mean DBH of trees in pre-tornado, pre-salvaged plots was 22.9 ± 0.74. Trees had a significantly lower mean DBH in plots that would be later salvaged (*t*-test; $p = 0.0015$). Most of that difference in pre-disturbance size structure was due to much greater stem abundances in the smallest size class (10–20 cm DBH; Figure 2a), in the plots that would later be salvaged.

The mean DBH six years after disturbance in salvaged plots was 19.4 ± 3.63, compared to the mean DBH of post-tornado, unsalvaged plots (25.5 ± 7.0). The average DBH was significantly higher for unsalvaged plots (*t*-test; $p = 0.037$). Unsalvaged plots have a tail of individuals with a large DBH, whereas the largest individual in a salvaged plot has a DBH in the 40–49.9 cm DBH category (Figure 2b).

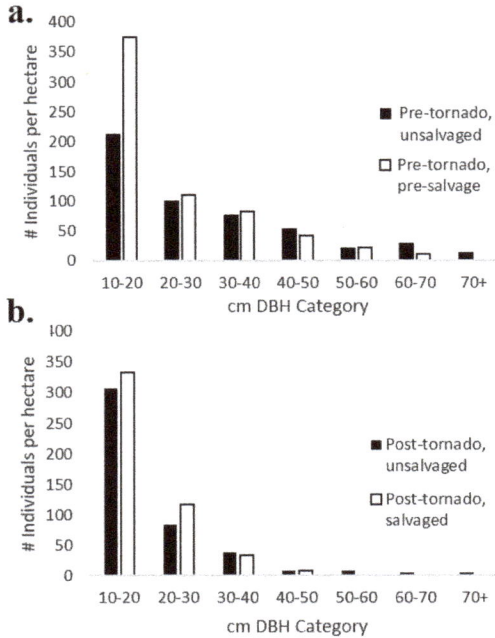

Figure 2. Frequency distributions among size (cm DBH) classes pre-tornado (**a**) and six years post-tornado (**b**) for salvaged and unsalvaged plots.

3.4. Tree Species Composition Ordinations

The final run of the nMDS had a stress of 0.24 (linear fit = 0.669; non-metric fit = 0.941). Prior to the tornado, species composition broadly overlapped between the plots that would be salvaged and those that would not. After the tornado, salvaged plots cluster away from the pre-disturbance plots, but overlap with the post-tornado, unsalvaged ordinal hull (Figure 3).

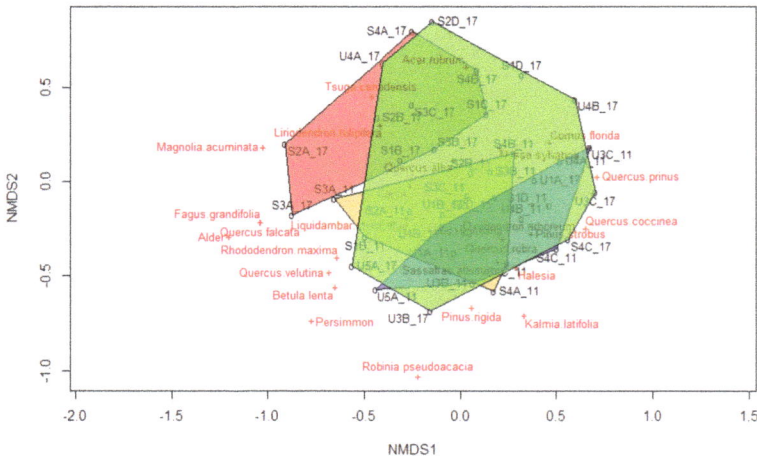

Figure 3. NMDS for trees pre-tornado, pre-salvage (**orange**); pre-tornado, unsalvaged (**blue**); six years post-tornado, post-salvage (**red**); and six years post-tornado, unsalvaged (**green**) plots.

3.5. Sapling Diversity and Density Parameters Six Years after Wind Disturbance and Salvage Logging

For saplings, 2017 Shannon diversity in salvaged and unsalvaged plots was 0.96 ± 0.48 and 1.32 ± 0.41, respectively; this difference was not statistically significant (*t*-test; $p = 0.11$). The total number of sapling species found in all unsalvaged plots was 18, compared to 17 for salvaged plots. A total of 201 saplings were sampled.

There was no significant Pearson correlation between plot basal area and sapling density in 2017 (Figure 4). For salvaged plots, the correlation coefficient was -0.11 ($t = -0.29$, dF = 7, $p = 0.78$). For unsalvaged plots, the correlation coefficient was -0.51 ($t = -1.67$, dF = 8, $p = 0.14$). There were no apparent relationships between sapling species composition, basal area, and condition.

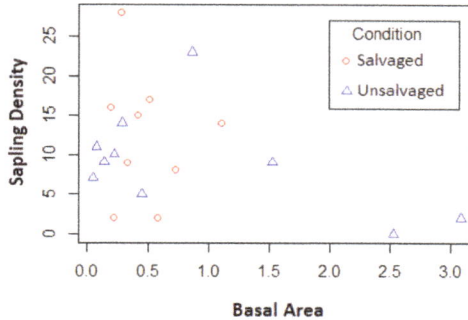

Figure 4. Scatterplot of plot (400 m^2) surviving tree basal areas vs sapling density in 2017 by condition. Salvaged plots had a Pearson's correlation of -0.11 ($p = 0.78$). Unsalvaged plots had a Pearson's correlation of -0.51 ($p = 0.14$).

3.6. Sapling Ordination

The nMDS for saplings in 2017 had a stress of 0.15 (linear fit = 0.871; non-metric fit = 0.979). Salvaged and unsalvaged plots separated in ordinal space, indicating different compositions (Figure 5). According to the nMDS, salvaged plots tended to have more *Acer rubrum* and *Pinus strobus*, while unsalvaged plots tended to have more *Sassafras albidum* and *Kamlia latifolia*. Early-successional species such as *Lirodendron tulipifera* or *Robinia pseudoacacia* were surprisingly rare.

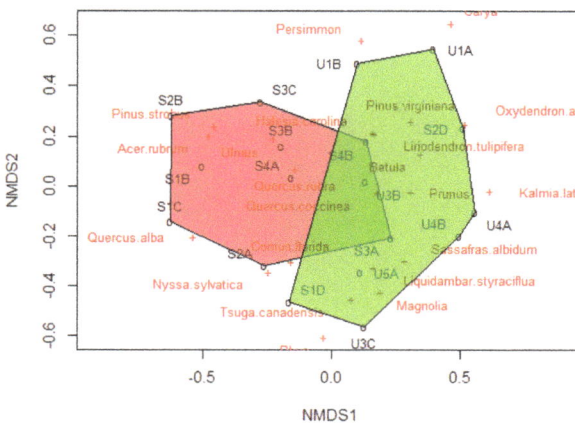

Figure 5. Non-metric multidimensional scaling (nMDS) of saplings in plots after tornado and salvage logging. Salvaged plots are indicated in **red**, and unsalvaged plots are **green**.

3.7. Sapling Species Accumulation Curve

Species accumulation curves showed that not only were more total species observed in unsalvaged plots (7), but that the curve for unsalvaged plots ascended more steeply (Figure 6).

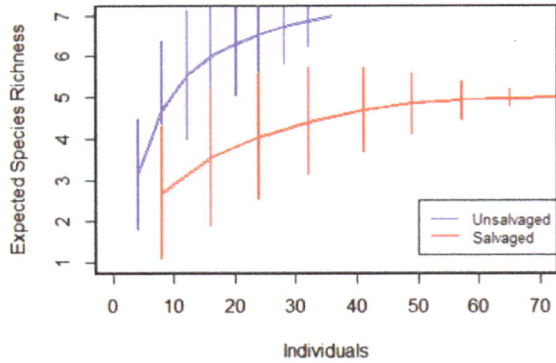

Figure 6. Species accumulation curve based on individual saplings in salvaged (**red**) and unsalvaged (**blue**) plots six years post-disturbance.

3.8. Tree and Sapling Species Basal Area Comparison

Combining 2017 basal area of saplings and trees (both pre-disturbance survivors and post-disturbance recruits) reveals that the species diversity and dominance in the four conditions as a whole are similar, with the exception of *P. strobus*, which is largely missing in the post-tornado post-salvage plots (Figure 7). While species that represented a small amount of basal area were grouped in the 'Other species' category, the total number of species for each condition were as follows: pre-tornado pre-salvage (22), pre-tornado unsalvaged (19), post-tornado post-salvage (24), post-tornado unsalvaged (26).

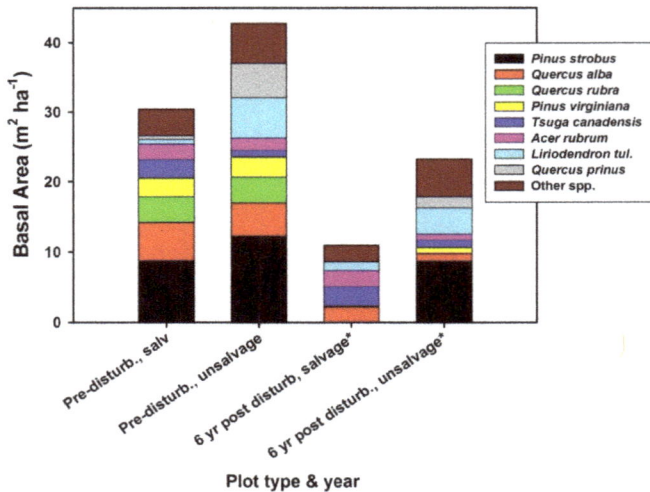

Figure 7. Stacked bars show the basal area (m^2ha^{-1}) per species per disturbance condition: pre-tornado pre-salvage, pre-tornado unsalvaged, post-tornado post-salvage, and post-tornado unsalvaged. Saplings were included in the post-tornado (2017) results(*).

4. Discussion

This study examined the effect of salvage logging on tree and sapling diversity, tree size structure, and sapling community composition. Our sampling allows a comparison of pre-disturbance, <1 year post-disturbance, and 6 years post-disturbance, a rarity for salvage logging studies [12]. Furthermore, although most conceptual frameworks relevant to compound disturbance research focus on multiple natural disturbances, such ideas must be confronted with findings from combined natural and anthropogenic disturbances. Wind followed by salvaging is an especially common natural + anthropogenic combination, and our findings add to the growing list of wind + salvage studies (e.g., [40]). Though the windthrow's destructive nature altered the tree size structure, the differences in size structure between plot types (salvaged and unsalvaged) remained consistent with the pattern found before the windthrow—plots that were unsalvaged continued to have a higher DBH than plots that were salvaged. However, an examination of the distribution of individuals in each size class shows that the drivers of these patterns have changed—before the windthrow, plots that would later be salvaged had more trees in the 10–20 cm DBH category. After the windthrow, this pattern was driven by the lack of large (>50 cm DBH) trees in salvaged plots. This pattern of tree size structure in salvaged vs. unsalvaged plots indicates the salvage logging operation removed large still-living trees, which were still present in the unsalvaged plots. We noted that there were fallen but still living trees six years post-windthrow in the unsalvaged plots (CAO, field observation). Therefore, when considering the effects of salvage logging, it is important to consider that trees that are standing or with minor damage may also be salvaged due to their size and value. This may have a counterproductive effect if encouraging a diverse, heterogenous forest is a management goal.

In partial support of our hypothesis, salvaged plots had greater abundance of mid-successional species (including *P. strobus*) when compared to unsalvaged plots, but very few pioneer species. It is possible that this is caused by the heterogeneous environment that resulted from the salvaging machinery. Of the salvaged plots, only one was directly located in the skid trails of the salvaging operation. Given the soil compaction and disturbance, we may expect, and did visually observe, more pioneer species in these skid trails outside of the plots. These findings are similar to those reported by Peterson and Leach [4], Nelson et al. [29], and Fidej et al. [41]. Therefore, the heterogeneous nature of salvage logging may result in modest overall impacts due to large areas that are lightly impacted, and small areas that are heavily impacted (e.g., by the skid trails) (c.f. [11]). Supporting this notion, Nelson et al. [29] found that tree diversity metrics did not differ overall between post-windthrow salvaged and unsalvaged plots over three years. However, these researchers found that pioneer species were more abundant in areas of high severity soil disturbance caused by the salvage logging.

These results provide support to the idea that salvage logging does not negatively impact species diversity, but that disturbance severity may affect species composition. Sapling community composition diverged in salvaged plots compared to unsalvaged plots. We found more mid-successional sapling species in salvaged plots, such as *P. strobus* and *A. rubrum*, whereas unsalvaged plots were characterized more by later successional species such as *T. canadensis*, *Kalmia latifolia* and *Carya* species. This may be a function of the increased soil disturbance and greater light availability as seen by reduced basal area in salvaged plots compared to unsalvaged. These results provide support to the Thorn et al. [12] meta-analysis that, broadly, salvage logging after a natural disturbance changes community composition but not diversity parameters even 6 years post-disturbance. In a Tennessee windthrow and subsequent salvage logging study, tree seedling diversity did not differ two years after disturbance [27]. The authors concluded that the disturbance and its interaction may not have been sufficiently severe to result in a lasting difference, but greater disturbance severity led to greater tree species composition change. Compared to Peterson and Leach [27] the present study characterized a more severe wind disturbance over a longer period of time, but we did not detect a difference in sapling diversity as a result of salvage logging. Royo et al. [11] found that salvage logging had effects on sapling diversity for the first several years after disturbance, but these effects were ephemeral. However, the authors determined that the soil disturbance caused by salvaging resulted in sapling compositional differences even five years after disturbance. These results are consistent with our own findings.

When examining the four conditions (pre-tornado pre-salvage and unsalvage, post-tornado salvage and unsalvage), we found that the relative dominances of species was consistent over time, with the exception of *P. strobus*. The most dominant species pre-tornado, its basal area decreased in the post-tornado post-salvage plots when compared to the unsalvaged plots. We believe that this further adds to the point made by our frequency diagrams that salvage logging affects trees that are both large and valuable, even if they are still standing or relatively undamaged. Therefore, in addition to the many *P. strobus* trees were killed by the windthrow, it is likely that living trees were also removed from the site through salvage logging. Another difference in species richness is apparent when examining the conditions as a whole: pre-tornado pre-salvage had a combined (tree + sapling) species richness of 22, while six years after the disturbance the same plots had a species richness of 24, an increase of 2 species. The pre-tornado unsalvaged area had a species richness of 19, but that increased to 26 six years after disturbance. The increase of seven species in unsalvaged plots compared to an increase of two species in salvaged plots suggests that salvage logging limit the accumulation of species, although the overall trend was not enough to cause 2017 diversity to differ between salvaged and unsalvaged plots.

It seems that this study, though spanning six years, is still too early to produce firm conclusions on the future canopy tree compositional changes caused by wind disturbance and salvage logging. Even six years post-disturbance, tree diversity, tree basal area, and total tree stems were lower than at <1 year post-disturbance, displaying the lag effect [42]. When we examined sapling density compared to surviving basal area, a metric to assess damage severity (Figure 3), we found that the pattern for unsalvaged plots resembles the bell shaped curve as predicted by the intermediate disturbance hypothesis [16]. The salvaged plots, which represent a combined disturbance, had a lower basal area overall and no apparent relationship between basal area and sapling density. However, despite expectations based on the IDH that sapling species diversity would be higher in unsalvaged (winthrow alone) than salvaged (two cumulative disturbances), there was only a small and nonsignificant trend towards more species in the unsalvaged plots. Our examination of sapling species composition six years post-disturbance showed only modest effects of disturbance on successional status of the sapling layer, as predicted by the cusp catastrophe model and Roberts model [3,15–17]. Interestingly, a study that examined sapling regeneration at 10 and 20 years post-disturbance found that site factors (pH and ground cover) and not disturbance type (windthrow or windthrow + salvage) controlled regeneration and sapling density [43]. Both salvaged and unsalvaged plots contained late successional species, but salvaged plots had greater abundance of mid-successional *Pinus strobus* and *Acer rubrum*. This partially supports the expectations of compositional change but falls short of the expected major compositional shift to pioneer species in salvaged areas.

The data available do not allow us to discern whether the saplings in our 2017 sample were composed of released advance regeneration or new recruits. Long-term Harvard Forest studies that examined the effects of a simulated hurricane found that recovery was initially comprised of sprouting from damaged trees and new seedlings, but the longer term (>20 years) patterns of composition were driven by surviving trees and release of trees already present pre-disturbance [44,45].

5. Conclusions

As this study and others have shown, it is crucial to monitor the impacts of combined disturbances >5 years post-disturbance. Ephemeral effects of salvage logging may be detected in the first years immediately after the disturbance, but they may be poor indicators of longer-term successional trajectories. Moreover, this study adds to a growing set of multi-year studies that consistently find roughly predictable effects of salvaging after wind disturbance but few examples of diversity reduction.

Forests **2019**, *10*, 129

Author Contributions: Conceptualization, C.O. and C.P; Methodology, C.P.; Software, C.P.; Validation, C.O. and C.P.; Formal analysis, C.O.; Investigation, C.O; Resources, C.P.; Data curation, C.O.; Writing—original draft preparation, C.O.; Writing—review and editing, C.O. and C.P.; Visualization, C.O. and C.P.; Supervision, C.P.; Project administration, C.P.; Funding acquisition, C.O. and C.P.

Funding: This research was funded by the National Science Foundation grants AGS-1141926 from Physical and Dynamic Meteorology and DEB-1143511 from Population and Community Ecology; NSF Graduate Research Fellowship Program, University of Georgia Graduate School, University of Georgia Department of Plant Biology through the Palfrey Award, Haines Family Field Botany Award, and Graduate Student Association Award.

Acknowledgments: We thank our field assistants who make this work possible: Rhett Parr, Ajay Patel, John Howard, Kathryn Thompson, and Sydney Mai. We thank the graduate students that began this project in 2012: Luke Snyder and Kaysandra Waldron, with assistance from Eli White, Trevor Sprague, Meredith Barrett, and Patrick Johnson. We are grateful for the insightful comments by Jacquelin Mohan, Daniel Markewitz, Doug Aubrey, and Chris Gough.

Conflicts of Interest: The authors declare no conflict of interest.

References

1. Cannon, J.B.; Hepinstall-Cymerman, J.; Godfrey, C.M.; Peterson, C.J. Landscape-scale characteristics of forest tornado damage in mountainous terrain. *Landsc. Ecol.* **2016**, *31*, 2097–2114. [CrossRef]

2. Paine, R.T.; Tegner, M.J.; Johnson, E.A. Compounded perturbations yield ecological surprises. *Ecosystems* **1998**, *1*, 535–545. [CrossRef]

3. Frelich, L.E. *Forest Dynamics and Disturbance Regimes: Studies from Evergreen-Deciduous Forests*; Cambridge University Press: Cambridge, UK, 2002.

4. Peterson, C.J.; Leach, A.D. Salvage logging after windthrow alters microsite diversity, abundance and environment, but not vegetation. *Forestry* **2008**, *81*, 361–376. [CrossRef]

5. Reyer, C.P.O.; Brouwers, N.; Rammig, A.; Brook, B.W.; Epila, J.; Grant, R.F.; Holmgren, M.; Langerwisch, F.; Leuzinger, S.; Lucht, W.; et al. Forest resilience and tipping points at different spatio-temporal scales: Approaches and challenges. *J. Ecol.* **2015**, *103*, 5–15. [CrossRef]

6. Williams, C.A.; Gu, H.; MacLean, R.; Masek, J.G.; Collatz, G.J. Disturbance and the carbon balance of US forests: A quantitative review of impacts from harvests, fires, insects, and droughts. *Glob. Planet. Change* **2016**, *143*, 66–80. [CrossRef]

7. Romme, W.H.; Everham, E.M.; Frelich, L.E.; Moritz, M.A.; Sparks, R.E. Are Large, infrequent disturbances qualitatively different from small, frequent disturbances? *Ecosystems* **1998**, *1*, 524–534. [CrossRef]

8. Everham, E.M.; Brokaw, N.V.L. Forest damage and recovery from catastrophic wind. *Bot. Rev.* **1996**, *62*, 113–185. [CrossRef]

9. Webb, S.L. Disturbance by wind in temperate-zone forests. In *Ecosystems of Disturbed Ground*; Walker, L.R., Ed.; Elsevier: Amsterdam, The Netherland, 1999; pp. 187–222.

10. Mitchell, S.J. Wind as a natural disturbance agent in forests: A synthesis. *Forestry* **2013**, *86*, 147–157. [CrossRef]

11. Royo, A.A.; Peterson, C.J.; Stanovick, J.S.; Carson, W.P. Evaluating the ecological impacts of salvage logging: Can natural and anthropogenic disturbances promote coexistence? *Ecology* **2016**, *97*, 1566–1582. [CrossRef]

12. Thorn, S.; Bässler, C.; Brandl, R.; Burton, P.J.; Cahall, R.; Campbell, J.L.; Castro, J.; Choi, C.Y.; Cobb, T.; Donato, D.C.; et al. Impacts of salvage logging on biodiversity: A meta-analysis. *J. Appl. Ecol.* **2018**, *55*, 279–289. [CrossRef]

13. Lindenmayer, D.B.; Noss, R.F. Salvage logging, ecosystem processes, and biodiversity conservation. *Conserv. Biol.* **2006**, *20*, 949–958. [CrossRef] [PubMed]

14. Connell, J.H. Diversity in tropical rain forests and coral reefs. *Science* **1978**, *199*, 1302–1310. [CrossRef] [PubMed]

15. Roberts, M.R. A conceptual model to characterize disturbance severity in forest harvests. *For. Ecol. Manage.* **2007**, *242*, 58–64. [CrossRef]

16. Roberts, M.R. Response of the herbaceous layer to natural disturbance in North American forests. *Can. J. Bot.* **2004**, *82*, 1273–1283. [CrossRef]

17. Frelich, L.E.; Reich, P.B. Neighborhood effects, disturbance severity, and community stability in forests. *Ecosystems* **1999**, *2*, 151–166. [CrossRef]

18. DellaSala, D.A.; Karr, J.R.; Schoennagel, T.; Perry, D.; Noss, R.F.; Lindenmayer, D.B.; Beschta, R.L.; Hutto, R.L.; Swanson, M.E.; Evans, J. Post-fire logging debate ignores many issues. *Science* **2006**, *312*, 1137–1137. [CrossRef] [PubMed]

19. McIver, J.D.; Starr, L. A literature review on the environmental effects of postfire logging. *West. J. Appl. For.* **2001**, *16*, 159–168.

20. Beschta, R.L.; Rhodes, J.J.; Kauffman, J.B.; Gresswell, R.E.; Minshall, G.W.; Karr, J.R.; Perry, D.A.; Hauer, F.R.; Frissell, C.A. Postfire management on forested public lands of the Western United States. (Special section: Wildfire and conservation in the western United States.). *Conserv. Biol.* **2004**, *18*, 957–967. [CrossRef]

21. Foster, D.R.; Orwig, D.A. Preemptive and salvage harvesting of New England forests: When doing nothing is a viable alternative. *Conserv. Biol.* **2006**, *20*, 959–970. [CrossRef]

22. Lindenmayer, D.B.; Foster, D.R.; Franklin, J.F.; Hunter, M.L.; Noss, R.F.; Schmiegelow, F.A.; Perry, D. Salvage harvesting policies after natural disturbance. *Science* **2004**, *303*, 1303. [CrossRef]

23. Donato, D.C.; Fontaine, J.B.; Campbell, J.L.; Robinsin, W.D.; Kauffman, J.B.; Law, B.E. Post-wildfire logging hinders regeneration and increases fire risk. *Science* **2006**, *311*, 352. [CrossRef] [PubMed]

24. Lindenmayer, D.B.; Ough, K. Salvage logging in the montane ash eucalypt forests of the Central Highlands of Victoria and its potential impacts on biodiversity. *Conserv. Biol.* **2006**, *20*, 1005–1015. [CrossRef] [PubMed]

25. Rumbaitis del Rio, C.M. Changes in understory composition following catastrophic windthrow and salvage logging in a subalpine forest ecosystem. *Can. J. For. Res.* **2006**, *36*, 2943–2954. [CrossRef]

26. Lain, E.J.; Haney, A.; Burris, J.M.; Burton, J. Response of vegetation and birds to severe wind disturbance and salvage logging in a southern boreal forest. *For. Ecol. Manage.* **2008**, *256*, 863–871. [CrossRef]

27. Peterson, C.J.; Leach, A.D. Limited salvage logging effects on forest regeneration after moderate-severity windthrow. *Ecol. Appl.* **2008**, *18*, 407–420. [CrossRef] [PubMed]

28. Fischer, A.; Fischer, H.S. Individual-based analysis of tree establishment and forest stand development within 25 years after wind throw. *Eur. J. For. Res.* **2012**, *131*, 493–501. [CrossRef]

29. Nelson, J.L.; Groninger, J.W.; Battaglia, L.L.; Ruffner, C.M. Bottomland hardwood forest recovery following tornado disturbance and salvage logging. *For. Ecol. Manage.* **2008**, *256*, 388–395. [CrossRef]

30. Palik, B.; Kastendick, D. Woody plant regeneration after blowdown, salvage logging, and prescribed fire in a northern Minnesota forest. *For. Ecol. Manage.* **2009**, *258*, 1323–1330. [CrossRef]

31. Lang, K.D.; Schulte, L.A.; Guntenspergen, G.R. Windthrow and salvage logging in an old-growth hemlock-northern hardwoods forest. *For. Ecol. Manage.* **2009**, *259*, 56–64. [CrossRef]

32. Elliott, K.J.; Hitchcock, S.L.; Krueger, L. Vegetation response to large scale disturbance in a southern appalachian forest: Hurricane Opal and salvage logging. *J. Torrey Bot. Soc.* **2002**, *129*, 48–59. [CrossRef]

33. D'Amato, A.W.; Fraver, S.; Palik, B.J.; Bradford, J.B.; Patty, L. Singular and interactive effects of blowdown, salvage logging, and wildfire in sub-boreal pine systems. *For. Ecol. Manage.* **2011**, *262*, 2070–2078. [CrossRef]

34. Bottero, A.; Garbarino, M.; Long, J.N.; Motta, R. The interacting ecological effects of large-scale disturbances and salvage logging on montane spruce forest regeneration in the western European Alps. *For. Ecol. Manage.* **2013**, *292*, 19–28. [CrossRef]

35. McDonald, J.R.; Kishor, C.M. *A Recommendation for an Enhanced Fujita Scale (EF-Scale)*; Wind Science and Engineering Center, Texas Tech University: Lubbock, TX, USA, 2006.

36. Web Soil Survey. Available online: https://websoilsurvey.sc.egov.usda.gov/ (accessed on 15 December 2018).

37. Diamond, H.J.; Karl, T.R.; Palecki, M.A.; Baker, C.B.; Bell, J.E.; Leeper, R.D.; Easterling, D.R.; Lawrimore, J.H.; Meyers, T.P.; Helfert, M.R.; et al. U.S. Climate reference network after one decade of operations status and assessment. *Bull. Am. Meteorol. Soc.* **2013**, *94*, 485–498. [CrossRef]

38. R Core Team. *R: A Language and Environment for Statistical Computing*; R Foundation for Statistical Computing: Vienna, Austria, 2017.

39. Oksanen, J.; Blanchet, F.G.; Kindt, R.; Legendre, P.; Minchin, P.R.; O'hara, R.B.; Simpson, G.L.; Solymos, P.; Stevens, M.H.; Wagner, H. Vegan: Community Ecology Package, R Package Version 2.5-2; 2018. Available online: https://CRAN.R-project.org/package=vegan (accessed on 2 February 2019).

40. Elliott, K.J.; Swank, W.T. Long-term changes in forest composition and diversity following early logging (1919–1923) and the decline of American chestnut (Castanea dentata). *Plant Ecol.* **2008**, *197*, 155–172. [CrossRef]

41. Fidej, G.; Rozman, A.; Nagel, T.A.; Dakskobler, I.; Diaci, J. Influence of salvage logging on forest recovery following intermediate severity canopy disturbances in mixed beech dominated forests of Slovenia. *iForest-Biogeosci. For.* **2016**, *9*, 430. [CrossRef]

42. Reich, P.B.; Bakken, P.; Carlson, D.; Frelich, L.E.; Friedman, S.K.; Reich, P.B.; Bakken, P.; Carlson, D.; Frelich, L.; Friedman, S.K.; et al. Influence of logging, fire, and forest type on biodiversity and productivity in southern boreal forests. *Ecology* **2001**, *82*, 2731–2748. [CrossRef]

43. Kramer, K.; Brang, P.; Bachofen, H.; Bugmann, H.; Wohlgemuth, T. Site factors are more important than salvage logging for tree regeneration after wind disturbance in Central European forests. *For. Ecol. Manage.* **2014**, *331*, 116–128. [CrossRef]

44. Plotkin, A.B.; Foster, D.; Carlson, J.; Magill, A. Survivors, not invaders, control forest development following simulated hurricane. *Ecology* **2013**, *94*, 414–423. [CrossRef] [PubMed]

45. Cooper-Ellis, S.; Foster, D.R.; Carlton, G.; Lezberg, A. Forest response to catastrophic wind: Results from an experimental hurricane. *Ecology* **1999**, *80*, 2683–2696. [CrossRef]

© 2019 by the authors. Licensee MDPI, Basel, Switzerland. This article is an open access article distributed under the terms and conditions of the Creative Commons Attribution (CC BY) license (http://creativecommons.org/licenses/by/4.0/).

MDPI

St. Alban-Anlage 66

4052 Basel

Switzerland

Tel. +41 61 683 77 34

Fax +41 61 302 89 18

www.mdpi.com

Forests Editorial Office

E-mail: forests@mdpi.com

www.mdpi.com/journal/forests

www.ingramcontent.com/pod-product-compliance
Lightning Source LLC
Chambersburg PA
CBHW051723210326
41597CB00032B/5581